Encyclopaedia of Mathematical Sciences

Volume 79

Editor-in-Chief: R.V. Gamkrelidze

Springer
Berlin
Heidelberg
New York
Barcelona
Budapest
Hong Kong
London
Milan
Paris
Santa Clara
Singapore
Tokyo

M. S. Agranovich · Yu. V. Egorov
M. A. Shubin (Eds.)

Partial Differential Equations IX

Elliptic Boundary Value Problems

Springer

Title of the Russian edition:
Itogi nauki i tekhniki, Sovremennye problemy matematiki,
Differentsial'nye uravneniya s chastnymi proizvodnymi 9,
Publisher VINITI, Moscow
(in preparation)

Mathematics Subject Classification (1991):
35J, 35P, 35S, 58G, 35Jxx

ISSN 0938-0396
ISBN 3-540-57044-6 Springer-Verlag Berlin Heidelberg New York

Typesetting: Camera-ready copy produced from the
translator's input files using a Springer TEX macro package.
SPIN: 10111578 41/3143 - 5 4 3 2 1 0 – Printed on acid-free paper

List of Editors, Authors and Translators

Editor-in-Chief

R. V. Gamkrelidze, Russian Academy of Sciences, Steklov Mathematical Institute, ul. Vavilova 42, 117966 Moscow; Institute for Scientific Information (VINITI), ul. Usievicha 20a, 125219 Moscow, Russia
e-mail: gam@ipsun.ras.ru

Consulting Editors

M. S. Agranovich, Moscow State Institute of Electronics and Mathematics (Technical University), 109028 Moscow, Russia, e-mail: msa@funcan.msk.su

Yu. V. Egorov, U.F.R. M.I.G., Université Paul Sabatier, 118, route de Narbonne, 31062 Toulouse Cedex, France, e-mail: egorov@mip.ups-tlse.fr

M. A. Shubin, Department of Mathematics, Northeastern University, Boston, MA 02115, USA, e-mail: shubin@neu.edu

Authors and Translators

M. S. Agranovich, Moscow State Institute of Electronics and Mathematics (Technical University), 109028 Moscow, Russia, e-mail: msa@funcan.msk.su

A. V. Brenner, Department of Mathematics, Technion, Technion City, 32000 Haifa, Israel, e-mail: brenner@techunix.technion.ac.il

B. A. Plamenevskij, Department of Mathematical Physics, Faculty of Physics, St. Petersburg State University, ul. Ulyanova 1, Stary Peterhof, 198904 St. Petersburg, Russia, e-mail: boris@plamen.niif.spb.su

Author

E. M. Shargorodsky, A. Razmadze Mathematical Institute, M. Aleksidze str. 1, 38093 Tbilisi, Republic of Georgia, e-mail: gmu@imath.acnet.ge and Department of Mathematics, King's College London, Strand, London, WC2R 2 LS, United Kingdom,
e-mail: sharg@mth.kcl.ac.uk

Contents

I. Elliptic Boundary Problems

M. S. Agranovich

Translated from the Russian
by the author

Contents

Preface

0.1. Scope of the Paper. This paper, somewhat delayed in the series "Fundamental Directions," is devoted to general linear elliptic boundary problems on a smooth compact manifold with boundary. The paper is intended for a large circle of readers. We hope the paper will be useful to many mathematicians with diverse scientific interests.

The main features of the theory we will discuss were formed in the 60's, beginning with the proof of the equivalence of the ellipticity conditions and the Fredholm property of the corresponding operator in the simplest Sobolev spaces. This was done on the basis of achievements of many mathematicians during the preceding decades. This basis was very extensive, but the results were incomplete. The elaboration of the general theory was stimulated by investigations of the index problem (see e.g. (Palais 1965) and (Fedosov 1990)) and went on under a strong influence of the microlocal analysis, beginning with the appearance of the calculus of pseudodifferential operators. During the last decades, new variants of the general theory appeared, and the old variants were enriched by new results.

At the present time, a number of books contain chapters devoted to the general theory of elliptic boundary problems, including (Hörmander 1963), (Agmon 1965a), (Berezanskij 1965), (Nečas 1967), (Lions and Magenes 1968), (Dieudonné 1978), (Triebel 1978), (Trèves 1980), (Chazarain and Piriou 1981), (Kumano-go 1981), (Taylor 1981), (Egorov 1984), and (Hörmander 1985a,b). In each of these books, certain fields and approaches are chosen for a systematic and detailed treatment, often with not simple proofs. Of course, we had no space for detailed proofs, but instead we had a possibility to try to give a general picture, without restriction to one approach, and to include results from a number of journal papers.

To decide what to include in this survey was a hard problem. Of course, the material was selected in accordance with scientific interests of the author. The paper is a continuation of the survey (Agranovich 1990a) devoted to elliptic operators on a closed manifold (i.e. on a compact manifold without boundary). It would be useful for the reader to look over that survey before reading the present paper. Like there, appreciable attention is given to the spectral properties of elliptic operators without obligatory assumption of selfadjointness; we relate the corresponding results to the general theory. There is more of variety in types of spectral problems on a manifold with boundary, though the results are often less complete. In some places we amplify the information contained in (Rozenblum et al. 1989).

Some of other surveys belonging to the same series "Fundamental Directions" abut on our paper. Two of them are contained in the present volume: (Brenner and Shargorodsky 1996) on pseudodifferential elliptic boundary problems, and (Plamenevskij 1996) on elliptic operators on nonsmooth manifolds. We also mention the surveys (Maz'ya 1988) on the method of boundary integral equations, (Levendorskij and Paneah 1990) on elliptic operators with degeneracies, (Fedosov 1990) on the index problem, (Dudnikov and Samborskij 1991), where overdetermined and underdetermined elliptic boundary problems are considered, and the paper of S.M. Vishik devoted to some problems from the global analysis, connected with the determinants of elliptic operators.

To simplify the treatment, we first consider main versions of the theory for elliptic boundary problems with a scalar unknown function, and we describe "rough" constructions at the level of principal symbols. After this, we pass to matrix boundary problems, to more complete constructions in terms of local complete symbols, and to other variants of major theorems. As in (Agranovich 1990a), discussing variants and generalizations, we generally take one step and do not consider combinations of generalizations. Our paper is not a text-book or a reference book, it is a survey. We also need to note that the degree of working out in detail is different in various places: sometimes we restrict ourselves to formulations, sometimes we explain simple things to assure continuity of our presentation.

At the request of the author, Sections 7.7 and 9.6 were written by Mikhailets, Section 7.9 by Roĭtberg, and Section 9.7 by Solomyak. Sections 7.5 and 7.6 were written in collaboration with Skubachevskij and Sternin, respectively. K. Naimark, Ya. Roĭtberg, and L. Dzhanlatyan have read the manuscript and have given a lot of useful suggestions. Many useful advices has been obtained from other mathematicians. The author is indebted to all of them for this friendly help. The almost final version of the English text was edited by Mr. M. Gross, New York, and his help was invaluable.

0.2. Some Assumptions and Notation. We denote by M a compact n-dimensional C^∞ manifold with boundary $\partial M = \Gamma$, connected for simplicity; $n \geq 2$. The subset $M \setminus \Gamma$ of M is denoted by M_+. The manifold M can be covered by coordinate neighborhoods O lying in M_+ and coordinate semi-neighborhoods O^+ adjacent to Γ. We permit ourselves to identify a point $x \in M$ with the corresponding tuple $(x^1, \ldots, x^n)$ of local coordinates. If $f(x)$ is a function on the manifold, then the corresponding function on a chart for O or O^+ will also be denoted by $f(x)$. The chart for O is a ball $\{x \in \mathbb{R}^n : |x| < r = r(O)\}$, and a chart for O^+ is a semi-ball $\{x \in \mathbb{R}^n : |x| < r(O^+), x^n \geq 0\}$, with the corresponding mapping.

All further assumptions are made without loss of generality. We do not need to use all of them everywhere, but each of them will often be convenient. We agree to consider M as a part of a closed n-dimensional C^∞ manifold M_0. The boundary Γ is an $(n-1)$-dimensional C^∞ submanifold of M_0, and we set

$M_- = M_0 \setminus M$. We assume that Γ has a neighborhood U on M_0 diffeomorphic to $\Gamma \times (-1,1)$. In U a unique coordinate $x_n \in (-1,1)$ is introduced; it is positive, equal to zero, and negative on $U \cap M_+$, on Γ, and on $U \cap M_-$, respectively. If $x' = (x^1, \dots, x^{n-1})$ are local coordinates on Γ, then (x', x^n) are local coordinates in the corresponding part of U. The set U is called a *tubular neighborhood* of Γ, and $U \times M$ is called a *collar* of M.

For functions $u \in C^\infty(M)$ we define the *trace of kth order* on the boundary Γ by the formula[1]

$$(\gamma_k u)(x) = D_n^k u(x', x^n)|_{x^n=0} . \tag{0.1}$$

We assume that positive C^∞ densities dx and dx' are given on M^0 and Γ, respectively. To simplify the representation of kernels of integral operators, we usually assume the local coordinates to be *consistent* with these densities. This means, for example, that if $f(x)$ is a function on M_0 with support lying in one coordinate neighborhood or semi-neighborhood, then its integral over M_0 is written in these coordinates in the form

$$\int f(x)\, dx = \int f(x^1, \dots, x^n)\, dx^1 \dots dx^n ,$$

without an additional factor $\rho(x)$ in the integrand. In particular, this assumption can be related to the coordinates (x', x^n) mentioned above, and then the local coordinates $x' = (x^1, \dots, x^{n-1})$ on Γ can be assumed to be consistent with dx'.

A Riemannian C^∞ metric $\{g_{j,k}\}$ can always be introduced on M_0, and if it is given, than it defines a density dx with $\rho(x) = (\det(g_{j,k}(x)))^{1/2}$. In local coordinates consistent with this density, $\det(g_{j,k}(x)) \equiv 1$. On Γ an induced metric arises, and it generates a density on Γ.

In particular, we will often consider the case in which M_+ is a bounded domain G in $\mathbb{R}^n$ with C^∞ boundary $\partial G = \Gamma$. In this case we have the usual Euclidean metric in G and the induced metric on Γ. The Cartesian coordinates in $\mathbb{R}^n$ can be used as local coordinates everywhere in G, and they are, of course, consistent with the corresponding density. In this situation we will often use the boundary operator $\gamma_0 \partial_\nu^k u$, where $\partial_\nu^k = \partial^k / \partial_\nu^k$ is the derivative of u of order k in the direction of the inner normal to Γ at a boundary point.

§1. Scalar Elliptic Boundary Problems

1.1. Formulation of Boundary Problems. We begin the consideration of the boundary problem

$$Au = f \quad \text{on} \quad M_+ , \quad B_j u = g_j \quad (j = 1, \dots, q) \quad \text{on} \quad \Gamma . \tag{1.1}$$

Here A is a partial differential operator of order $m = 2q$ on M, and B_j are boundary partial differential operators of orders r_j. The coefficients of all

[1] $D_j = i^{-1}\partial_j$, where $\partial_j = \partial / \partial x^j$.

differential operators are assumed to be C^∞ on M and, in general, complex-valued. In local coordinates in O or near Γ we have[2]

$$A = A(x, D) = \sum_{|\alpha| \leq m} a_\alpha(x) D^\alpha \,, \tag{1.2}$$

where $a_\alpha(x) \in C^\infty$, and in local coordinates near Γ we have

$$B_j = B_j(x, D) = \sum_{|\beta| \leq r_j} \gamma_0 b_{j,\beta}(x) D^\beta \,, \tag{1.3}$$

where $b_{j,\beta} \in C^\infty$. If $M = \overline{G}$, where G is a domain in $\mathbb{R}^n$, then A and B may be written globally in coordinates of $\mathbb{R}^n$. From the outset we assume that A has even order $2q$, and that the number of boundary conditions is equal to half of this order, since this will follow from the ellipticity conditions (see Sect. 1.2). It is possible to assume that the coefficients of the boundary operators are given only for $x \in \Gamma$, but our assumption that they are given near Γ does not restrict generality. At the beginning we assume that $u(x) \in C^\infty(M)$ in (1.1), and we associate the operator

$$\mathcal{A} = (A, B_1, \ldots, B_q) \tag{1.4}$$

with our boundary problem. This operator acts from $C^\infty(M)$ to $C^\infty(M) \times [C^\infty(\Gamma)]^q$.

1.2. Ellipticity Conditions. The boundary problem (1.1) is called *elliptic* if the following three conditions are satisfied.

1. The operator A is elliptic on M, i.e. its *principal symbol*, locally given by the formula[3]

$$a_0(x, \xi) = \sum_{|\alpha| = m} a_\alpha(x) \xi^\alpha \,, \tag{1.5}$$

is different from zero everywhere on the cotangent bundle $T^*M \setminus 0$.

The following two conditions relate to an arbitrary fixed point on Γ. We write A and B_j in local coordinates near Γ (see Sect. 0.2). Assume that the given point on Γ has coordinates $(x', 0)$.

2. Consider the equation

$$a_0(x', 0, \xi', \zeta) = 0 \tag{1.6}$$

with respect to ζ. In its left-hand side, the coefficient in the term of highest degree is nonzero in view of the ellipticity of A. This equation must have the same number (q) of roots in the upper and the lower halfplanes

$$\mathbb{C}_+ = \{\zeta : \operatorname{Im} \zeta > 0\} \quad \text{and} \quad \mathbb{C}_- = \{\zeta : \operatorname{Im} \zeta < 0\} \tag{1.7}$$

[2] $D^\alpha = D_1^{\alpha_1} \ldots D_n^{\alpha_n}$; $|\alpha| = \alpha_1 + \ldots + \alpha_n$.
[3] $\xi^\alpha = \xi_1^{\alpha_1} \ldots \xi_n^{\alpha_n}$.

for any real $\xi' \neq 0$. This is the condition of *proper ellipticity* of A. From this condition it follows, of course, that the order of A must be even.

For $n > 2$ proper ellipticity follows from ellipticity. Indeed, denote by $n_+(\xi')$ and $n_-(\xi')$ the number of the roots of equation (1.7) in $\mathbb{C}_+$ and $\mathbb{C}_-$, respectively, for $\xi' \neq 0$. Then we have $n_\pm(\xi') = n_\mp(-\xi')$, since the polynomial $a_0(x,\xi)$ is homogeneous, and $n_\pm(\xi') = n_\pm(-\xi')$, since it is possible to connect ξ' and $-\xi'$ by a continuous curve in $\mathbb{R}_{n-1}$ not passing through the origin. Hence $n_+(\xi') = n_-(\xi')$.

In the case $n = 2$, $D_1 + iD_2$ is the simplest example of an elliptic operator of odd order.

For $n = 2$ the condition of proper ellipticity is satisfied everywhere if it is satisfied at one point of the boundary. This follows from the connectedness of M.

3. Consider the following boundary problem on the halfline:

$$a_0(x', 0, \xi', D_n)v(t) = 0 \qquad (t = x_n > 0)\,, \tag{1.8}$$

$$b_{j,0}(x', 0, \xi', D_n)v(t)|_{t=0} = h_j \qquad (j = 1, \dots, q)\,, \tag{1.9}$$

where $b_{j,0}$ is the principal symbol of the boundary operator B_j:

$$b_{j,0}(x,\xi) = \sum_{|\beta|=r_j} b_{j,\beta}(x)\xi_\beta\,. \tag{1.10}$$

The boundary problem (1.8), (1.9) is obtained from the boundary problem (1.1) with $f = 0$ by "freezing" the coefficients at a boundary point, omitting the lower order terms, and using the formal Fourier transform $F'_{x'\to\xi'}$.[4]

Denote by $\mathfrak{M}(\xi')$ the space of solutions $v(t)$ of equation (1.8) with $|v(t)| \to 0$ (exponentially) as $t \to \infty$. In view of proper ellipticity, $\dim \mathfrak{M}(\xi')$ is equal to q for $\xi' \neq 0$. For these ξ' and any complex numbers h_j, the boundary problem (1.8), (1.9) is required to have a unique solution in $\mathfrak{M}(\xi')$. This condition is called the *Shapiro–Lopatinskij condition* at the given point of the boundary.[5] As a consequence, the number of boundary operators must be equal to q, the dimension of $\mathfrak{M}(\xi')$.

Thus, the boundary problem (1.1) is elliptic if A is elliptic on M, and properly elliptic, and if thc Shapiro–Lopatinskij condition is satisfied at any point of the boundary. The operator (1.4) will also be called *elliptic* if the boundary problem (1.1) is elliptic.

Note that the ellipticity conditions are invariant with respect to coordinate transformations that preserve the flat part of the boundary. [6] In particular this relates to the Shapiro-Lopatinskij condition. More generally, the ellipticity

---

[4] $v(\xi', t) = F'[u(x', t)] = \int e^{-ix'\cdot\xi'} u(x', t)\, dx'$.

[5] The expressions "complementing condition", "ellipticity of $\{B_j\}$ with respect to A", and "A is covered by $\{B_j\}$" are also used.

[6] Here we do not assume that the coordinates are consistent with the density.

conditions are invariant with respect to diffeomorphisms of the manifold M. This follows, e.g., from the equivalence of the ellipticity conditions and the a priori estimate for solutions in Sobolev norms (see Sect. 2.2 below).

Of course, the invariance of the first condition of ellipticity follows also from the sense of the principal symbol as a function on the cotangent bundle T^*M (see (Agranovich 1990a, Sect. 2.1). The same for the Shapiro–Lopatinskij condition can be deduced from the possibility of a reduction of an elliptic boundary problem to an elliptic system on the boundary (see Sect. 5.2).

In the case $M = \overline{G}$, where G is a bounded domain in $\mathbb{R}^n$, the Shapiro–Lopatinskij condition at a point x_0 of the boundary may be verified in the coordinate system associated with x_0. This means that the coordinate system is obtained from the original Cartesian system in $\mathbb{R}^n$ by the shift of the origin to x_0 and a subsequent rotation around the origin, such that x^n-axis takes the direction of the inner normal to Γ at x_0.

1.3. Other Forms of the Shapiro–Lopatinskij Condition. Here, to simplify the notation, we agree to write $a_0(\xi', \zeta)$ and $b_{j,0}(\xi', \zeta)$ instead of $a_0(x', 0, \xi', \zeta)$ and $b_{j,0}(x', 0, \xi', \zeta)$, respectively. Let $\zeta_1(\xi'), \ldots, \zeta_q(\xi')$ be all the roots of the equation $a_0(\xi', \zeta) = 0$ in $\mathbb{C}_+$ $(0 \neq \xi' \in \mathbb{R}_{n-1})$. Set

$$a_0^+(\xi', \zeta) = (\zeta - \zeta_1(\xi')) \ldots (\zeta - \zeta_q(\xi')) \,. \tag{1.11}$$

Let us divide the polynomial $b_{j,0}(\xi', \zeta)$ (in ζ) by $a_0^+(\xi', \zeta)$ with remainder, and denote the remainder by $\widetilde{b}_{j,0}(\xi', \zeta)$ $(j = 1, \ldots, q)$. It is a polynomial in ζ of degree not greater than $q - 1$:

$$\widetilde{b}_{j,0}(\xi', \zeta) = \sum_{l=1}^{q} \beta_{j,l}(\xi')\zeta^{l-1} \,. \tag{1.12}$$

The boundary problem (1.8), (1.9) in $\mathfrak{M}(\xi')$ is equivalent to the boundary problem

$$a_0^+(\xi', D_n)v(t) = 0 \qquad (t > 0) \,, \tag{1.13}$$

$$\widetilde{b}_{j,0}(\xi', D_n)v(t)|_{t=0} = \widetilde{h}_j \qquad (j = 1, \ldots, q) \,. \tag{1.14}$$

The last boundary problem has a unique solution for any $\widetilde{h}_j$ if and only if conditions (1.14) uniquely determine the Cauchy data

$$v(0),\ D_n v(0), \ldots, D_n^{q-1} v(0) \tag{1.15}$$

and vice versa. From this remark we obtain the following formulation of the Shapiro–Lopatinskij condition: for any $\xi' \in \mathbb{R}_{n-1}$, the polynomials $\widetilde{b}_{j,0}(\xi', \zeta)$ $(j = 1, \ldots, q)$ are linearly independent. An equivalent formulation is: the polynomials $b_{j,0}(\xi', \zeta)$ $(j = 1, \ldots, q)$ are linearly independent modulo $a_0^+(\xi', \zeta)$ for $0 \neq \xi' \in \mathbb{R}_{n-1}$, i.e. no nontrivial linear combination $\sum \alpha_j b_{j,0}(\xi', \zeta)$ is divisible by $a_0^+(\xi', \zeta)$ without remainder.

If $v_l(\xi', t)$ $(l = 1, \ldots, q)$ is a basis in $\mathfrak{M}(\xi')$, then we can write the Shapiro–Lopatinskij condition in the form of the inequality

$$L(\xi') = \det L_{j,l}(\xi') \neq 0 , \tag{1.16}$$

where

$$L_{j,l}(\xi') = b_{j,0}(\xi', D_n) v_l(\xi', t)|_{t=0} \qquad (j, l = 1, \ldots, q) . \tag{1.17}$$

If the roots $\zeta_1(\xi'), \ldots, \zeta_q(\xi')$ are pairwise distinct, then we can choose a basis in the form $e^{i\zeta_l(\xi')t}$ $(l = 1, \ldots, q)$. However, when ξ' varies, multiple roots can appear, therefore it is more convenient to take $v_l(\xi', t)$ in the form of contour integrals. Namely, setting

$$a_0^+(\xi', \zeta) = \sum_{k=0}^{q} c_k(\xi') \zeta^{q-k} , \tag{1.18}$$

we introduce the functions

$$v_l(\xi', t) = \frac{1}{2\pi i} \oint_{\Gamma_+} \frac{e^{i\zeta t} a_l^+(\xi', \zeta)}{a_0^+(\xi', \zeta)} \, d\zeta \qquad (l = 1, \ldots, q) , \tag{1.19}$$

where

$$a_l^+(\xi', \zeta) = \sum_{k=0}^{q-l} c_k(\xi') \zeta^{q-l-k} \qquad (l = 1, \ldots, q) ; \tag{1.20}$$

the contour Γ_+ lies in $\mathbb{C}_+$, surrounds all roots of the polynomial $a_l^+(\xi', \zeta)$ lying there, and has a positive orientation with respect to the domain bounded by it. It is easy to check the relations

$$D_n^{j-1} v_l(\xi', t)|_{t=0} = \delta_{j,l} \qquad (j, l = 1, \ldots, q) , \tag{1.21}$$

from which it is clear that $\{v_l\}$ is a basis in $\mathfrak{M}(\xi')$. The matrix $(L_{j,l}(\xi'))$ with elements (1.17), where $\{v_l\}$ is the basis (1.19), is called the *Lopatinskij matrix* of the boundary problem (1.8)–(1.9). It is easy to see that this matrix coincides with the matrix of the coefficients $\beta_{j,l}(\xi')$ of polynomials (1.12). The determinant $L(\xi')$ of the Lopatinskij matrix is called the *Lopatinskij determinant* of the boundary problem at the given point on Γ.

1.4. Examples. Completely Elliptic Boundary Conditions. The simplest example of an elliptic boundary problem for an elliptic operator of order $2q$ is the *Dirichlet problem*

$$Au = f \quad \text{on} \quad M_+ , \quad \gamma_{j-1} u = g_j \quad (j = 1, \ldots, q) \quad \text{on} \quad \Gamma . \tag{1.22}$$

The Shapiro–Lopatinskij condition is here obviously satisfied, since the polynomials ζ^{j-1} $(j = 1, \ldots, q)$ are linearly independent.

The system of boundary operators $B_1, \ldots, B_q$ is called *completely elliptic* (or absolutely elliptic) if the boundary problem (1.1) is elliptic for any properly elliptic operator A of order $2q$. Such systems have been considered in

(Hörmander 1958), (Agmon 1962), and (Senator 1967). For complete ellipticity it is necessary and sufficient that, at any point of the boundary, each nontrivial linear combination of the polynomials $b_{j,0}(x', 0, \xi', \zeta)$ have at most $q-1$ roots in the upper halfplane for every $\xi' \in \mathbb{R}_{n-1} \setminus 0$. Of course, the system of the boundary operators of the Dirichlet problem is an example of such a system. In the second of the papers indicated above, the following more general example of a completely elliptic boundary problem is mentioned:

$$B_j = \gamma_0 \partial_\mu^{s+j-1} \qquad (j = 1, \ldots, q) ,$$

where $0 \leq s \leq q$ and ∂_μ is the derivative in a non-tangential smoothly variable direction $\mu = \mu(x)$. In particular, in the case $q = 1$ the boundary operator γ_1 of the *Neumann problem*

$$Au = f \quad \text{on} \quad M_+ , \qquad \gamma_1 u = g \quad \text{on} \quad \Gamma, \tag{1.23}$$

is completely elliptic. It is also clear that any completely elliptic system admits the addition of arbitrary lower order terms.

We add that if $n \geq 3$, then for the classical oblique derivative problem

$$\Delta u = f \quad \text{in} \quad G, \qquad \gamma_0 \partial_\mu u = g, \tag{1.24}$$

where ∂_μ is the derivative in a direction $\mu(x)$, the Shapiro–Lopatinskij condition is not satisfied at those, and only those, points where this direction is tangent to the boundary.

1.5. Uniformly Elliptic Boundary Problems in a Halfspace. Here and below we denote by $\mathbb{R}^n_+$ the halfspace $\{x \in \mathbb{R}^n : x^n > 0\}$. Consider the boundary problem

$$Au = f \quad \text{in} \quad \mathbb{R}^n_+ , \quad B_j u = g_j \quad (j = 1, \ldots, q) \quad \text{on} \quad \mathbb{R}^{n-1} = \partial\mathbb{R}^n_+ . \tag{1.25}$$

Here A is a differential operator of the form (1.2) in $\overline{\mathbb{R}^n_+}$ with coefficients belonging to $B^\infty(\overline{\mathbb{R}^n_+})$ (the space of C^∞ functions on $\overline{\mathbb{R}^n_+} = \{x \in \mathbb{R}^n : x^n \geq 0\}$ bounded along with their derivatives up to any fixed order); the B_j are boundary operators of the form (1.3) with coefficients belonging to $B^\infty(\overline{\mathbb{R}^n_+})$ or $B^\infty(\mathbb{R}^{n-1})$ (the space of C^∞ functions on $\mathbb{R}^{n-1}$ bounded with their derivatives up to any fixed order). For the time being we assume that $u(x) \in C^\infty(\overline{\mathbb{R}^n_+})$.

The ellipticity conditions for (1.25) are formulated in an obvious way. This boundary problem is called *uniformly elliptic* if

$$|a_0(x, \xi)| \geq C_1 > 0 \quad \text{for} \quad x \in \mathbb{R}^n_+ , \ |\xi| = 1$$

and

$$|L(x', 0, \xi')| \geq C_2 > 0 \quad \text{for} \quad x' \in \mathbb{R}^{n-1}, \ |\xi'| = 1 .$$

Here a_0 is the principal symbol of A, and L is the Lopatinskij determinant. We will touch on uniformly elliptic boundary problems, in passing, in §2.

A boundary problem in $\mathbb{R}^n_+$ with constant $a_\alpha(x) = a_\alpha$ that can be distinct from zero only for $|\alpha| = m$, and constant $b_{j,\beta}(x) = b_{j,\beta}$ that can be distinct from zero only for $|\beta| = r_j$, is called a *model boundary problem*. If it is uniformly elliptic, then it is called an elliptic model boundary problem.

§2. Elliptic Boundary Problems in Sobolev Spaces

2.1. Sobolev Spaces H_s. Here we recall some elementary definitions and assertions related to the Sobolev L_2-spaces in $\mathbb{R}^n_+$ and on a manifold with boundary, in addition to the appropriate information concerning the analogous spaces in $\mathbb{R}^n$ and on a closed manifold, which was included in (Agranovich 1990a). See also Sect. 7.9 below.

a. The space $H_s(\mathbb{R}^n_+)$ $(s \geq 0)$ can be defined as a completion of $C_0^\infty(\overline{\mathbb{R}^n_+})$ with respect to the norm

$$\|u\|_s = \|u\|_{s,\mathbb{R}^n_+} = \left\{ \sum_{|\alpha| \leq s} \int_{\mathbb{R}^n_+} |D^\alpha u(x)|^2 dx \right\}^{1/2} \tag{2.1}$$

if s is an integer, and with respect to the norm

$$\begin{aligned} \|u\|_s &= \|u\|_{s,\mathbb{R}^n_+} \\ &= \left\{ \|u\|^2_{[s],\mathbb{R}^n_+} + \sum_{|\alpha|=[s]} \int_{\mathbb{R}^n_+} \int_{\mathbb{R}^n_+} \frac{|D^\alpha u(x) - D^\alpha u(y)|^2}{|x-y|^{n+2(s-[s])}} \, dx\, dy \right\}^{1/2} \end{aligned} \tag{2.2}$$

if s is noninteger and $[s]$ is its integral part. These norms can be expressed in terms of the Fourier transform $F'[u(x)] = (F'u)(\xi', x^n)$. All $H_s(\mathbb{R}^n_+)$ $(s \geq 0)$ are Hilbert spaces; the corresponding scalar products can easily be written. In particular, $H_0(\mathbb{R}^n_+)$ coincides with $L_2(\mathbb{R}^n_+)$. For $|\alpha| \leq s$ the derivatives $D^\alpha u(x)$ of a function $u(x) \in H_s(\mathbb{R}^n_+)$, in the sense of distributions, belong to $H_{s-|\alpha|}(\mathbb{R}^n_+)$, and the norms (2.1) and (2.2) retain their meaning for functions in $H_s(\mathbb{R}^n_+)$. Moreover, D^α is a bounded operator from $H_s(\mathbb{R}^n_+)$ to $H_{s-|\alpha|}(\mathbb{R}^n_+)$ if $|\alpha| \leq s$.

If u is a function defined on $\mathbb{R}^n$, we denote by $\mathcal{R}u$ its restriction to $\mathbb{R}^n_+$. Obviously, $\mathcal{R}$ is a bounded operator from $H_s(\mathbb{R}^n)$ to $H_s(\mathbb{R}^n_+)$. We now construct a bounded *extension operator* $\mathcal{E}$ following (Seeley 1964): if u is a function on $\mathbb{R}^n_+$, then $\mathcal{E}u$ will be a function on $\mathbb{R}^n$ coinciding with u on $\mathbb{R}^n_+$. It can be shown that there exist two sequences $\{a_k\}_0^\infty$ and $\{b_k\}_0^\infty$ of real numbers such that $b_k > 0$,

$$\sum_k |a_k|\, b_k^l < \infty, \quad \text{and} \quad \sum_k a_k(-b_k)^l = 1 \qquad (l = 0, 1, \ldots)\,. \tag{2.3}$$

Namely, if we take $b_k = 2^{-k}$ and define $a_{k,N}$ by the equations

$$\sum_{k=0}^{N} a_{k,N}(-b_k)^l = 1 \qquad (l = 0, \ldots, N) ,$$

then, as can be checked, there exist limits $a_k = \lim_{N\to\infty} a_{k,N}$ with properties (2.3). Let $\theta_0(t)$ be a function from $C_0^\infty(\overline{\mathbb{R}}_+)$ equal to 1 near 0. We set

$$\mathcal{E}u(x) = \mathcal{E}_{\mathbb{R}^n_+}u(x) = \begin{cases} u(x) & \text{for } x^n \geq 0 , \\ \sum_{k=0}^\infty a_k\theta_0(-x^n/b_k)u(x', -b_kx^n) & \text{for } x^n < 0 . \end{cases} \tag{2.4}$$

The last series consists of a finite number of terms for any $x^n < 0$. Seeley shows that if $u \in C_0^\infty(\overline{\mathbb{R}^n_+})$, then $\mathcal{E}u \in C_0^\infty(\mathbb{R}^n)$ and

$$\|\mathcal{E}u\|_{s,\mathbb{R}^n} \leq C_s\|u\|_{s,\mathbb{R}^n_+} \tag{2.5}$$

for all s. Hence $\mathcal{E}$ has an extension to a bounded operator from $H_s(\mathbb{R}^n_+)$ to $H_s(\mathbb{R}^n)$ for all s.

It follows that $H_s(\mathbb{R}^n_+)$ $(s \geq 0)$ may be defined as the space of restrictions u of functions $v \in H_s(\mathbb{R}^n)$ to $\mathbb{R}^n_+$, with norm $\inf \|v\|_{s,\mathbb{R}^n}$, where inf is taken over all $v \in H_s(\mathbb{R}^n)$ equal to u in $\mathbb{R}^n_+$ (almost everywhere). This norm is equivalent to (2.1) or (2.2) for integer and noninteger s, respectively.

This permits us to derive some properties of the spaces $H_s(\mathbb{R}^n_+)$ from the corresponding properties of the spaces $H_s(\mathbb{R}^n)$. First, we obtain the embedding theorem: for $s > l + (n/2)$ with $l \in \mathbb{Z}_+$, the space $H_s(\mathbb{R}^n_+)$ is embedded in the space $B^{(l)}(\overline{\mathbb{R}^n_+})$ of functions that are continuous and bounded in $\overline{\mathbb{R}^n_+}$ along with their derivatives up to order l. The norm in $B^{(l)}(\overline{\mathbb{R}^n_+})$ is defined in an obvious way.

Furthermore, we obtain the *interpolation inequality*: if $0 < s < s'$, then

$$\|u\|_{s,\mathbb{R}^n_+} \leq \varepsilon\|u\|_{s',\mathbb{R}^n_+} + C_\varepsilon\|u\|_{0,\mathbb{R}^n_+} \tag{2.6}$$

for any $\varepsilon > 0$.

b. By means of the Fourier transform it can be easily checked that for functions from $C_0^\infty(\mathbb{R}^n)$ (or $S(\mathbb{R}^n)$) the following inequality is true:

$$\|u(x', 0)\|_{s-\frac{1}{2},\mathbb{R}_{n-1}} \leq C'_s\|u(x)\|_{s,\mathbb{R}^n} \quad (s > 1/2) . \tag{2.7}$$

It can be shown by examples that such an inequality is not true for $s \leq 1/2$. Using (2.7), we can define the trace $u(x', 0) \in H_{s-\frac{1}{2}}(\mathbb{R}^{n-1})$ of a function $u \in H_s(\mathbb{R}^n)$ on the hyperplane $\mathbb{R}^{n-1} = \{x \in \mathbb{R}^n : x^n = 0\}$ as a limit in $H_{s-\frac{1}{2}}(\mathbb{R}^{n-1})$ of the traces $u_\nu(x', 0)$ of functions $u_\nu(x) \in C_0^\infty(\mathbb{R}^n)$ converging to $u(x)$ in $H_s(\mathbb{R}^n)$ as $\nu \to \infty$. Inequality (2.7) remains true for $u \in H_s(\mathbb{R}^n)$.

Using the operator $\mathcal{E}$, we see that for functions $u(x) \in C^\infty(\overline{\mathbb{R}^n_+})$ we have

$$\|u(x', 0)\|_{s-\frac{1}{2},\mathbb{R}^{n-1}} \leq C''_s\|u(x)\|_{s,\mathbb{R}^n_+} \qquad (s > 1/2) . \tag{2.8}$$

For $s > 1/2$, we now can define the boundary value, or the trace,

$$(\gamma_0 u)(x') = u(x', 0) \tag{2.9}$$

of a function $u(x) \in H_s(\mathbb{R}^n_+)$. This boundary value belongs to $H_{s-\frac{1}{2}}(\mathbb{R}^{n-1})$, and inequality (2.8) remains true for functions $u(x) \in H_s(\mathbb{R}^n_+)$.

From this it follows that for $s > q + (1/2)$ with $q \in \mathbb{N}$ the operator

$$\gamma^{(q)} = (\gamma_0, \ldots, \gamma_q)', \qquad (\gamma_j u)(x') = (D_n^j u)(x', 0)\ , \tag{2.10}$$

is defined on $H_s(\mathbb{R}^n_+)$ and is bounded from this space to $\prod_0^q H_{s-j-\frac{1}{2}}(\mathbb{R}^{n-1})$. This operator and the similar operator from $H_s(\mathbb{R}^n)$ to $\prod_0^q H_{s-j-\frac{1}{2}}(\mathbb{R}^{n-1})$ have right inverses. More precisely, let $\psi_k(t)$ $(k = 1, \ldots, q)$ be functions in $C_0^\infty(\mathbb{R})$ such that

$$(D_n^j \psi_k)(0) = \delta_k^j \qquad (j,\, k = 0, \ldots, q)\ . \tag{2.11}$$

For $g_k \in H_{s-k-\frac{1}{2}}(\mathbb{R}^{n-1})$ $(k = 0, \ldots, q)$, set

$$\begin{aligned} v_q(x) &= E_q(g_1, \ldots, g_q) \\ &= \sum_{k=0}^{q} F'^{-1}_{\xi' \mapsto x'} [\psi_k(x^n(1 + |\xi'|^2)^{1/2})(F' g_k)(\xi')(1 + |\xi'|^2)^{-k/2}]\ . \end{aligned} \tag{2.12}$$

Then E_q is a bounded operator from $\prod_0^q H_{s-k-\frac{1}{2}}(\mathbb{R}^{n-1})$ to $H_s(\mathbb{R}^n)$, the function $v_q(x)$ belongs to C^∞ for $x^n \neq 0$, and

$$\lim_{x^n \to 0} D_n^k v_q(x) = g_k(x') \qquad (k = 0, \ldots, q) \tag{2.13}$$

in the sense of convergence in $H_{s-k-\frac{1}{2}}(\mathbb{R}^{n-1})$, where $s \geq q + (1/2)$. See (Slobodetskij 1958). A different construction of an operator with such properties for any real s is indicated in (Seeley 1966), where more general spaces of functions are considered.

c. If $\varphi(x)$ is a function from $C_0^\infty(\mathbb{R}^n)$ or $C_0^\infty(\overline{\mathbb{R}^n_+})$, then the operator of multiplication by this function is bounded in $H_s(\mathbb{R}^n)$ or in $H_s(\mathbb{R}^n_+)$, respectively. This enables us to introduce the Sobolev norms on a C^∞ manifold using a partition of unity. Assuming the manifold M to be compact, we fix a finite covering of M by coordinate neighborhoods O and semi-neighborhoods O^+, and a partition of unity

$$\sum_1^K \varphi_k(x) \equiv 1 \tag{2.14}$$

on M subordinated to the covering. Let the support of φ_k lie in O_k for $1 \leq k \leq K'$ and in O_k^+ for $K' < k \leq K$. It is convenient to assume that these O_k^+ lie on a collar of the manifold M. Now for $u \in C^\infty(M)$ and $s \geq 0$ we set

$$\|u\|_s = \|u\|_{s,M} = \left\{ \sum_1^{K'} \|\varphi_k u\|^2_{s,\mathbb{R}^n} + \sum_{K'+1}^{K} \|\varphi_k u\|^2_{s,\mathbb{R}^n_+} \right\}^{1/2} , \tag{2.15}$$

where all norms are calculated in local coordinates. The space $H_s(M)$ is defined as the completion of $C^\infty(M)$ with respect to this norm. The norms corresponding to different coverings, choices of local coordinates, and partitions of unity are equivalent. Note that instead of (2.14) it suffices to assume that $\sum \varphi_k(x) > 0$ on M. If (2.14) is valid and if the local coordinates are consistent with the given density dx on M (see Sect. 0.2), then the scalar product $(u, v) = (u, v)_{0,M}$ corresponding to the norm $\|\cdot\|_{0,M}$ coincides with the integral of $u \cdot \overline{v}$ with respect to this density:

$$(u, v) = \int_M u \cdot \overline{v}\, dx . \tag{2.16}$$

All our assertions concerning the spaces $H_s(\mathbb{R}^n)$, $H_s(\mathbb{R}^n_+)$, and $H_s(\mathbb{R}^{n-1})$ can easily be carried over to the spaces $H_s(M_0)$, $H_s(M)$, and $H_s(\Gamma)$. We will identify the spaces $H_\infty(M) = \bigcap H_s(M)$ and $H_\infty(\Gamma)$ with $C^\infty(M)$ and $C^\infty(\Gamma)$, respectively. The analogs of the assertions in Subsect. 2.1b are formulated using an operator of the form (2.4) for M and M_0 constructed in a tubular neighborhood of the boundary. In addition, for $s < s'$ the embedding $H_{s'}(M) \subset H_s(M)$ is compact. This follows from the compactness of the embedding $H_{s'}(M_0) \subset H_s(M_0)$.

If $M = \overline{G}$, where G is a bounded domain in $\mathbb{R}^n$ with smooth boundary, then the norms in $H_s(M) = H_s(G)$ can be defined by formulas (2.1) and (2.2) with integration over G instead of $\mathbb{R}^n_+$.

d. In (Agranovich 1990a) we formulated analogs of Agmon's theorems on integral representation of operators of sufficiently low negative order on a closed manifold. In (Agmon 1965) such theorems were proved for operators in a domain of $\mathbb{R}^n$. Similar theorems hold for operators on a manifold M.

Theorem 2.1.1. *Let T be a bounded operator in $H_0(M)$, and let T^* be its adjoint. Assume that T and T^* are also bounded operators from $H_0(M)$ to $H_s(M)$, where $s > n$. Then T is an integral operator of the form*

$$Tu(x) = \int_M K(x, y) u(y)\, dy , \tag{2.17}$$

where the kernel $K(x, y)$ is continuous on $M \times M$. In addition,

$$|K(x, y)| \le C(\|T\|_{0,s} + \|T^*\|_{0,s})^{\frac{n}{2s}} \|T\|_{0,0}^{1-\frac{n}{2s}} , \tag{2.18}$$

where $C = C(n, s, M)$.

Here and below $\|T\|_{0,t}$ is the norm of the operator $T\colon H_0(M) \to H_t(M)$.

e. In (Agranovich 1990a) we presented an exact estimate of s-numbers of an operator T of negative order[7] on a closed manifold. A similar theorem holds for operators on M:

Theorem 2.1.2. *Let T be a bounded operator from $H_s(M)$ to $H_{s+t}(M)$ for some $t > 0$, $s \geq 0$. Then for T as an operator in $H_s(M)$ we have*

$$s_\nu(T) = O(\nu^{-t/n}) . \tag{2.19}$$

In order to derive this estimate, we represent T in the form $\mathcal{R}(\mathcal{E}T\mathcal{R})\mathcal{E}$, where $\mathcal{E}$ is an operator of a smooth extension of functions on M to functions on M_0 (an analog of the operator (2.4) for M and M_0) and $\mathcal{R}$ is the operator of the restriction of functions from M_0 to M. Now it suffices to apply the above-mentioned estimate for operators on a closed manifold to $\mathcal{E}T\mathcal{R}$.

In particular, if $t > n/2$, then T is a Hilbert–Schmidt operator, i.e. $\sum s_\nu^2(T) < \infty$. For $s = 0$ this is equivalent to the possibility of representing T in the form (2.17) with $K \in L_2(M \times M)$. If $t > n$, then T is a trace class operator, i.e. $\sum s_\nu(T) < \infty$. In the last case T has a trace (the sum of the absolutely convergent series consisting of all eigenvalues, counted with multiplicities). The following assertion is well known (see e.g. (Duistermaat 1981)).

Theorem 2.1.3. *Let (2.17) be a trace class operator in $H_0(M)$, and let its kernel $K(x, y)$ be continuous on $M \times M$. Then*

$$\operatorname{tr} T = \int_M K(x, x)\, dx . \tag{2.20}$$

In particular, this formula holds under the assumptions of Theorem 2.1.1.

f. In Sects. 5.4, 6.5, and 7.1 we will need Sobolev spaces $H_s(M_\pm)$ of negative order. Here we follow e.g. (Triebel 1978) and define $H_s(M_\pm)$ as the space of restrictions u to $M_\pm$ of all $w \in H_s(M_0)$, with $\|u\|_{s,M_\pm} = \inf \|w\|_{s,M_0}$, where the infimum is taken over all $w \in H_s(M_0)$ such that $w = u$ in $M_\pm$ a.e. If $M_+ = G$ is a bounded domain in $\mathbb{R}^n$ with a smooth boundary, then $\mathbb{R}^n$ is used instead of M_0. For $s < 0$ such that $s - \frac{1}{2}$ is not an integer, an equivalent definition is $H_s(M_\pm) = \left[\overset{\circ}{H}_{|s|}(M_\pm)\right]^*$ (the dual space with respect to the extension of the scalar product in $M_\pm$), where $\overset{\circ}{H}_{|s|}(M_\pm)$ is the completion of $C_0^\infty(M_\pm)$ in $H_{|s|}(M)$ (see Triebel 1978). Somewhat different definition will be accepted in Sect. 7.9 (written by Ya.A. Roĭtberg).

In Sects. 7.2 and 9.8 we will need Sobolev spaces $H_s(G)$ with $s = 1, 2 \ldots$ in bounded domains $G \subset \mathbb{R}^n$ 1) with arbitrary boundaries, 2) with Lipschitz boundaries (see the definition in Sect. 7.2). Here we follow (Agmon 1965)

---

[7] I.e. of eigenvalues of the nonnegative operator $(T^*T)^{1/2}$ numbered in the nonincreasing order, taking multiplicities into account.

or (Adams 1975) and define $H_s(G)$ as the completion of $\{u \in C^{(s)}(G) : \|u\|_{s,G} < \infty\}$ with respect to the norm $\| \ \|_{s,G}$ of the form (2.1) with G instead of $\mathbb{R}^n_+$. In the second case this definition is equivalent to the definition of $H_s(G)$ as the space of restrictions u of functions $w \in H_s(\mathbb{R}^n)$ to G (with $\|u\|_{s,G} = \inf \|w\|_{s,\mathbb{R}^n}$, $w = u$ a.e. in G), since in this case there exists a bounded extension operator $\mathcal{E} : H_s(G) \to H_s(\mathbb{R}^n)$ (with $\mathcal{E}u = u$ a.e. in G) constructed by Calderón. See e.g. (Adams 1975) and references therein.

2.2. A Priori Estimate.

a. Let us return to the boundary problem (1.1). Assume that

$$s \geq m = 2q \quad \text{and} \quad s > r_j + \frac{1}{2} \quad (j = 1, \ldots, q) . \tag{2.21}$$

Set

$$H_{\{s-r_j-\frac{1}{2}\}}(\Gamma) = H_{s-r_1-\frac{1}{2}}(\Gamma) \times \ldots \times H_{s-r_q-\frac{1}{2}}(\Gamma) \tag{2.22}$$

and

$$H_s(M, \Gamma) = H_{s-m}(M) \times H_{\{s-r_j-\frac{1}{2}\}}(\Gamma) . \tag{2.23}$$

From what has been said in the previous section, it is clear that the operator (1.4) corresponding to this boundary problem is bounded from $H_s(M)$ to $H_s(M, \Gamma)$, i.e. the inequality

$$\|Au\|_{s-m,M} + \sum_{j=1}^{q} \|B_j\|_{s-r_j-\frac{1}{2},\Gamma} \leq C_s \|u\|_{s,M} \tag{2.24}$$

holds.

Theorem 2.2.1. *Assume that the boundary problem* (1.1) *is elliptic, and s satisfies inequalities* (2.21)*. Then the inequality*

$$\|u\|_{s,M} \leq C'_s \left\{ \|Au\|_{s-m,M} + \sum_{j=1}^{q} \|B_j\|_{s-r_j-\frac{1}{2},\Gamma} + \|u\|_{0,M} \right\} \tag{2.25}$$

holds for functions $u \in H_s(M)$.

We indicate the main steps of a possible proof; technically it apparently is the simplest one. It suffices to obtain (2.25) for functions $u \in C^\infty(M)$.

1. Consider a model elliptic boundary problem in $\mathbb{R}^n_+$. For the function $v(\xi', x^n) = F^{-1}_{x' \mapsto \xi'} u$, where $u \in C_0^\infty(\overline{\mathbb{R}^n_+})$, explicit formulas can be derived. Namely, we set

$$v_1(\xi', x^n) = F^{-1}_{\xi_n \mapsto x^n} a_0^{-1}(\xi) F_{x \mapsto \xi}(\mathcal{E}_{\mathbb{R}^n_+} u) , \qquad v_2 = v - v_1$$

and obtain a boundary problem of the form (1.8), (1.9) for v_2 on the halfline $\mathbb{R}_+ = \{x^n : x^n > 0\}$. For $\xi' \neq 0$ the solution is expressed using the basis (1.19) and the matrix inverse to the Lopatinskij matrix (cf. formulas (2.33) and (2.34) below). Assume for simplicity that s is an integer, and consider the seminorm

$$\left(\sum_{|\alpha|=s}\int_{\mathbb{R}_+^n}|D^\alpha u(x)|^2dx\right)^{1/2}.$$

Taking into account the homogeneity properties of the functions in the expression for v_2, we can estimate this seminorm in terms of $\|Au\|_{s-m,\mathbb{R}_+^n}$ and $\|B_ju\|_{s-r_j-\frac{1}{2},\mathbb{R}^{n-1}}$. Adding $\|u\|_{0,\mathbb{R}_+^n}$ to both sides, we obtain the inequality

$$\|u\|_{s,\mathbb{R}_+^n}\leq C_s''\left\{\|Au\|_{s-m,\mathbb{R}_+^n}+\sum_{j=1}^q\|B_ju\|_{s-r_j-\frac{1}{2},\mathbb{R}^{n-1}}+\|u\|_{0,\mathbb{R}_+^n}\right\}. \tag{2.26}$$

2. Consider a boundary problem in $\mathbb{R}_+^n$ that is elliptic at the point $x=0$ (ellipticity at a point is defined in the obvious way). Inequality (2.26) extends to solutions of this boundary problem supported in a sufficiently small semi-neighborhood $O^+=\{x:|x|<\delta,\ x^n\geq 0\}$ of the origin. The properties of the Sobolev norms are used, including the interpolation inequality (2.6). The lower order terms may be arbitrary; however, δ depends on the given higher order terms (they must be almost constant in O^+).

3. The final result is obtained by means of a sufficiently fine partition of unity on M.

The "localization method" used in this proof (see also Sect. 2.3 below) is typical of many papers on elliptic equations, see e.g. (Agmon et al. 1959, 1964).

Remark 2.2.2. If uniqueness holds for the boundary problem (1.1), then we may drop the term $\|u\|_{0,M}$ in the right-hand side of (2.25).

Theorem 2.2.3. *Assume that inequality* (2.25) *is valid for functions $u(x)$ in $C^\infty(M)$ and some s satisfying inequalities* (2.21). *Then the boundary problem* (1.1) *is elliptic.*

The ellipticity of the operator A at interior points $x_0\in M_+$ follows from the corresponding theorem for pseudodifferential operators in $\mathbb{R}^n$ (see Theorem 1.8.7 in (Agranovich 1990a)). To verify the ellipticity conditions at a boundary point $x_0\in\Gamma$, we use the way of the proof of the a priori estimate but in the opposite direction, and show that an estimate of the form (2.25) in $\mathbb{R}_+^n$ holds for the appropriate model boundary problem on functions supported in a sufficiently small semi-neighborhood of this point. After this, the ellipticity of the model boundary problem is verified by contradiction, and for this appropriate sequences of functions $u(x)$ are constructed.

Remark 2.2.4. Assertions similar to those stated above are valid for uniformly elliptic boundary problems in $\mathbb{R}_+^n$.

Remark 2.2.5. Using Theorems 2.2.1 and 2.2.3, we can easily verify the ellipticity of a composition of elliptic boundary problems. Namely, consider, besides (1.1), the second boundary problem

$$\widetilde{A}v = f \quad \text{on} \quad M_+ , \quad \widetilde{B}_j v = \widetilde{g}_j \quad (j = 1, \dots, \widetilde{q}) \quad \text{on} \quad \Gamma . \tag{2.27}$$

Setting $v = Au$, we obtain the boundary problem

$$\widetilde{A}Au = f \quad \text{on} \quad M_+ , \tag{2.28}$$

$$B_j u = g_j \quad (j = 1, \dots, q) \quad \text{and} \quad \widetilde{B}_j Au = \widetilde{g}_j \quad (j = 1, \dots, \widetilde{q}) \quad \text{on} \quad \Gamma . \tag{2.29}$$

This boundary problem is elliptic if the boundary problems (1.1) and (2.27) are elliptic.

2.3. Right Parametrix.

Theorem 2.3.1. *Assume that the boundary problem* (1.1) *is elliptic, and that s satisfies conditions* (2.21). *Then there exists a right parametrix $\mathcal{B}$ for the operator $\mathcal{A}$ corresponding to this boundary problem. More precisely,*

$$\mathcal{A}\mathcal{B} = \mathcal{I} + \mathcal{T} , \tag{2.30}$$

where $\mathcal{I}$ is the identity operator in $H_s(M, \Gamma)$ and $\mathcal{T}$ is a bounded operator from this space to $H_{s+1}(M, \Gamma)$.

Here $\mathcal{B}$ is a "rough" right parametrix (cf. Theorem 2.4.5 below). As in the proof of the a priori estimate, in order to construct $\mathcal{B}$ we make three steps.

1. First we construct a right parametrix $\mathcal{B}_0^+$ for the operator $\mathcal{A}_0$ corresponding to a model elliptic boundary problem. We write it in the form

$$a_0(D)u(x) = f(x) \quad \text{in} \quad \mathbb{R}^n_+ , \quad b_{j,0}(D)u(x)|_{x^n=0} = g_j(x') \quad (j = 1, \dots, q) . \tag{2.31}$$

Here $a_0(\xi)$ and $b_{j,0}(\xi)$ are homogeneous polynomials of order m and r_j, respectively. We set

$$\mathcal{B}_0^+(f, g) = u_0 + u_1 , \tag{2.32}$$

where

$$u_0(x) = F^{-1} \frac{|\xi|^m}{1 + |\xi|^m} \frac{1}{a_0(\xi)} F \mathcal{E}_{\mathbb{R}^n_+} f \tag{2.33}$$

and

$$u_1(x) = F'^{-1} \sum_{j,k=1}^{q} \frac{|\xi'|^{r_j+1}}{1 + |\xi'|^{r_j+1}} L^{k,j}(\xi') F' [g_j - b_{j,0}(D)u_0|_{x^n=0}](\xi') v_k(\xi', x^n) \tag{2.34}$$

(see (1.19)). The function $u_0(x)$ is an "almost solution" of the equation $Au = f$. The fraction before $1/a_0(\xi)$ "cancels" the singularity of this function at $\xi = 0$ and is equal to $1 + O(|\xi|^{-m})$ at infinity. The right-hand side in (2.33) is assumed to be restricted to $\mathbb{R}^n_+$. In (2.34), $L^{k,j}(\xi')$ are the elements of the matrix inverse to the Lopatinskij matrix. The fractions before them cancel the singularities of the functions $L^{j,k}v_k$ at $\xi' = 0$ and are equal to $1 + O(|\xi'|^{-r_j - 1})$. It is easy to verify that $\mathcal{B}_0^+$ is a bounded operator from $H_s(\mathbb{R}^n_+, \mathbb{R}^{n-1})$ into $H_s(\mathbb{R}^n_+)$, and that $\mathcal{A}_0\mathcal{B}_0^+ = \mathcal{I}_0 + \mathcal{T}_0$, where $\mathcal{I}_0$ is the identity operator in $H_s(\mathbb{R}^n_+, \mathbb{R}^{n-1})$ and $\mathcal{T}_0$ is a bounded operator from this space to $H_{s+1}(\mathbb{R}^n_+, \mathbb{R}^{n-1})$ (these spaces are defined by a formula similar to (2.23)).

2. For a boundary problem in $\mathbb{R}^n_+$ elliptic at the point $x = 0$, a local right parametrix is constructed, i.e. an operator $\mathcal{B}_+$ such that $\varphi\mathcal{A}\mathcal{B}^+ = \varphi(\mathcal{I}_0 + \mathcal{T}_1)$ if φ is a function from $C_0^\infty(\overline{\mathbb{R}^n_+})$ supported in sufficiently small semi-neighborhood of the origin. Here $\mathcal{A}$ is the operator corresponding to the boundary problem and $\mathcal{T}_1$ is a bounded operator from $H_s(\mathbb{R}^n_+, \mathbb{R}^{n-1})$ to $H_{s+1}(\mathbb{R}^n_+, \mathbb{R}^{n-1})$.

3. Finally, we construct a right parametrix for the operator $\mathcal{A}$ corresponding to the original boundary problem, in the form

$$\mathcal{B} = \sum_1^{K'} \psi_k \mathcal{B}_k(\varphi_k \,\cdot\,) + \sum_{K'+1}^{K} \psi_k \mathcal{B}_k^+(\varphi_k \,\cdot\,)\,. \tag{2.35}$$

Here $\psi_k \in C^\infty(M)$, and $\{\varphi_k\}$ is a sufficiently fine partition of unity defined in Sect. 2.1; in the same notation as there, $\operatorname{supp}\psi_k$ lies in O_k for $k \le K'$ and in O_k^+ for $k > K'$; in addition, $\psi_k = 1$ in a neighborhood of $\operatorname{supp}\varphi_k$ if $k \le K'$, and in the intersection of M with such a neighborhood if $k > K'$. If $k \le K'$, then $\mathcal{B}_k$ is a parametrix for A in O_k: in local coordinates we can set

$$\mathcal{B}_k = F^{-1}_{\xi\mapsto x} \frac{|\xi|^m}{1 + |\xi|^m} \frac{1}{a_0(x,\xi)} F_{x\mapsto\xi}\,;$$

and if $k \ge K'$, then $\mathcal{B}_k^+$ is a local parametrix for $\mathcal{A}$ in O_k^+ constructed in the second step of the proof. For details, see e.g. (Agranovich 1965), where more general boundary problems containing pseudodifferential operators are considered.

We also note that the parametrix $\mathcal{B}$ may be assumed to be independent of s, at least on any segment $[s_0, s_1]$, and that a theorem similar to Theorem 2.3.1 holds for uniformly elliptic boundary problems in $\mathbb{R}^n_+$.

2.4. The Fredholm Property of Elliptic Operators, and Smoothness of Solutions.

a. We recall the definition. Let X_1 and X_2 be Banach spaces and A be a bounded operator from X_1 to X_2. Then A is called a *Fredholm operator* if the following three conditions hold: 1) The kernel $\operatorname{Ker} A = \{u \in X_1 : Au = 0\}$ is finite-dimensional, 2) the range $R(A)$ is closed in X_2, and 3) the cokernel $\operatorname{Coker} A = X_2/R(A)$ is finite-dimensional. The difference

$$\kappa(A) = \dim \operatorname{Ker} A - \dim \operatorname{Coker} A \tag{2.36}$$

is called the *index* of the Fredholm operator A.

Theorem 2.4.1. *Let* $\mathcal{A}: H_s(M) \to H_s(M,\Gamma)$ *be the operator that corresponds to the boundary problem* (1.1). *Then* $\mathcal{A}$ *is a Fredholm operator if and only if this boundary problem is elliptic. Here* s *is assumed to satisfy inequalities* (2.21).

Indeed, as we have seen in Sect. 2.2, ellipticity of the boundary problem is equivalent to validity of a priori estimate (2.25). Conditions 1) and 2) follow from it, and conditions 2) and 3) follow from the existence of a right parametrix. Here it is important that the embedding $H_{s'}(M) \subset H_s(M)$ is compact for $s' > s$: in view of this, in particular, in (2.30) $\mathcal{T}$ is a compact operator in $H_s(M,\Gamma)$. It remains to note that if $\mathcal{A}$ is Fredholm, then the estimate (2.25) is true. Here we use some abstract theorems on Fredholm operators; they are stated, e.g., in (Agranovich 1990a, Sect. 2.3).

Additionally, we find that the right parametrix is also a *left parametrix*:

$$\mathcal{B}\mathcal{A} = I + T\,, \tag{2.37}$$

where I is the identity operator in $H_s(M)$ and T is a bounded operator from $H_s(M)$ to $H_{s+1}(M)$. Using this left parametrix, we obtain the assertion on improved smoothness of solutions:

Theorem 2.4.2. *Let the boundary problem* (1.1) *be elliptic,* $u \in H_s(M)$, *and* $\mathcal{A}u \in H_{s'}(M)$, *where* s *satisfies inequalities* (2.21) *and* $s' > s$. *Then* $u \in H_{s'}(M)$.

In particular, $\operatorname{Ker}\mathcal{A}$ *belongs to* $H_\infty(M)$, *i.e., in essence, it belongs to* $C^\infty(M)$, *and does not depend on* s.

A similar result is true for uniformly elliptic boundary problems in $\mathbb{R}^n_+$.

Theorem 2.4.3. *Let the boundary problem* (1.1) *be elliptic. Then for all* s *satisfying inequalities* (2.21) *the index of the operator* $\mathcal{A}: H_s(M) \to H_s(M,\Gamma)$ *corresponding to this boundary problem does not depend on* s.

To verify this, it suffices to prove that $\dim \operatorname{Coker}\mathcal{A}$ does not depend on s. For this we apply the following known assertion (see e.g. Lemma 2.1 in (Gohberg and Kreĭn, 1957)).[8]

Lemma 2.4.4. *Let* X *be a Banach space, and let* $X = X_1 \dotplus Q$, *where* X_1 *and* Q *are subspaces in* X *and* Q *is finite-dimensional. Furthermore, let* X_∞ *be a linear submanifold dense in* X. *Then* $X = X_1 \dotplus Q_1$, *where* Q_1 *is a finite-dimensional subspace in* X *contained in* X_∞.

In view of this lemma,

$$H_s(M,\Gamma) = \mathcal{A}H_s(M) \dotplus Q_1\,, \tag{2.38}$$

[8] Any subspace is assumed, by definition, to be closed. We use $\dotplus$ to denote the direct sum.

where Q_1 is a finite-dimensional subspace in $H_s(M,\Gamma)$ lying in $H_\infty(M,\Gamma) = \bigcap H_t(M,\Gamma)$. Now let $s' > s$; consider the intersections of both parts in (2.38) with $H_{s'}(M,\Gamma)$. Using Theorem 2.4.2, we obtain

$$H_{s'}(M,\Gamma) = \mathcal{A}H_{s'}(M) \dot{+} Q_1 ,$$

so that $Q_1 =$ Coker A does not change when we pass from s to s'.

Further discussion of some questions connected with the index is postponed to Sect. 7.8.

c. Theorems 2.4.1 and 2.4.2 above are proved by means of a rough parametrix for $\mathcal{A}$. However, these theorems imply the existence of a *precise parametrix* $\mathcal{B}_\infty$, with properties indicated in the following theorem.

Theorem 2.4.5. *Assume that the boundary problem* (1.1) *is elliptic and that* s_0 *satisfies conditions* (2.21). *Then there exists an operator* $\mathcal{B}_\infty$ *such that it is bounded from* $H_s(M,\Gamma)$ *to* $H_s(M)$ *for all* $s \geq s_0$, *and*

$$\mathcal{A}\mathcal{B}_\infty = \mathcal{I} + \mathcal{T}_\infty , \qquad \mathcal{B}_\infty\mathcal{A} = I + T_\infty , \tag{2.39}$$

where $\mathcal{I}$ *and* I *are the identity operators in* $H_s(M,\Gamma)$ *and* $H_s(M)$, *while* $\mathcal{T}_\infty$ *and* T_∞ *are finite-dimensional operators in these spaces with ranges lying in* $H_\infty(M,\Gamma)$ *and* $H_\infty(M)$, *respectively.*

Indeed, let us fix a direct complement $(\operatorname{Ker}\mathcal{A})^\bullet$ of $\operatorname{Ker}\mathcal{A}$ in $H_{s_0}(M)$ and a direct complement $(R(\mathcal{A}))^\bullet$ of $R(\mathcal{A})$ in $H_{s_0}(M,\Gamma)$ lying in $H_\infty(M,\Gamma)$ (see (2.38)). The operator $\mathcal{A}$: $(\operatorname{Ker}\mathcal{A})^\bullet \to R(\mathcal{A})$ is invertible by Banach's theorem. We set $\mathcal{B}_\infty = \mathcal{A}^{-1}$ on $R(\mathcal{A})$ and $\mathcal{B}_\infty = 0$ on $(R(\mathcal{A}))^\bullet$ and extend $\mathcal{B}_\infty$ to $H_{s_0}(M,\Gamma)$ by linearity. We obtain (2.39) for $s = s_0$. It remains to use Theorem 2.4.2.

We will not investigate the operator $\mathcal{B}_\infty$. However, in Sect. 8.1 we describe a precise approximation of the resolvent of the operator A_B defined below in (2.41), and in Sect. 5.2 we examine a precise approximation of the parametrix for the boundary problem $Au = 0$ on M_+, $B_ju = g_j$ $(j = 1,\ldots,q)$ on Γ.

d. Consider the following elliptic boundary problem with homogeneous boundary conditions:

$$Au = f \quad \text{on} \quad M_+ , \quad B_ju = 0 \quad (j = 1,\ldots,q) \quad \text{on} \quad \Gamma , \tag{2.40}$$

where $r_j = \operatorname{ord} B_j < m$. With this boundary problem we can associate a bounded operator acting from $H_{s+m}(M)$ to $H_s(M)$ for nonnegative s. The same operator but considered as an unbounded operator in $H_s(M)$ will be denoted by A_B:

$$\begin{gathered} A_Bu = Au \quad \text{for } u \in D(A_B) , \\ D(A_B) = \{u \in H_{s+m}(M) : B_ju = 0 \ (j = 1,\ldots,q) \text{ on } \Gamma\} . \end{gathered} \tag{2.41}$$

This is one of the main objects in the spectral theory for elliptic boundary problems. We will discuss some functions of this operator in §8 and its spectral properties in Sects. 9.1–9.3. Here we only present some elementary preliminary assertions.

Proposition 2.4.6. *The operator A_B is closed.*

Indeed, if $u_k \in H_{s+m}(M)$ $(k = 1, 2, \ldots)$, $B_j u_k = 0$ $(j = 1, \ldots, q)$, $u_k \to u$ and $A_B u_k \to f$ in $H_s(M)$, then $\{u_k\}$ converges in $H_{s+m}(M)$ in view of the a priori estimate; from this it easily follows that $u \in D(A_B)$ and $A_B u = f$.

Proposition 2.4.7. *If the resolvent $R_{A_B}(\lambda) = (A - \lambda I)^{-1}$ exists for some λ, it is compact, so that A_B is an operator with a discrete spectrum* (see Sect. 2.5 in (Agranovich 1990a)). *In addition, in this case the eigenfunctions and the associated functions belong to $H_\infty(M)$, and hence the spectrum does not depend on s.*

The compactness of the resolvent follows from the compactness of the embedding $H_{s+m}(M) \subset H_s(M)$, and the smoothness of the root functions follows from Theorem 2.4.2.

We will supplement these remarks at the ends of Subsects. 3.2b and 4.3b.

§3. Ellipticity with Parameter

3.1. Definitions and Examples. Consider the boundary problem

$$A(\lambda)u = f \quad \text{on} \quad M_+ , \quad B_j(\lambda)u = g_j \quad (j = 1, \ldots, q) \quad \text{on} \quad \Gamma . \tag{3.1}$$

Here all operators are polynomials in λ, $\lambda \in \mathbb{C}$, and λ has the "weight" τ with respect to differentiation, $\tau \in \mathbb{N}$:

$$A(\lambda) = \sum_{0 \le l\tau \le m} \lambda^l A_{m-l\tau} \quad \text{and} \quad B_j(\lambda) = \sum_{0 \le l\tau \le r_j} \lambda^l B_{j,r_j - l\tau} , \tag{3.2}$$

where A_s is a differential operator on M of order s, $m = 2q$, and $B_{j,s}$ is a boundary operator on Γ of the same structure as in (1.3) but of order s instead of r_j. Our smoothness assumptions are the same as in Sect. 1.1.

Let us fix a closed angle (angular domain) $\mathcal{L}$ on the complex plane with vertex at the origin, and assume that $\lambda \in \mathcal{L}$.

a. We call the boundary problem (3.1) *elliptic with parameter* in $\mathcal{L}$ if the following conditions hold.

1. The operator $A(\lambda)$ is elliptic with parameter on M. This means that

$$a_0(x, \xi, \lambda) = \sum \lambda^l a_{m-l\tau,0}(x, \xi) \neq 0 \quad \text{for } (x, \xi) \in T^*M, \ \lambda \in \mathcal{L}, \ (\xi, \lambda) \neq 0 , \tag{3.3}$$

where $a_{s,0}(x,\xi)$ is the principal symbol of A_s. It follows, in particular, that $\rho = m/\tau$ is an integer (this can be easily checked by setting $\xi = 0$).

2. Fixing an arbitrary point on Γ, we write the operators $A(\lambda)$ and $B_j(\lambda)$ in local coordinates on a collar of M. Let $(x', 0)$ be the coordinates of this point. Consider the following boundary problem on the ray $\mathbb{R}_+$:

$$a_0(x', 0, \xi', D_n, \lambda)v(t) = 0 \qquad (t = x_n > 0) , \tag{3.4}$$

$$b_{j,0}(x', 0, \xi', D_n, \lambda)v(t)|_{t=0} = h_j \qquad (j = 1, \ldots, q) . \tag{3.5}$$

Here

$$b_{j,0}(x, \xi, \lambda) = \sum \lambda^l b_{j,r_j - l\tau,0}(x, \xi) , \tag{3.6}$$

where $b_{j,s,0}$ is the principal symbol of the boundary operator $B_{j,s}$; this principal symbol is homogeneous in ξ of degree s. It is required that this boundary problem have a unique solution $v(t)$ with $|v(t)| \to 0$ as $t \to \infty$, for any $\xi' \in \mathbb{R}_{n-1}$, $\lambda \in \mathcal{L}$, $(\xi', \lambda) \neq 0$, and any numbers h_j.

More precisely, this is the ellipticity with parameter *of weight* τ in $\mathcal{L}$. However we assume τ to be fixed and will not mention the weight.

From the conditions of ellipticity with parameter it follows that our boundary problem (3.1) is elliptic for any fixed λ. The condition of proper ellipticity holds automatically (for $n \geq 2$): the equation $a_0(x', 0, \xi', \zeta, \lambda) = 0$ with respect to ζ has q roots in the upper halfplane and the same number of roots in the lower halfplane. This can be easily checked by using the possibility of joining the points $(\xi', 0)$ and $(-\xi', 0)$, with any $\xi' \neq 0$, in the set $\{(\xi', \lambda) \in \mathbb{R}_{n-1} \times \mathcal{L} : (\xi', \lambda) \neq 0\}$ by a continuous curve.

In $\mathbb{R}^n_+$ it is also possible to consider a boundary problem with a parameter:

$$A(\lambda)u = f \text{ in } \mathbb{R}^n_+ , \quad B_j(\lambda)u = g_j \ (j = 1, \ldots, q) \text{ on } \mathbb{R}^{n-1} = \partial\mathbb{R}^n_+ . \tag{3.7}$$

Assume that the coefficients in the operators A_s and $B_{j,s}$ belong to $B^\infty(\overline{\mathbb{R}^n_+})$ and $B^\infty(\mathbb{R}^{n-1})$, respectively. The conditions of ellipticity with parameter are formulated in an obvious way. Ellipticity with parameter in $\mathcal{L}$ is said to be *uniform* if $|a_0(x, \xi, \lambda)|$ is bounded by a positive constant from below for $x \in \overline{\mathbb{R}^n_+}$, $(\xi, \lambda) \in \mathbb{R}_n \times \mathcal{L}$, $|\xi|^2 + |\lambda|^{2/\tau} = 1$ and if a similar condition for the Lopatinskij determinant is valid for $x^n = 0$, $(\xi', \lambda) \in \mathbb{R}_{n-1} \times \mathcal{L}$, $|\xi'|^2 + |\lambda|^{2/\tau} = 1$.

In a similar way, we can formulate the conditions of ellipticity with parameter of a boundary problem (on M or in $\mathbb{R}^n_+$) at a fixed boundary point and the definition of a model boundary problem elliptic with parameter in $\mathbb{R}^n_+$.

The angle $\mathcal{L}$ can coincide, in particular, with a ray. However, from ellipticity with parameter in the angle $\mathcal{L} = \{\alpha \leq \arg\lambda \leq \beta\} \cup \{0\}$ it obviously follows that ellipticity with parameter holds in a somewhat larger angle $\mathcal{L}' = \{\alpha - \varepsilon \leq \arg\lambda \leq \beta + \varepsilon\} \cup \{0\}$, $\varepsilon > 0$. In other words, the set of directions of the rays along which the given boundary problem is elliptic with parameter is open.

b. Examples. 1. Let the operator $A(\lambda)$ of order $m = 2q$ on M be elliptic with parameter in $\mathcal{L}$, and let $B_1, \ldots, B_q$ be a completely elliptic system of boundary operators not depending on λ. Then the boundary problem

$$A(\lambda)u = f \quad \text{on} \quad M_+ \,, \quad B_j u = g_j \quad (j = 1, \ldots, q) \quad \text{on} \quad \Gamma \tag{3.8}$$

is elliptic with parameter in $\mathcal{L}$. In particular, this is true in the case of the Dirichlet boundary conditions.

2. The boundary problem

$$\Delta u + \ldots + \lambda^2 u = f \quad \text{in} \quad G \,, \quad \gamma_0 \partial_\nu u - \lambda \gamma_0 u = g \quad \text{on} \quad \Gamma \tag{3.9}$$

is elliptic with parameter outside arbitrary narrow angular neighborhoods of the rays $\mathbb{R}_\pm$. This can be verified directly. Here and in Examples 3–5, ∂_ν is the derivative along the inner normal, and the dots denote lower order terms.

3. The boundary problem

$$\Delta u + \ldots + \lambda u = f \quad \text{in} \quad G \,, \quad \gamma_0 \partial_\nu u - \alpha \lambda \gamma_0 u = g \quad \text{on} \quad \Gamma \,, \tag{3.10}$$

where $\alpha \neq 0$, is not elliptic with parameter: here λ has different weights in the equation and in the boundary condition.

4. The boundary problem

$$\Delta u = f \quad \text{in} \quad G \,, \quad \gamma_0 \partial_\nu u - \lambda \gamma_0 u = g \quad \text{on} \quad \Gamma \tag{3.11}$$

is also not elliptic with parameter: here, for any angle $\mathcal{L}$, the first condition of ellipticity with parameter is violated at $\xi = 0$. However, this boundary problem with $f = 0$ can be reduced to a pseudodifferential equation on Γ that is elliptic with parameter and even selfadjoint, so that, taking $f = 0$ and $g = 0$, we have a well posed spectral problem (see Sects. 5.3 and 9.4 below).

5. The boundary problem

$$\Delta u + \ldots + \lambda u = f \quad \text{in} \quad G \,, \quad \gamma_0 \partial_\nu^3 u = g \quad \text{on} \quad \Gamma \tag{3.12}$$

is elliptic with parameter outside an arbitrary narrow angular neighborhood of the ray $\mathbb{R}_+$. We see that in a boundary problem elliptic with parameter the boundary condition can have higher order than the elliptic equation.

6. Here we briefly discuss the relation between boundary problems elliptic with parameter and nonstationary boundary problems.

Consider the following mixed boundary problem:

$$A(\partial_t) w(t, x) = f(t, x) \quad \text{in} \quad \mathbb{R}_+ \times M_+ \,, \tag{3.13}$$

$$B_j(\partial_t) w(t, x) = g_j(t, x) \quad (j = 1, \ldots, q) \quad \text{on} \quad \mathbb{R}_+ \times \Gamma \,, \tag{3.14}$$

$$\partial_t^k w(t, x) = 0 \quad (k = 0, \ldots, \rho - 1) \quad \text{on} \quad M_+ \quad \text{for} \quad t = 0 \,. \tag{3.15}$$

Here $\partial_t = \partial/\partial_t$; $A(\lambda)$ and $B_j(\lambda)$ have the same form as in (3.1). For simplicity we assume that the coefficients in A_s and $B_{j,s}$ do not depend on the time t and that the right-hand sides in the initial conditions (3.15) are equal to zero. The formal Laplace transform

$$w(t,x) \mapsto \int_0^\infty e^{-\lambda t} w(t,x)\,dt = u(x,\lambda)$$

converts this boundary problem into a stationary boundary problem having the form (3.1). If the last boundary problem is elliptic with parameter in the right halfplane $\{\lambda : |\arg\lambda| \le \pi/2\}$, then the boundary problem (3.13)–(3.15) can be called *parabolic*. In this case τ is even: $\tau = 2b, b \in \mathbb{N}$. The Dirichlet problem for the heat equation can serve as an example:

$$\begin{aligned} &\partial_t w(t,x) - \Delta_x w(t,x) = f(t,x) \quad \text{in} \quad R_+ \times G\,, \\ &w(t,x) = g(t,x) \quad \text{on} \quad \mathbb{R}_+ \times \Gamma\,, \quad w(0,x) = 0\,. \end{aligned} \tag{3.16}$$

Here $\tau = 2$, and the corresponding stationary boundary problem is elliptic with parameter outside an arbitrary narrow angular neighborhood of $\mathbb{R}_-$.

3.2. Unique Solvability.

a. We can consider the boundary problem (3.1) in the same Sobolev spaces as before. However it is convenient to include the parameter in the norms, setting, for positive s,

$$\begin{aligned} |||u|||_{s,M} &= \{\|u\|_{s,M}^2 + |\lambda|^{2s/\tau}\|u\|_{0,M}^2\}^{1/2}\,, \\ |||u|||_{s,\Gamma} &= \{\|u\|_{s,\Gamma}^2 + |\lambda|^{2s/\tau}\|u\|_{0,\Gamma}^2\}^{1/2}\,. \end{aligned} \tag{3.17}$$

For a fixed λ, these norms are equivalent to the previous norms $\|u\|_{s,M}$ and $\|u\|_{s,\Gamma}$. Using the norms (3.17), we have

$$|||A(\lambda)u|||_{s-m,M} + \sum_{j=1}^q |||B_j(\lambda)u|||_{s-r_j-\frac{1}{2},\Gamma} \le C_s |||u|||_{s,M} \tag{3.18}$$

with a constant C_s not depending on u and λ, without any assumptions of ellipticity with parameter, or even ellipticity.

Theorem 3.2.1. *Let the boundary problem* (3.1) *be elliptic with parameter in* $\mathcal{L}$, *and let inequalities* (2.21) *be satisfied for* s. *Then there exists a number* h_0 *such that for any fixed* $\lambda \in \mathcal{L}$ *with* $|\lambda| \ge h_0$ *this boundary problem has one and only one solution* $u \in H_s(M)$ *for any* $f \in H_{s-m}(M)$ *and* $g_j \in H_{s-r_j-\frac{1}{2}}(\Gamma)$, $j = 1, \ldots, q$. *In addition, the following inequality holds*:

$$|||u|||_{s,M} \le C_s' \left\{ |||A(\lambda)u|||_{s-m,M} + \sum_{j=1}^q |||B_j(\lambda)u|||_{s-r_j-\frac{1}{2},\Gamma} \right\}, \tag{3.19}$$

where C_s' *is independent of* u *and* s.

Conversely, if this inequality holds for some s satisfying conditions (2.21), *then the boundary problem is elliptic with parameter in* $\mathcal{L}$.

This theorem can be proved, essentially, by using a plan similar to that described in §2.

Namely, we first prove an inequality of the form (3.19) for a model boundary problem elliptic with parameter for $0 \neq \lambda \in \mathcal{L}$. Next we prove it for a boundary problem in $\mathbb{R}^n_+$ elliptic with parameter at $x = 0$, with variable coefficients and lower order terms, on functions supported in a sufficiently small semi-neighborhood of the origin, and for $\lambda \in \mathcal{L}$ with sufficiently large modulus. Passing to our boundary problem on M, we obtain an estimate for functions with support in a sufficiently small semi-neighborhood of a boundary point. A similar estimate in a neighborhood O of an interior point follows from the results for operators $A(\lambda)$ elliptic with parameter in $\mathbb{R}^n$ and on a closed manifold. Now we obtain (3.19) by means of a sufficiently fine partition of unity on M. From (3.19) we obtain uniqueness for the boundary problem (3.1).

In addition, we construct not only the right parametrix (see Sect. 2.3) but also the right inverse operator. For a model boundary problem in a halfspace it is defined by the formula

$$R_0(\lambda)(f,g) = u_0 + u_1 \tag{3.20}$$

for $0 \notin \mathcal{L}$, where

$$u_0(x) = F^{-1}_{\xi \mapsto x} a_0^{-1}(\xi,\lambda) F_{x \mapsto \xi} \mathcal{E}_{\mathbb{R}^n_+} f \tag{3.21}$$

and

$$u_1(x) = F'^{-1}_{\xi' \mapsto x'} \sum_{j,k=1}^{q} L^{k,j}(\xi',\lambda) F'_{x' \mapsto \xi'} [g_j - b_{j,0}(D,\lambda) u_0|_{x^n=0}](\xi',\lambda) v_k(\xi',\lambda,x^n) \,. \tag{3.22}$$

Here $\{v_k\}$ is the analog of the basis (1.19) for the boundary problem (3.4)–(3.5) on a ray, and $(L^{j,k}(\xi',\lambda)) = (L_{j,k}(\xi,\lambda))^{-1}$, where $(L_{j,k}(\xi,\lambda))$ is the analog of the Lopatinskij matrix for this boundary problem. At the next two steps we use the invertibility of an operator sufficiently close in norm to an invertible operator; this can be done for $\lambda \in \mathcal{L}$ with sufficiently large modulus. Having the right inverse, we see that the boundary problem is solvable. In view of uniqueness, the right inverse operator is also the left inverse.

Finally, to deduce ellipticity with parameter from inequality (3.19), we pass to a model boundary problem; for it, ellipticity with parameter is verified by contradiction, using appropriate sequences of functions.

For boundary problems in $\mathbb{R}^n$ that are uniformly elliptic with parameter in $\mathcal{L}$, we also obtain a theorem on unique solvability for $\lambda \in \mathcal{L}$ with sufficiently large modulus. An a priori estimate of the form (3.19) is valid for such boundary problems, with $\mathbb{R}^n_+$ and $\mathbb{R}_{n-1}$ instead of M and Γ, respectively.

b. From Theorem 3.2.1 it is possible to deduce a theorem on unique solvability of parabolic mixed boundary problems of the form (3.13)–(3.15), and even more general, with coefficients depending on t, and also in a finite cylinder $[0,T] \times M$ instead of $\mathbb{R}_+ \times M$. See (Agranovich and Vishik 1964).

Theorem 3.2.1 also finds important applications to various spectral problems; this will be discussed in Sect. 9.3. The simplest of these problems relates to the operator A_B (see 2.4d). In particular, it is clear that the condition of ellipticity with parameter in an angle of a boundary problem of the form (2.40) with $A - \lambda I$ instead of A guarantees the discreteness of the spectrum of this operator and the absence of eigenvalues in this angle sufficiently far from the origin. A construction of a precise parametrix for the operator $A_B - \lambda I$ in this case will be described in Sect. 8.1.

§4. Adjoint Elliptic Boundary Problems

In essence, here we follow papers (Aronszajn and Milgram 1953) and (Schechter 1959a); see also (Lions and Magenes 1968, Chapter 2) and (Berezanskij 1965). Using the symbols γ_k introduced in Sect. 0.2, we rewrite the boundary operators in the form

$$B_j u = \sum_{k=0}^{r_j} B_{j,k} \gamma_k u \, , \tag{4.1}$$

where the $B_{j,k}$ are differential operators on Γ of orders $r_j - k$ with infinitely smooth coefficients.

4.1. Normal Boundary Conditions and Dirichlet Systems.

a. The system (4.1) of boundary operators is called *normal* if

$$r_j \neq r_k \quad \text{for} \quad j \neq k \quad \text{and} \quad B_{j,r_j} \neq 0 \quad \text{everywhere on} \quad \Gamma \tag{4.2}$$

$(j,k = 1,\dots,q)$. Here the B_{j,r_j} are functions from $C^\infty(\Gamma)$, and the last inequality means that the boundary Γ is noncharacteristic with respect to each of the boundary operators B_j. The boundary problem (1.1) with a normal system of boundary operators is also called *normal.*

In this definition there are no restrictions on the orders r_j of boundary operators and no assumptions of ellipticity. Note that if the boundary problem (1.1) with $A - \lambda I$ instead of A is elliptic with parameter in an angle $\mathcal{L}$, then the system of boundary operators in this boundary problem is normal. This can be seen from Condition 2 in Subsect. 3.1a at $\xi' = 0$. The orders of boundary operators in such a problem need not be less than the order of the elliptic equation (see Example 5 in Subsect. 3.1b).

However now we will additionally assume that

$$r_j < m = 2q \, . \tag{4.3}$$

Condition (2.21) reduces to

$$s \geq m\,. \tag{4.4}$$

b. The system

$$B_1, \dots, B_m \tag{4.5}$$

of m boundary operators of the form (4.1) of orders $r_j < m$ is called the *Dirichlet system* of order m if it satisfies conditions (4.2) $(j, k = 1, \dots, m)$.

The system

$$\gamma_0, \gamma_1, \dots, \gamma_{m-1} \tag{4.6}$$

can serve as an example, and here arbitrary lower order terms may be added (beginning with γ_1).

It is not difficult to describe the general form of the Dirichlet system. Let $\Phi = (\Phi_{j,k})$ be an $m \times m$ matrix, in which the elements $\Phi_{j,k}$ below the main diagonal are differential operators of orders $j-k$ on Γ with C^∞ coefficients, the elements $\Phi_{j,j}$ on the main diagonal are C^∞ functions on Γ that are different from zero everywhere, and the elements $\Phi_{j,k}$ above the main diagonal are zero. Such a matrix will be called *admissible.* Denote by B and γ the columns $(B_1, \dots, B_m)'$ and $(\gamma_0, \dots, \gamma_{m-1})'$, respectively.

Proposition 4.1.1. *The system* (4.5) *with* $\operatorname{ord} B_j = j - 1$ *is a Dirichlet system if and only if* $B = \Phi\gamma$, *where* Φ *is an admissible matrix.*

Obviously, admissible matrices form a group, and therefore we obtain the following two corollaries. 1) If B and $\widetilde{B}$ are two columns of boundary operators of orders $0, \dots, m-1$ forming Dirichlet systems, then $B = \Psi\widetilde{B}$, where Ψ is an admissible matrix. 2) If $\widetilde{B}$ corresponds to a Dirichlet system and Ψ is an admissible matrix, then $B = \Psi\widetilde{B}$ corresponds to a Dirichlet system.

Proposition 4.1.2. *Let* (4.5) *be a Dirichlet system,* $\operatorname{ord} B_j = r_j$, *and let* g_j *be functions from* $H_{m-r_j-\frac{1}{2}}(\Gamma)$ $(j = 1, \dots, m)$. *Then there exists a function* $v \in H_m(M)$ *satisfying the conditions*

$$B_j v = g_j \quad (j = 1, \dots, m) \quad \text{on} \quad \Gamma \tag{4.7}$$

and

$$\|v\|_{m,M} \leq C \sum_1^m \|g_j\|_{m-r_j-\frac{1}{2},\Gamma}\,, \tag{4.8}$$

where C *does not depend on* g_j.

To verify this, in view of the previous proposition it suffices to consider the case $B_j = \gamma_{j-1}$ $(j = 1, \dots, m)$, and in this case we have in essential described the construction in Subsect. 2.1b. Going from $\mathbb{R}^n_+$ and $\mathbb{R}^{n-1}$ to M and Γ, we use a partition of unity on a collar of the manifold.

Obviously any normal system consisting of k boundary operators, where $k < m$, is a part of a Dirichlet system of order m, and the additional normal operators of needed orders can easily be chosen.

4.2. Green's Formula and Formally Adjoint Boundary Problems.

a. If A is the Laplace operator Δ in the domain G, then the following Green's formula is well known:

$$(\Delta u, v)_G - (u, \Delta v)_G = (\gamma_0 u, \gamma_0 \partial_\nu v)_\Gamma - (\gamma_0 \partial_\nu u, \gamma_0 v)_\Gamma \,. \tag{4.9}$$

Here $\gamma_1 = \gamma_0 \partial_\nu$, where ∂_ν is the derivative along the inner normal. In this section we describe a generalization of this formula.

We write the elliptic differential operator A on a collar of the manifold M in the form

$$A = \sum_{j=0}^{m} A_j D_n^j \,. \tag{4.10}$$

Here A_j is a differential operator of order $m - j$ on Γ with coefficients depending on x^n; $m = 2q$. The operator A^* formally adjoint to A has the form

$$A^* = \sum_{j=0}^{m} D_n^j (A_j^* \cdot) \,, \tag{4.11}$$

where A_j^* is the differential operator on Γ formally adjoint to A_j. Using integration by parts, we obtain (a special) Green's formula

$$(Au, v)_M - (u, A^* v)_M = \sum_{l=0}^{m-1} (\gamma_l u, L_{m-l} v)_\Gamma \,, \tag{4.12}$$

where L_{m-l} is a differential operator of order $m - 1 - l$:

$$L_{m-l} = \sum_{j=l+1}^{m} (-1)^{j-l} \gamma_{j-l-1} A_j^* \,. \tag{4.13}$$

In view of the ellipticity of A, $A_m^* = \overline{A_m}$ is a function different from zero everywhere. Thus we see that $L_1, \ldots, L_m$ is a Dirichlet system. This result can be generalized:

Theorem 4.2.1. *Let A be an elliptic differential operator of order m on M, and let $B_1, \ldots, B_m$ be a Dirichlet system. Then there exists another Dirichlet system $C_1, \ldots, C_m$ such that*

$$(Au, v)_M - (u, A^* v)_M = \sum_{j=q+1}^{m} (B_j u, C_{m+1-j} v)_\Gamma - \sum_{j=1}^{q} (B_j u, C_{m+1-j} v)_\Gamma \tag{4.14}$$

for any $u, v \in C^\infty(M)$.

To prove this, we obviously need to use Proposition 4.1.1. We write the right-hand side in (4.12) in the form $(\gamma u, Lv)_\Gamma$, where $L = (L_m, \ldots, L_1)'$ and $(\cdot, \cdot)_\Gamma$ is the scalar product of vector-valued functions on Γ. Setting $\gamma = \Psi B$, where Ψ is an admissible matrix, we see that the right-hand side in (4.12) can be written in the form $(Bu, \Psi^* Lv)$, and here $\Psi^* L = (\widetilde{L}_m, \ldots, \widetilde{L}_1)'$, where $\widetilde{L}_1, \ldots, \widetilde{L}_m$ is a Dirichlet system. It remains to set $C_j = \widetilde{L}_j$ for $j \leq q$ and $C_j = -\widetilde{L}_j$ for $j > q$.

Formula (4.14) remains valid for functions $u, v \in H_m(M)$.

b. Let (1.1) be a normal boundary problem. Choose operators $B_{q+1}, \ldots, B_m$ (in one of possible ways) so that $B_1, \ldots, B_m$ is a Dirichlet system. Let $C_1, \ldots, C_m$ be the Dirichlet system indicated in Theorem 4.2.1. Consider the boundary problem

$$A^* v = \widetilde{f} \quad \text{on} \quad M_+ \,, \quad C_j v = \widetilde{g}_j \quad (j = 1, \ldots, q) \quad \text{on} \quad \Gamma \,. \tag{4.15}$$

The boundary problems (1.1) and (4.15) are called *formally adjoint*. Formula (4.14) is called the *Green formula* for these boundary problems. If u and v are functions from $H_m(M)$ satisfying homogeneous boundary conditions

$$B_1 u = \ldots = B_q u = 0 \quad \text{on} \quad \Gamma \tag{4.16}$$

and

$$C_1 v = \ldots = C_q v = 0 \quad \text{on} \quad \Gamma \,, \tag{4.17}$$

then it follows from (4.14) that

$$(Au, v)_M = (u, A^* v)_M \,. \tag{4.18}$$

Using Proposition 4.1.2, we can easily verify that if (4.18) is true for some function $v \in H_m(M)$ and any function $u \in H_m(M)$ satisfying conditions (4.16), then v satisfies conditions (4.17).

In the case of the Dirichlet problem for A, i.e. when $B_j = \gamma_{j-1}$ ($j = 1, \ldots, q$), it is always possible to take the Dirichlet problem for A^* as the formally adjoint boundary problem. The appropriate $B_q, \ldots, B_m$ and $C_q, \ldots, C_m$ can be found.

Two normal systems of boundary operators, $\{B_j\}_1^q$ and $\{T_j\}_1^q$, are called *equivalent* if conditions (4.16) follow from the conditions

$$T_1 u = \ldots = T_q u = 0 \quad \text{on} \quad \Gamma \,, \tag{4.19}$$

and vice versa.

Two equivalent systems are connected in a way similar to that for two Dirichlet systems:

Proposition 4.2.2. *Let $\{B_j\}_1^q$ and $\{T_j\}_1^q$ be normal systems of boundary operators of orders r_j and s_j, respectively, where $r_1 < \ldots < r_q$ and $s_1 < \ldots < s_q$. Then the systems $\{B_j\}$ and $\{T_j\}$ are equivalent if and only if $r_j = s_j$ for all j and $B = \Xi T$, where $B = (B_1, \ldots, B_q)'$, $T = (T_1, \ldots, T_q)'$, and Ξ is a $q \times q$ matrix of the following structure: its elements $\Xi_{j,k}$ are differential operators of orders $r_j - r_k$ on Γ with infinitely smooth coefficients for $j > k$,* *infinitely smooth functions on Γ different from* 0 *everywhere for $j = k$, and* 0 *for $j < k$.*

This can be verified by including the given systems in Dirichlet systems and using Propositions 4.1.1 and 4.1.2.

If the boundary problem (1.1) is elliptic and if $\{B_j\}_1^q$ and $\{T_j\}_1^q$ are equivalent normal systems of boundary operators, then the boundary problem

$$Au = f \quad \text{on} \quad M_+ \,, \quad T_j u = g_j \quad (j = 1, \ldots, q) \quad \text{on} \quad \Gamma \tag{4.20}$$

is also elliptic. This can be easily verified using the equivalence of ellipticity and a priori estimate (see Sect. 2.2).

d. Let us return to the Green formula. We have seen that the boundary problem (4.15) formally adjoint to the normal elliptic boundary problem (1.1) is not unique. However, from Theorem 4.2.1 and Proposition 4.2.2 it is easy to deduce

Proposition 4.2.3. *Two normal boundary problems,* (4.15) *and*

$$A^* v = \widehat{f} \quad \text{on} \quad M_+ \,, \quad S_j v = \widehat{g}_j \quad (j = 1, \ldots, q) \quad \text{on} \quad \Gamma \,, \tag{4.21}$$

are formally adjoint to the same normal elliptic boundary problem (1.1) *if and only if the systems $\{C_j\}$ and $\{S_j\}$ are equivalent.*

e. The following theorem admits an algebraic proof, see (Schechter 1959a); another proof can be found in (Lions and Magenes 1968, Ch. 2, Sect. 2.5).

Theorem 4.2.4. *Let* (1.1) *be a normal elliptic boundary problem. Then the boundary problem* (4.15) *formally adjoint to* (1.1) *is also elliptic.*

4.3. Range of the Operator Corresponding to a Normal Elliptic Boundary Problem.

a. Let (1.1) be a normal elliptic boundary problem, and let $s \geq m$. As we know, the range of the corresponding operator $\mathcal{A}\colon H_s(M) \to H_s(M, \Gamma)$ is closed and has a finite-dimensional direct complement (see Sect. 2.4). For the boundary problems under consideration, this range admits a more precise description. Assume that the system $\{B_j\}_1^q$ is included in the Dirichlet system $\{B_j\}_1^m$ and we have Green's formula (4.14).

Theorem 4.3.1. *Let $(f, g) \in H_s(M, \Gamma)$. Then $(f, g) \in R(\mathcal{A})$ it and only if*

$$(f, v)_M + \sum_{j=1}^{q} (g_j, C_{m+1-j} v)_\Gamma = 0 \tag{4.22}$$

for any $v \in \operatorname{Ker} \mathcal{A}^{(*)}$, where $\mathcal{A}^{(*)}$ is the operator of the form (1.4) *corresponding to the formally adjoint boundary problem* (4.15).

The necessity of condition (4.22) for the inclusion of (f, g) into $R(\mathcal{A})$ follows immediately from Green's formula (4.14). The proof of the sufficiency is somewhat more complicated. For this it is necessary to consider the operator $\mathcal{A}^*$ adjoint to $\mathcal{A}$ and to verify the smoothness of the elements of $\operatorname{Ker}\mathcal{A}^*$. Here we do not present the appropriate statement in terms of Sobolev spaces on M with negative indices; see (Schechter 1960b), (Lions and Magenes 1968), (Roĭtberg 1991), and also Sect. 7.9 below.

b. Theorem 4.3.1 has some useful consequences for the operator A_B (see Subsect. 2.4d). We will now consider it as acting in $H_0(M)$. The boundary problem (1.1) is assumed to be elliptic. Consider the boundary problem (4.15) formally adjoint to (1.1), with homogeneous boundary conditions. The corresponding operator is (cf. Subsect. 2.4d)

$$\begin{aligned} &A_C^* v = A^* v \,, \\ &v \in D(A_C^*) = \{v \in H_m(M) : \quad C_j v = 0 \quad (j = 1, \dots, q) \quad \text{on} \quad \Gamma\} \,. \end{aligned} \tag{4.23}$$

Proposition 4.3.2. *The operator A_C^* is adjoint to A_B in $H_0(M)$.*

Here we have to verify the following assertion: if

$$(Au, v)_M = (u, h)_M$$

for all $u \in D(A_B)$ (see (2.41) in the case $s = 0$) and some v and h in $H_0(M)$, then $v \in D(A_C^*)$ and $A^* v = h$. Applying Theorem 4.3.1 to the boundary problem (4.15), we see that there exists $w \in D(A_C^*)$ such that $A^* w = h$. Now $(Au, v - w)_M = 0$; applying the same theorem to the initial boundary problem, we find that $v - w \in \operatorname{Ker} A_C^*$, and from this our assertion follows.

If the boundary problem formally adjoint to (1.1) coincides with (1.1), it is called *formally selfadjoint.* Equivalent assumptions are: $A = A^*$, and in the boundary problem (4.15) formally adjoint to (1.1) the system $\{C_j\}_1^q$ is equivalent to $\{B_j\}_1^q$. As an example, any normal boundary problem (1.1) with $A = A^*$ and $\operatorname{ord} B_j = j - 1$ $(j = 1, \dots, q)$ can be used (see Proposition 4.2.2), including the Dirichlet problem.

From Proposition 4.3.2 we obtain

Corollary 4.3.3. *Let* (1.1) *be a formally selfadjoint elliptic boundary problem. Then A_B is a selfadjoint operator in $H_0(M)$.*

c. We have some additional remarks:

Consider the boundary problem

$$(A - \lambda)u = f \quad \text{on} \quad M_+ \,, \quad B_j u = g_j \quad (j = 1, \dots, q) \quad \text{on} \quad \Gamma \,, \tag{4.24}$$

assuming that it is formally selfadjoint for $\lambda = 0$ (or, what is the same, for real λ), or at least close to a formally selfadjoint boundary problem:

$$A = A_0 + \dots \,, \quad B_j = B_{j,0} + \dots \,, \tag{4.25}$$

where

$$A_0 u = f \quad \text{on} \quad M\,, \quad B_{j,0}u = g_j \quad (j = 1, \dots, q) \quad \text{on} \quad \Gamma \tag{4.26}$$

is a formally selfadjoint boundary problem and the dots denote lower order terms. Let $\mathcal{L}$ be any closed angle on the complex plane with vertex at the origin and no other common points with the real axis.

Remark 4.3.4. Under these assumptions, the boundary problem (4.24) is elliptic with parameter in $\mathcal{L}$ (of weight $\tau = m$).

It suffices to verify this assertion for $A = A_0$, $B_j = B_{j,0}$, since the condition of ellipticity with parameter does not involve lower order terms. In this case the norm of the resolvent $R_{A_B} = (A_B - \lambda I)^{-1}$ in $H_0(M)$ is equal to $1/d(\lambda)$, where $d(\lambda)$ is the distance between λ and the spectrum of A_B. Using the a priori estimate for the boundary problem (1.1), we obtain

$$|||u|||_{m,M} \leq C\|(A - \lambda I)u\|_{0,M} \tag{4.27}$$

for $0 \neq \lambda \in \mathcal{L}$, $u \in D(A_B)$, where C is independent of u and λ. Now it suffices to show that from here we can derive the estimate

$$|||u|||_{m,M} \leq C'(\|(A - \lambda I)u\|_{0,M} + \sum |||B_j u|||_{m-r_j-\frac{1}{2},\Gamma}) \quad \text{for} \quad \lambda \in \mathcal{L} \tag{4.28}$$

with sufficiently large $|\lambda|$ and $u \in H_m(M)$. We include the system $\{B_j\}_1^q$ in a Dirichlet system $\{B_j\}_1^m$ and set $g_j = B_j u = 0$ $(j = q+1, \dots, m)$ on Γ. Let $\operatorname{ord} B_j = r_j$. Now it remains to prove the following variant of Proposition 4.1.2:

Proposition 4.3.5. *For any* $g_j \in H_{m-r_j-\frac{1}{2}}(\Gamma)$ $(j = 1, \dots, m)$ *and* λ, $|\lambda| \geq \lambda_0$, *there exists a function* $v_\lambda \in H_m(G)$ *such that*

$$B_j v_\lambda = g_j \quad (j = 1, \dots, m) \quad \text{on} \quad \Gamma \tag{4.29}$$

and

$$|||v|||_{m,M} \leq C \sum_{j=1}^{m} |||g_j|||_{m-r_j-\frac{1}{2},\Gamma}\,, \tag{4.30}$$

where C *does not depend on* g_j *and* λ.

Using Proposition 4.1.1, we need only check this in the case $B_j = \gamma_{j-1}$. In this case we can use the construction of Subsect. 2.1b, replacing $1 + |\xi'|^2$ by $|\lambda|^{1/q} + |\xi'|^2$. For $\mathbb{R}^n_+$ and $\mathbb{R}^{n-1}$ instead of M and Γ, the function v_λ is given by the formula

$$v_\lambda(x) = \sum_{j=1}^{m} F'^{-1}_{\xi' \mapsto x'}[\psi_j(x^n(|\lambda|^{1/q} + |\xi'|^2)^{1/2})(F'g_j)(\xi')(|\lambda|^{1/q} + |\xi'|^2)^{-(j-1)/2}]\,.$$

Passing to M and Γ, we use a partition of unity on a collar of the manifold.

Remark 4.3.6. Let the normal elliptic boundary problem (1.1) be formally selfadjoint, so that A_B is a selfadjoint operator. Then it is bounded from below (i.e. its spectrum is bounded from below) if and only if the boundary problem (4.24) is elliptic with parameter along $\mathbb{R}_-$ (or, what is the same, outside an arbitrarily narrow angular neighborhood of $\mathbb{R}_+$).

Indeed, here the part "if" of the assertion is evident, and the part "only if" is verified by means of estimate (4.28), which is true for $\lambda \in \mathbb{R}_-$ with sufficiently large modulus if A_B has no eigenvalues there.

Positiveness of the principal symbol of A is insufficient for boundedness of A_B from below. An example can be found in (Agmon 1961b) (see also (Agmon 1962, p. 134)): it is a boundary problem for the biharmonic operator Δ^2.

If the boundary problem (1.1) is only close to a selfadjoint boundary problem, then ellipticity with parameter along $\mathbb{R}_-$ remains sufficient for boundedness from below.

Remark 4.3.7. The composition of two normal or normal elliptic boundary problems is a normal or normal elliptic boundary problem, respectively. Hence, if the operators A_B and $\widetilde{A}_{\widetilde{B}}$ correspond to normal elliptic boundary problems with homogeneous boundary conditions, then the operator $\widetilde{A}_{\widetilde{B}} A_B$ corresponds to a normal elliptic boundary problem with homogeneous boundary conditions. In particular, if the operator A_B corresponds to a normal elliptic boundary problem with homogeneous boundary conditions, then the selfadjoint nonnegative operator $(A_B)^* A_B$ corresponds to a normal elliptic boundary problem with homogeneous boundary conditions. Of course, this boundary problem is formally selfadjoint.

§5. Reduction of Elliptic Boundary Problems to Equations on the Boundary

In classical mathematical physics, methods of using potentials are well known. The inhomogeneous elliptic equation $Au = f$ is reduced to the homogeneous equation $Av = 0$ by means of the substitution $u = v + u_0$, where u_0 is a particular solution of the inhomogeneous equation having the form of a volume potential. The elliptic boundary problem for the homogeneous equation can be reduced (by various methods) to integral or integro-differential (actually elliptic pseudodifferential) equations on Γ. These reductions are used to prove the Fredholm property and to determine solvability conditions for concrete boundary problems, as well as to solve these problems numerically. This is possible if a fundamental solution for the given equation is known, since then we can construct the needed potentials explicitly.

Similar reductions are possible in the general theory of elliptic boundary problems. In general, they have only a theoretical nature, and we can only hope to calculate the symbols of required pseudodifferential operators. In some papers and monographs (cited below) this approach is used to establish the Fredholm property of operators corresponding to elliptic and even more general boundary problems for elliptic equations.

In Sect. 5.1 we present a description of the range $R(A)$ of an elliptic operator A on M, considered without boundary conditions, in the cases of uniqueness and non-uniqueness for the Cauchy problem for the adjoint equation. Then we construct the right inverse of A, which permits us to reduce the boundary problem (1.1) to the case of the homogeneous equation $Au = 0$.

In Sect. 5.2 we explain the method of reducing of an elliptic boundary problem for the homogeneous elliptic equation $Au = 0$ to an elliptic system of pseudodifferential equations on Γ with respect to, say, the "Dirichlet data." Here we follow the paper (Vainberg and Grushin 1967) (where nonelliptic boundary problems were considered) and the monograph (Kumano-go 1981).

In Sect. 5.3 we discuss the solvability of boundary problems with a parameter only in boundary conditions, following (Panich 1966, 1973); later on we return to such boundary problems in Sect. 9.4.

In Sect. 5.4 we consider the "Cauchy data" for homogeneous elliptic equations and the Calderón projectors. These projectors are related to the decomposition of the space of vector-valued functions on the boundary into the direct sum of the subspaces of Cauchy data for the equation $Au = 0$ on M_+ and on M_-, and to the reconstruction of the solution in terms of its Cauchy data. First we consider one possible example that seems to illustrate the situation quite clear, and then we present general results, following, in general, the papers (Calderón 1963) and (Seeley 1966). Boundary problems for elliptic equations can be reduced to equations on the boundary with respect to Cauchy data. See also the books (Dieudonné 1978), (Trèves 1980), (Chazarain and Piriou 1981), (Taylor 1981), (Hörmander 1985a), and the paper (Costabel and Wendland 1986).

We consider as known the results on the Fredholm property of elliptic boundary problems (outlined in the preceding sections), and we do not deal with nonelliptic boundary problems. However the material we discuss contains additional information.

5.1. Transition to Homogeneous Elliptic Equation.

a. Here we consider the equation

$$Au = f \quad \text{on} \quad M_+ \,, \tag{5.1}$$

where A is an elliptic differential operator of order $m = 2q$, without boundary conditions. Simultaneously, we consider the homogeneous Cauchy problem for the formally adjoint equation:

$$A^* v = 0 \quad \text{on} \quad M_+ \,, \quad \gamma_0 v = \ldots = \gamma_{m-1} v = 0 \quad \text{on} \quad \Gamma \,. \tag{5.2}$$

Denote by $R(A)$ the range of the operator $A\colon H_m(M) \to H_0(M)$, and by $N_0(A^*)$ the subspace of solutions of the Cauchy problem (5.2) in $H_m(M)$. The notations $R(A^*)$ and $N_0(A)$ will have a similar sense. The following proposition contains a description of $R(A)$ and $R(A^*)$.

Proposition 5.1.1. *The subspace $N_0(A^*)$ is finite-dimensional and is contained in $C^\infty(M)$. The range $R(A)$ coincides with the orthogonal complement of $N_0(A^*)$ in $H_0(M)$ and is, in particular, closed. Analogous assertions hold for $N_0(A)$ and $R(A^*)$.*

The easiest way to prove this in the case of scalar-valued functions, which we consider at present, is to use what we know about elliptic boundary problems. The first assertion immediately follows from the fact that $N_0(A^*)$ is contained in the space of solutions of the homogeneous Dirichlet problem for A^*. The orthogonality of $N_0(A^*)$ and $R(A)$ is clear from Green's formula (4.12). Now let f be a function in $H_0(M)$ orthogonal to $N_0(A^*)$. We set $C_j = \gamma_{j-1}$ $(j = 1, \ldots, 2m)$ and choose B_j $(j = 1, \ldots, 2m)$ such that Green's formula (4.14) is valid. We take a basis $v_1, \ldots, v_r$ in $N_0(A^*)$ and complete it by functions $v_{r+1}, \ldots, v_p$ such that $v_1, \ldots, v_p$ is a basis in the space of solutions of the homogeneous Dirichlet problem for A^*. The system of vectors

$$w_k = (\gamma_{m-1} v_k, \ldots, \gamma_q v_k)' \qquad (k = r+1, \ldots, p) \tag{5.3}$$

is linearly independent; we assume, without loss of generality, that it is orthonormal in $[H_0(\Gamma)]^q$. Set

$$(g_1, \ldots, g_q)' = -\sum_{k=r+1}^{p} (f, v_k)_M w_k \; ; \tag{5.4}$$

then the boundary problem

$$Au = f \quad \text{on} \quad M_+ \, , \quad B_j u = g_j \quad (j = 1, \ldots, q) \quad \text{on} \quad \Gamma \tag{5.5}$$

is solvable, by virtue of Theorem 4.3.1. Thus, $f \in R(A)$.

b. Corollary 5.1.2. *$R(A) = H_m(M)$ if and only if the Cauchy problem* (5.2) *has no nontrivial solutions.*

An extensive literature is devoted to the question of conditions for uniqueness for the Cauchy problem. This question has been investigated in a local setting: assuming that the equation $Au = 0$ is satisfied in the domain G near a boundary point, and that the Cauchy data γu on Γ are also equal to zero near this point, one asks when it follows that $u = 0$ near the same point. Instead of ellipticity, it is often assumed only that the boundary Γ is noncharacteristic with respect to A. We present some simple corollaries of local uniqueness theorems:

Theorem 5.1.3. *Let A be an elliptic operator of order $m = 2q$ on M. Then $N_0(A) = N_0(A^*) = \{0\}$ if $q = 1$ and also if M_0 and the coefficients in A are real-analytic.*

The local uniqueness theorem, from which the first assertion follows, was apparently obtained in its full generality in (Aronszajn 1957); some papers with more special results appeared earlier. The second assertion follows from Holmgren's classical theorem, which is presented e.g. in (Hörmander 1963, Ch. V). We also mention the local uniqueness theorem due to Calderón (Calderón 1958), in which the Cauchy problem in $\mathbb{R}^n_+$ is considered under the assumption that the equation $a_0(x', 0, \xi', \zeta) = 0$ has no multiple roots ζ for $\xi' \neq 0$. Further information can be found e.g. in (Hörmander 1963 and 1983b). On the other hand, there are examples of elliptic equations $Au = 0$ having solutions with compact support: see (Pliš 1954)) and also (Hörmander 1983b) and references therein.

c. Now we discuss the passage to the homogeneous equation $Au = 0$ in the boundary problem (1.1), in general without any assumption about uniqueness for the Cauchy problem (5.2).

Proposition 5.1.4. *There exists a linear unbounded operator*

$$A^{(-1)}: R(A) \to H_m(M)$$

with the following properties: it is a right inverse of A, and from $f \in R(A) \cap H_s(M)$ $(s \geq 0)$ it follows that $A^{(-1)}f \in H_{s+m}(M)$.

Having the exact right inverse of A, we can reduce the boundary problem (1.1) to the form

$$Au = 0 \quad \text{on} \quad M_+ \,, \quad B_j u = g_j \quad (j = 1, \ldots, q) \quad \text{on} \quad \Gamma \tag{5.6}$$

by the substitution

$$u \mapsto u + A^{(-1)}f \,, \quad g_j \mapsto g_j - B_j A^{(-1)} f \quad (j = 1, \ldots, q) \,. \tag{5.7}$$

As for the proof of Proposition 5.1.4, we first note that the existence of a right inverse of A follows from Banach's theorem on inverse operator: it suffices to apply it to $A: (\operatorname{Ker} A)^\bullet \to R(A)$, where $(\operatorname{Ker} A)^\bullet$ is a direct complement of $\operatorname{Ker} A$ in $H_m(M)$. However to prove the assertion on smoothness, we need some additional argument.

Denote by Ψ the operator $f \mapsto g$ defined by (5.4). Here $g \in C^\infty(\Gamma)$. Let $\mathcal{A}_B$ be the operator corresponding to the boundary problem (5.5) and sending $u \in H_m(M)$ to (f, g). Let $(\operatorname{Ker} \mathcal{A}_B)^\bullet$ be a direct complement of $\operatorname{Ker} \mathcal{A}_B$, for example the orthogonal complement with respect to the scalar product in $H_0(M)$. By Banach's theorem, there exists the operator $\mathcal{A}_B^{-1}$ inverse to $\mathcal{A}_B: (\operatorname{Ker} \mathcal{A}_B)^\bullet \to R(\mathcal{A}_B)$ (a right inverse of $\mathcal{A}_B: H_m(M) \to H_m(M, \Gamma)$). Set

$$A^{(-1)} f = \mathcal{A}_B^{-1}(f, \Psi f) \,. \tag{5.8}$$

It remains to use Theorem 2.4.2 on improved smoothness of solutions for the boundary problem (5.5).

Without using elliptic boundary problems, it is easy to construct the operator $A^{(-1)}$ if we assume that the operator A has an elliptic invertible extension on the closed manifold M_0; this extension may be a differential operator or at least a pseudodifferential operator that is differential near Γ. Retaining the notation A for this extension, we set

$$A^{(-1)}f = \mathcal{R}A^{-1}\mathcal{E}^{(0)}f\,, \tag{5.9}$$

where $\mathcal{R}$ is the operator of restriction of functions defined on M_0 to M, and $\mathcal{E}^{(0)}$ is the operator of extension of functions from M to M_0 by setting them equal to zero outside M. The operator $\mathcal{R}A^{-1}\mathcal{E}^{(0)}$ is bounded from $H_s(M)$ to $H_{s+m}(M)$. This can be verified by using the transmission property of the operator A^{-1} (see the definition in (Hörmander 1985a) or in the paper of Brenner and Shargorodsky in this volume).

Since $\operatorname{ord} A^{-1} < 0$, this pseudodifferential operator admits the representation

$$A^{-1}w(x) = \int_{M_0} \Phi(x,y)w(y)\,dy\,, \tag{5.10}$$

where the kernel $\Phi(x,y)$ is infinitely smooth for $x \neq y$ and has an integrable singularity at $x = y$; in local coordinates its asymptotics for $y \to x$ can be calculated (see (Seeley 1965) or the survey (Agranovich 1990a)). The function $\Phi(x,y)$ satisfies the equation

$$A(x, D_x)\Phi(x,y) = \delta_y(x) \tag{5.11}$$

in x in the sense of distributions on M_0, where $\delta_y(x)$ is the delta-function supported at the point y. In other words, $\Phi(x,y)$ is a *fundamental solution* for A. Thus, formula (5.9) can be rewritten in the form

$$(A^{(-1)}f)(x) = \int_M \Phi(x,y)f(y)\,dy \qquad (x \in M)\,, \tag{5.12}$$

where Φ is a fundamental solution for A.

5.2. Transition to Elliptic Equations on the Boundary.

a. Above we denoted by $\mathcal{A}_D$ the operator from $H_m(M)$ to $H_m(M,\Gamma)$ corresponding to the Dirichlet problem (1.22). As we know, this boundary problem is always elliptic. The following theorem is due to Browder (Browder 1961).

Theorem 5.2.1. *The index $\kappa(\mathcal{A}_D)$ is equal to zero.*

For the proof, we introduce the operator $\mathcal{A}_D^*$ corresponding to the Dirichlet problem formally adjoint to (1.22), and the operator $\overline{\mathcal{A}}_D$ corresponding to the Dirichlet problem for the operator $\overline{A}$ that is the complex adjoint of A (in the local representation (1.2) of A all the coefficients are replaced by the complex conjugate numbers). The assertion follows from the evident relations

$$\kappa(\mathcal{A}_D) = -\kappa(\mathcal{A}_D^*)\,, \quad \kappa(\mathcal{A}_D) = \kappa(\overline{\mathcal{A}}_D)\,, \quad \text{and} \quad \kappa(\mathcal{A}_D^*) = \kappa(\overline{\mathcal{A}}_D)\,.$$

Let $s \geq m = 2q$. Introduce the spaces

$$K(A,s) = \{u \in H_s(M): \ Au = 0 \ \text{ on } M_+\} \tag{5.13}$$

and

$$H_{(s)}(\Gamma) = H_{\{s-j+\frac{1}{2}\}}(\Gamma) = H_{s-\frac{1}{2}}(\Gamma) \times \ldots \times H_{s-(q-1)-\frac{1}{2}}(\Gamma) \tag{5.14}$$

and the operator

$$D_A u = (\gamma_0 u, \ldots, \gamma_{q-1} u)' ; \tag{5.15}$$

it is defined, in particular, on functions $u \in K(A,s)$ and associates with each of them the column of its Dirichlet data. Obviously, $K(A,s)$ is a (closed) subspace in $H_s(M)$, and D_A is a bounded operator from $K(A,s)$ to $H_{(s)}(\Gamma)$.

Theorem 5.2.2. *The operator D_A is Fredholm,*

$$\operatorname{Ker} D_A = \operatorname{Ker} \mathcal{A}_D , \quad \textit{and} \quad \kappa(D_A) = \dim N_0(A^*) . \tag{5.16}$$

In particular, D_A is invertible if and only if the homogeneous Dirichlet problem for A and the homogeneous Cauchy problem for A^* have no nontrivial solutions.

The Fredholm property of D_A follows from the Fredholm property of $\mathcal{A}_D$. The first relation in (5.16) is obvious, and from Theorem 4.3.1 it is easy to deduce that the number of linearly independent conditions of solvability of the equation $D_A u = g$ is equal to $\dim \operatorname{Ker} \mathcal{A}_D^* - \dim N_0(A^*)$.

b. Define the *Poisson operator $\mathcal{P} = \mathcal{P}_D = \mathcal{P}_{D_A}$ of the Dirichlet problem* for the equation $Au = 0$ on M_+ as a precise two-sided parametrix for D_A, i.e. as a bounded operator from $H_{(s)}(\Gamma)$ to $K(A,s)$ $(s \geq m)$ such that

$$D_A \cdot \mathcal{P}_{D_A} = \mathcal{I}_1 + \mathcal{T}_1 \quad \text{and} \quad \mathcal{P}_{D_A} \cdot D_A = \mathcal{I}_2 + \mathcal{T}_2 , \tag{5.17}$$

where $\mathcal{I}_1$ and $\mathcal{I}_2$ are the identity operators in $H_{(s)}(\Gamma)$ and $K(A,s)$, and $\mathcal{T}_1$ and $\mathcal{T}_2$ are finite-dimensional operators in these spaces with ranges lying in $[H_\infty(\Gamma)]^q$ and $\{u \in H_\infty(M) : Au = 0 \text{ on } M\}$, respectively. The existence of such a parametrix is verified as in the proof of Theorem 2.4.5. If the operator D_A is invertible, then we set $\mathcal{P}_{D_A} = (D_A)^{-1}$. We now describe the construction of some approximation $\mathcal{P}_N = \mathcal{P}_{D_A,N}$ of $\mathcal{P}$ with an accuracy that grows unboundedly as $N \to \infty$.

Consider first the Dirichlet problem in a halfspace,

$$Au = 0 \quad \text{in} \quad \mathbb{R}^n_+ , \quad D_n^{j-1} u(x)|_{x^n=0} = g_j(x') \quad (j = 1, \ldots, q) , \tag{5.18}$$

assuming that the operator A is (properly) elliptic in $\overline{\mathbb{R}^n_+}$ and its coefficients do not depend on x for sufficiently large $|x|$. We intend to construct a right parametrix for this boundary problem; the function $v = \mathcal{P}_N g$, where $g = (g_1, \ldots, g_q)'$, will be a good approximation of the solution for large N and small $x^n > 0$.

We set

$$\mathcal{P}_N = (\mathcal{P}_{N,1}, \ldots, \mathcal{P}_{N,q}) , \tag{5.19}$$

where the scalar operators $\mathcal{P}_{N,l}$ are defined by the formula

$$(\mathcal{P}_{N,l}h)(x) = \sum_{s=0}^{N} \frac{1}{(2\pi)^{n-1}} \int_{\mathbb{R}_{n-1}} e^{ix'\cdot\xi'} p_{s,l}(x,\xi')\theta(\xi')(F'h)(\xi')\, d\xi' \tag{5.20}$$

on functions $h(x') \in C_0^\infty(\mathbb{R}^{n-1})$. Here $\theta(\xi')$ is a function from $C_0^\infty(\mathbb{R}_{n-1})$ equal to 1 for $|\xi'| \geq 1$ and 0 for small $|\xi'|$. We assume the function $p_{s,l}(x,\xi')$ to be positive homogeneous in $((x^n)^{-1}, \xi')$ of degree $1-l-s$; from its construction (described below) it will be clear that it is infinitely smooth for $x^n > 0$, $\xi' \neq 0$, and its modulus tends to zero (exponentially) as $x^n \to +\infty$:

$$|p_{s,l}(x,\xi')| \leq C_{s,l}(1+|\xi'|)^{1-l-s} e^{-cx^n|\xi'|} , \tag{5.21}$$

where the positive constants $C_{s,l}$ and c do not depend on (x,ξ'). We do not write out the analogous estimates for the derivatives of $p_{s,l}$.

To define $p_{s,l}(x,\xi')$, we replace all coefficients in the complete symbol $a(x,\xi)$ of A by their Taylor expansions in powers of x^n, and denote by $a^{(r)}(x,\xi)$ the sum of the terms in $a(x,\xi)$ that are positive homogeneous in $((x^n)^{-1}, \xi')$ of degree $m-r$. We require that

$$\sum_{r\geq 0} a^{(r)}(x,\xi',D_n) \circ \sum_{s\geq 0} p_{s,l}(x,\xi') = 0 \qquad (x^n > 0\, ;\, l = 1, \ldots, q) , \tag{5.22}$$

$$D_n^{j-1} \sum_{s\geq 0} p_{s,l}(x,\xi')|_{x^n=0} = \delta_l^j \qquad (j,l = 1, \ldots, q) , \tag{5.23}$$

$$p_{s,l}(x,\xi') \to 0 \qquad (x^n \to +\infty) . \tag{5.24}$$

In (5.22) $\circ$ is the composition of symbols depending on (x',ξ'); x^n is considered as a parameter. This actually is a consequence of the structure of the integrals in (5.20). Explicitly, relations (5.22) are written in the form

$$\sum_{\alpha';r;s} \frac{1}{\alpha'!}[\partial_{\xi'}^{\alpha'} a^{(r)}](x,\xi',D_n) D_{x'}^{\alpha'} p_{s,l}(x,\xi') = 0 \qquad (x^n > 0) . \tag{5.25}$$

Now, fixing l, we separate the terms with $|\alpha'| + r + s = 0, 1, \ldots$ in (5.25), and the terms homogeneous of degree $j-l-s$ with respect to ξ' in (5.23), $s = 0, 1, \ldots$, regarding $\delta_j^j = 1$ as homogeneous functions of degree 0. We obtain a sequence of boundary problems on $\mathbb{R}_+$ for successive definition of the functions $p_{0,l}, p_{1,l}, \ldots$; the first boundary problem has the form

$$a_0(x',0,\xi',D_n) p_{0,l}(x,\xi') = 0 \qquad (x^n > 0) , \tag{5.26}$$

$$D_n^{j-1} p_{0,l}(x,\xi')|_{x^n=0} = \delta_l^j \qquad (j = 1, \ldots, q) , \tag{5.27}$$

$$p_{0,l}(x,\xi') \to 0 \qquad (x^n \to +\infty) . \tag{5.28}$$

In the next boundary problems $p_{0,l}$ is replaced by $p_{1,l}$, etc. The right-hand side appears in (5.26), and δ_l^j on the right in (5.27) is replaced by zero, but the operators on the left remain the same.

The functions $p_{0,l}(x,\xi')$ form the standard basis in the space of decreasing solutions of the boundary problem (5.26)–(5.27); in Sect. 1.3 they are denoted by $v_l(\xi', x^n)$ and are expressed by integrals (1.19); but now they additionally depend on x'. From these formulas we see, in particular, that for $p_{0,l}(x,\xi')$ inequality (5.21) holds. In the sequel, i.e. in defining $p_{1,l}$ etc., we consider nonhomogeneous boundary problems, but their solutions are also written in the form of contour integrals, for which the necessary estimates are obtained.

Returning to the Dirichlet problem on M, we define the desired approximation $\mathcal{P}_N = \mathcal{P}_{D,N}$ of the Poisson operator $\mathcal{P} = \mathcal{P}_D$ by the formula

$$\mathcal{P}_N = \sum_{k=K'+1}^{K} \psi_k \mathcal{P}_N^{(k)}(\varphi_k \cdot) . \tag{5.29}$$

Here $\mathcal{P}_N^{(k)}$ are the operators (5.19) defined in local coordinates from Sect. 0.2 near the boundary; $\{\varphi_k\}$ and $\{\psi_k\}$ are the same systems of functions as in Sect. 2.1.

The following theorem can be proved by means of these approximations.

Theorem 5.2.3. *Let $B = (B_1, \dots, B_q)'$ be the column of boundary operators* (1.3). *Then the composition*

$$B\mathcal{P} = (B_1, \dots, B_q)'(\mathcal{P}_1, \dots, \mathcal{P}_q) \tag{5.30}$$

is a $q \times q$ matrix consisting of polyhomogeneous pseudodifferential operators $B_j\mathcal{P}_k$ of orders $r_j - k + 1$ on Γ; the matrix of the principal symbols of these operators coincides with the Lopatinskij matrix of the boundary problem (1.8)–(1.9).

Thus, if we assume that s satisfies conditions (2.21), then the substitution $u = \mathcal{P}_D w$ reduces the boundary problem (5.6) to the system

$$B\mathcal{P}_D w = g , \tag{5.31}$$

and this system is elliptic (in the Douglis–Nirenberg sense, of type $\{r_j, 1-k\}$) if and only if the boundary problem (5.6) is elliptic. The system (5.31) is equivalent to the boundary problem (5.6) if the operator $\mathcal{P}_D$ is invertible. In the general case we have the equivalence modulo finite-dimensional subspaces.

c. More generally, instead of the Dirichlet problem we can use any other elliptic boundary problem.

Assuming the operator A on M to be properly elliptic, and s to satisfy conditions (2.21), we introduce the operator

$$B_A u = (B_1 u, \dots, B_q u)' \tag{5.32}$$

from $K(A,s)$ to $H_{\{s-r_j-\frac{1}{2}\}}(\Gamma)$ (see (2.22)). This operator is bounded, and its Fredholm property is equivalent to the ellipticity of the boundary problem (1.1). The following theorem is a simple generalization of Theorem 5.2.2:

Theorem 5.2.4. *Let the boundary problem* (1.1) *be elliptic. Then*

$$\operatorname{Ker} B_A = \operatorname{Ker} \mathcal{A}_B \quad \textit{and} \quad \kappa(B_A) = \kappa(\mathcal{A}_B) + \dim N_0(A^*)\ . \tag{5.33}$$

In particular, the operator B_A is invertible if and only if $\operatorname{Ker} \mathcal{A}_B = \{0\}$ *and* $\kappa(\mathcal{A}_B) = -\dim N_0(A^*)$.

We define the *Poisson operator* $\mathcal{P}_B = \mathcal{P}_{B_A}$ *of the boundary problem* (5.6) as a precise two-sided parametrix for B_A, i.e. as a bounded operator from $H_{\{s-r_j-\frac{1}{2}\}}(\Gamma)$ to $K(A,s)$ such that

$$B_A \cdot \mathcal{P}_{B_A} = \mathcal{I}_1 + \mathcal{T}_1 \quad \text{and} \quad \mathcal{P}_{B_A} \cdot B_A = \mathcal{I}_2 + \mathcal{T}_2\ , \tag{5.34}$$

where $\mathcal{I}_1$ and $\mathcal{I}_2$ are identity operators in $H_{\{s-r_j-\frac{1}{2}\}}(\Gamma)$ and $K(A,s)$, respectively, and $\mathcal{T}_1$ and $\mathcal{T}_2$ are finite-dimensional operators in these spaces with ranges lying in $[H_\infty(\Gamma)]^q$ and $\{u \in H_\infty(M) : Au = 0 \text{ on } M_+\}$, respectively. If the operator B_A is invertible, then we set $\mathcal{P}_{B_A} = B_A^{-1}$. Note that if Q is a precise parametrix for $B\mathcal{P}_D$, then

$$\mathcal{P}_B = \mathcal{P}_D Q \tag{5.35}$$

at least modulo addition of an operator bounded from $H_{\{s-r_j-\frac{1}{2}\}}(\Gamma)$ to $K(A, s+\tau)$ for any $\tau > 0$.

The approximation $\mathcal{P}_{B,N}$ of the operator $\mathcal{P}_B$ can be constructed 1) similarly to $\mathcal{P}_{D,N}$, or 2) in the form $\mathcal{P}_{B,N} = \mathcal{P}_{D,N}Q$ or $\mathcal{P}_{D,N}Q_N$, where Q_N is an approximation of Q constructed by means of the N first terms in the expansion of the local complete symbol of Q (it is not difficult to make this assertion more precise, but we do not dwell on this).

Now let the second elliptic boundary problem be given:

$$Au = 0 \quad \text{on} \quad M_+\ , \quad C_j u = g_j \quad (j = 1, \dots, q) \quad \text{on} \quad \Gamma\ , \tag{5.36}$$

where $\operatorname{ord} C_j = s_j$. The substitution $u = \mathcal{P}_B w$ reduces this boundary problem to the system

$$C\mathcal{P}_B w = g\ , \tag{5.37}$$

which is elliptic in the Douglis–Nirenberg sense, of type $\{s_j, -r_k\}$. This system is equivalent to the boundary problem (5.36) if the operator B_A is invertible. The principal symbol of the operator on the left in (5.37) is equal to

$$L_C(x', \xi') \cdot [L_B(x', \xi')]^{-1},$$

where L_B and L_C are the Lopatinskij matrices for the boundary problems (5.6) and (5.36).

5.3. Boundary Problems with Parameter only in Boundary Conditions. Consider the boundary problem

$$Au = f \quad \text{on} \quad M_+ , \quad \sum_{l=0}^{N} \lambda^{N-l} B_{j,l} u = g_j \quad (j = 1, \dots, q) \quad \text{on} \quad \Gamma . \quad (5.38)$$

Here A is a properly elliptic differential operator of order $m = 2q$ on M; $N \in \mathbb{N}$. The boundary operators are written in a form different from that in (3.1): in (5.38) $\operatorname{ord} B_{j,l} = \rho_j + l$, where $\rho_j \in \mathbb{Z}_+$. For simplicity, we assume that the weight of the parameter is equal to 1. The number s is subjected to the conditions

$$s \geq m , \quad s > \max \left(\rho_j + N + \frac{1}{2} \right) . \quad (5.39)$$

Let $\mathcal{L}$ be a closed angle on the complex plane with vertex at the origin. Following Panich, we call the boundary problem (5.38) (with a parameter only in boundary conditions) *elliptic with parameter in* $\mathcal{L}$ if the following two conditions are fulfilled:

1) The boundary problem

$$Au = 0 \quad \text{on} \quad M_+ , \quad B_{j,0} u = g_j \quad (j = 1, \dots, q) \quad \text{on} \quad \Gamma \quad (5.40)$$

is uniquely solvable in $H_s(M)$ for any $g \in H_{\{s-\rho_j-\frac{1}{2}\}}(\Gamma)$, i.e. the corresponding operator

$$B_{0,A} = (B_{1,0}, \dots, B_{q,0})' \colon K(A, s) \to H_{\{s-\rho_j-\frac{1}{2}\}}(\Gamma) \quad (5.41)$$

is invertible.

2) For any point on Γ, the boundary problem on the ray

$$a_0(x', 0, \xi', D_n) v(t) = 0 \qquad (t = x^n > 0) , \quad (5.42)$$

$$\sum_{l=0}^{N} \lambda^{N-l} b_{j,l,0}(x', 0, \xi', D_n) v(t)|_{t=0} = h_j \qquad (j = 1, \dots, q) \quad (5.43)$$

has one and only one solution $v(t)$ in the space of solutions $\mathfrak{M}(\xi')$ to equation (5.42) with $|v(t)| \to 0$ as $t \to +\infty$, for any $(\xi', \lambda) \in \mathbb{R}_{n-1} \times \mathcal{L}$, $(\xi', \lambda) \neq 0$, and any complex numbers h_j. Here a_0 and $b_{j,l,0}$ are the principal symbols of the operators A and $B_{j,0}$ written in local coordinates of Sect. 0.2; the boundary point under consideration has the coordinates $(x', 0)$.

Under condition 1) we can use the Poisson operator $\mathcal{P}_{B_{0,A}}$ inverse to $B_{0,A}$ and reduce the boundary problem (5.40) to a system of pseudodifferential equations on Γ depending on the parameter polynomially:

$$\sum_{l=0}^{N} \lambda^{N-l} C^{(l)} w = g , \quad (5.44)$$

where $C^{(l)}$ is a matrix pseudodifferential operator with rows $B_{j,l}\mathcal{P}_{B_{0,A}}$; its type is $\{\rho_j + l, -\rho_k\}$, and $C^{(0)} = I$. This system is equivalent to the boundary problem. Let $c_0^{(l)}(x', \xi')$ be the principal symbol of $C^{(l)}$. Under condition 2), as can be seen from what has been said at the end of Sect. 5.2,

$$\det \sum_{l=0}^{N} \lambda^{N-l} c_0^{(l)}(x', \xi') \neq 0 \qquad (\lambda \in \mathcal{L},\ (x', \xi') \in T^*\Gamma \setminus 0)\ , \tag{5.45}$$

i.e. the system (5.44) is elliptic with parameter in $\mathcal{L}$ in the Douglis–Nirenberg sense (see (Agranovich 1990a, Subsect. 4.3e)). Hence, it is uniquely solvable for $\lambda \in \mathcal{L}$ with sufficiently large modulus; in addition, the a priori estimate

$$\sum_{k=0}^{q} \|w_k\|_{\sigma+N-\rho_k,\Gamma} + |\lambda|^N \sum_{k=1}^{q} \|w_k\|_{\sigma-\rho_k,\Gamma} \leq C_\sigma \sum_{j=1}^{q} \|g_j\|_{\sigma-\rho_j,\Gamma} \tag{5.46}$$

holds ($\sigma \in \mathbb{R}$), where C_σ does not depend on w and λ. Setting $\sigma = s - N - \frac{1}{2}$, we obtain the following result:

Theorem 5.3.1. *Assume that the boundary problem* (5.38) *is elliptic with parameter in $\mathcal{L}$, and that s satisfies conditions* (5.39). *Then for $f = 0$ and $\lambda \in \mathcal{L}$ with sufficiently large modulus, this boundary problem is uniquely solvable, and the a priori estimate*

$$\|u\|_{s,M} + |\lambda|^N \|u\|_{s-N,M} \leq C_s' \sum_{j=1}^{q} \|g_j\|_{s-\rho_j-N-\frac{1}{2},\Gamma} \tag{5.47}$$

holds with a constant C_s' not depending on u and λ.

From this and from Corollary 5.1.2 we obtain

Corollary 5.3.2. *If, in addition to the assumptions of Theorem* 5.3.1, *$s \geq m + N$ and the Cauchy problem* (5.2) *has no nontrivial solutions, then the boundary problem* (5.38) *is uniquely solvable for $\lambda \in \mathcal{L}$ with sufficiently large modulus, and the a priori estimate*

$$\|u\|_{s,M} + |\lambda|^N \|u\|_{s-N,M}$$
$$\leq C_s'' \Big\{ \|f\|_{s-m,M} + |\lambda|^N \|f\|_{s-N-m,M} + \sum_{j=1}^{q} \|g_j\|_{s-\rho_j-N-\frac{1}{2},\Gamma} \Big\} \tag{5.48}$$

holds with C_s'' not depending on u and λ.

Example 5.3.3. Let $b(x)$ be a complex-valued function belonging to $C^\infty(\Gamma)$ and different from zero everywhere, with $|\arg b(x)| \leq \alpha < \pi$. Then the boundary problem

$$\Delta u = f \quad \text{in} \quad G\,, \quad b(x)\gamma_0\partial_\nu u(x) + \lambda\gamma_0 u(x) = g(x) \quad \text{on} \quad \Gamma \tag{5.49}$$

is elliptic with parameter in the angle $\{\lambda : |\arg \lambda| \geq \alpha + \varepsilon\}$ for arbitrarily small $\varepsilon > 0$, and is uniquely solvable for λ in this angle with sufficiently large modulus, with the estimate

$$\|u\|_{s,G} + |\lambda| \|u\|_{s-1,G} \leq C_s(\|f\|_{s-2,G} + |\lambda| \|f\|_{s-3,G} + \|g\|_{s-\frac{3}{2},\Gamma}) \qquad (s \geq 3) .$$

Here we use the unique solvability of the Dirichlet problem $\Delta u = f$ in G, $\gamma_0 u = g$ on Γ ($u \in H_2(G)$). The last fact is well known; recall that it follows from the equality to zero of the index of the corresponding operator (see Theorem 5.2.1) and from uniqueness. In turn, uniqueness is verified by means of integration by parts in the case $f = 0$, $g = 0$:

$$0 = (\Delta u, u)_G = -\int_G \sum |\partial_j u|^2 dx , \tag{5.50}$$

which, along with the boundary condition $\gamma_0 u = 0$, implies that $u = 0$. Condition 2) can be verified directly.

5.4. Cauchy Data and Calderón Projectors.

a. Example 5.4.1. Consider the *Helmholtz equation*

$$\Delta u + k^2 u = 0 \tag{5.51}$$

in $\mathbb{R}^3$ outside a smooth bounded closed surface Γ that divides its complement into the inner part G_+ and the outer part G_-. For simplicity, let $k > 0$. The solution in G_- is subjected to the *radiation condition* at infinity:

$$\frac{\partial u(x)}{\partial r} + iku(x) = o\Big(\frac{1}{r}\Big) \quad \text{for} \quad r = |x| \to \infty . \tag{5.52}$$

Here we will construct the so-called *Calderón projectors* for the equation (5.51). By means of these projectors, any sufficiently smooth vector-valued function $\varphi = (\varphi_1, \varphi_2)'$ on Γ can be decomposed into the sum of the Cauchy data of a solution to the equation (5.51) in G_+, and the Cauchy data of a solution to this equation in G_- with the radiation condition at infinity.

The fundamental solution $\Phi(x)$ of the Helmholtz equation (i.e. the solution of the equation $\Delta\Phi + k^2\Phi = \delta(x)$) satisfying condition (5.52) has the form

$$\Phi(x) = -\frac{1}{4\pi|x|} e^{-ik|x|} . \tag{5.53}$$

We now recall the integral formulas for solutions to the Helmholtz equation (see e.g. (Colton and Kress 1983)). These formulas are valid for solutions in G_+ belonging to $H_2(G)$ and solutions in G_- satisfying the radiation condition and locally belonging to H_2 near Γ (inside G_+ and G_- they belong to C^∞). Let $\chi_\pm(x)$ be functions equal to 1 in $G_\pm$ and 0 in $G_\mp$. To simplify the notation, we here denote by $u^\pm$ the boundary values of solutions u in G_+ and G_- and by $\partial_\nu u^\pm$ the values on Γ of their derivatives in the direction of the inner normal to Γ. We have

$$\chi_+(x)u(x) = \int_\Gamma [\Phi(x-y)\partial_\nu u^+(y) - (\partial_{\nu,y}\Phi(x-y))u^+(y)]\, dS_y\,, \qquad (5.54')$$

$$\chi_-(x)u(x) = -\int_\Gamma [\Phi(x-y)\partial_\nu u^-(y) - (\partial_{\nu,y}\Phi(x-y))u^-(y)]\, dS_y \qquad (5.54'')$$

for $x \notin \Gamma$. Here dS_y is the area element on Γ, and $\partial_{\nu,y}\Phi(x-y)$ is the derivative of $\Phi(x-y)$ along the inner normal at y. The integrals over Γ in the right-hand side with Φ and $\partial_{\nu,y}\Phi$ in the integrand are the well-known single layer potential and double layer potential, respectively, for the Helmholtz equation. Using properties of these potentials and passing to the limit as $x \to \Gamma$ from G_+ in (5.54′) and from G_- in (5.54″), we obtain

$$u^+ = \Big(\frac{1}{2}I - T_2\Big)u^+ + T_1\partial_\nu u^+\,, \qquad (5.55')$$

$$u^- = \Big(\frac{1}{2}I + T_2\Big)u^- - T_1\partial_\nu u^- \qquad (5.55'')$$

on Γ. Here [9]

$$\begin{aligned} T_1 v(x) &= \int_\Gamma \Phi(x-y)v(y)\, dS_y \qquad (x\in\Gamma)\,, \\ T_2 v(x) &= \int_\Gamma (\partial_{\nu,y}\Phi(x-y))v(y)\, dS_y \qquad (x\in\Gamma)\,. \end{aligned} \qquad (5.56)$$

Taking into account the asymptotics of the kernels of these integral operators for $y \to x$, we can verify that they are polyhomogeneous pseudodifferential operators on Γ. More precisely, T_1 is an elliptic pseudodifferential operator of order -1, and T_2 is a pseudodifferential operator of order not greater than -1. In addition it is easy to verify that the operator T_1 has an inverse of order 1 for the given k if and only if the homogeneous Dirichlet problem

$$\Delta u + k^2 u = 0 \quad \text{in} \quad G_+\,, \quad u^+ = 0 \quad \text{on} \quad \Gamma \qquad (5.57)$$

has no nontrivial solutions, i.e. k^2 is not an eigenvalue of $-\Delta_D$, where Δ_D is the "Dirichlet Laplacian" (see (Agranovich 1977)). For simplicity we assume that this last condition holds, so we exclude a countable set $\{k\}$, $k \to \infty$, from consideration.

Now we differentiate relations (5.54′) for $x \in G_+$ and (5.54″) for $x \in G_-$ along the inner normal and again pass to the limit as $x \to \Gamma$. We obtain

$$\partial_\nu u^+ = -T_4 u^+ + \Big(\frac{1}{2}I + T_3\Big)\partial_\nu u^+\,, \qquad (5.58')$$

$$\partial_\nu u^- = T_4 u^- + \Big(\frac{1}{2}I - T_3\Big)\partial_\nu u^- \qquad (5.58'')$$

on Γ, where

[9] In (Agranovich 1990a, Sect. 2.2) the operator T_1 is denoted by A. In (Agranovich 1977) the operators T_1 and $-T_2$ are denoted by A and B, respectively.

$$\begin{aligned}T_3v(x) &= \int_\Gamma \partial_{\nu,x}\Phi(x-y)v(y)\,dS_y \qquad (x\in\Gamma)\,,\\ T_4v(x) &= \partial_{\nu,x}\int_\Gamma (\partial_{\nu,y}\Phi(x-y))v(y)\,dS_y \qquad (x\in\Gamma)\,,\end{aligned} \tag{5.59}$$

and $\partial_{\nu,x}\Phi(x-y)$ is the derivative of $\Phi(x-y)$ along the inner normal at x. Again, here T_3 is a pseudodifferential operator of order not greater than -1; obviously, T_2 and T_3 are mutually transposed operators, i.e.

$$\int_\Gamma T_3\varphi\cdot\psi\,dS = \int_\Gamma \varphi\cdot T_2\psi\,dS\,. \tag{5.60}$$

To clarify the meaning of T_4, we define $w(x)$ as the solution of the Dirichlet problem

$$\Delta w + k^2 w = 0 \quad \text{in} \quad G\,, \quad \partial_\nu w^+ = v \quad \text{on} \quad \Gamma\,. \tag{5.61}$$

Under our assumption concerning k, this solution is defined uniquely, at least for $v\in H_{3/2}(\Gamma)$. From a formula of the form (5.54′) we obtain

$$\int_\Gamma (\partial_{\nu,y}\Phi(x-y))v(y)\,dS_y = \int_\Gamma \Phi(x-y)\partial_\nu w^+(y)\,dS_y - \chi_+(x)w(x)$$

for $x\notin\Gamma$. It follows that there exist

$$\partial^+_{\nu,x}\int_\Gamma \partial_{\nu,y}\Phi(x-y))v(y)\,dS_y = \int_\Gamma \partial_{\nu,x}\Phi(x-y)\partial_\nu w^+(y)\,dS_y - \frac{1}{2}\partial_\nu w^+(x)\,. \tag{5.62}$$

Furthermore, from (5.55′) we obtain

$$\Big(\frac{1}{2}I + T_2\Big)v = T_1\partial_\nu w^+\,,$$

From the last two formulas we obtain

$$\begin{aligned}T_4 &= T_3T_1^{-1}\Big(\frac{1}{2}I+T_2\Big) - \frac{1}{2}T_1^{-1}\Big(\frac{1}{2}I+T_2\Big)\\ &= -\frac{1}{4}T_1^{-1} + T_3T_1^{-1}T_2 = -\frac{1}{4}T_1^{-1} + T_2'T_1^{-1}T_2\,,\end{aligned} \tag{5.63}$$

since $T_2T_1 = T_1T_3$ (see e.g. (Agranovich 1977), p. 365) and $T_3' = T_2$ in view of (5.60). From (5.63) we see that T_4, along with T_1^{-1}, is an elliptic pseudodifferential operator of the first order.[10] Now we set (only in this subsection)

$$\gamma^\pm u = (u^\pm, -i\partial_\nu u^\pm)' \tag{5.64}$$

[10]Like T_1, the operator T_4 possesses the property $T_4^* = \overline{T_4}$ (i.e. $T_4^*v = \overline{T_4\overline{v}}$). From this it is easy to deduce that, just as T_1, the operator T_4 is "infinitely close to a selfadjoint operator:" $\operatorname{Im} T_4 = (T_4 - T_4^*)/(2i)$ has the order $-\infty$. Here we have an additional example for Sect. 6.2 in (Agranovich 1990a).

and rewrite formulas (5.55) and (5.58) in the form

$$\gamma^{\pm}u = P^{\pm}\gamma^{\pm}u\,, \tag{5.65}$$

where

$$P^{\pm} = \begin{pmatrix} \frac{1}{2}I \mp T_2 & \pm iT_1 \\ \pm iT_4 & \frac{1}{2}I \pm T_3 \end{pmatrix}. \tag{5.66}$$

The matrix pseudodifferential operators $P^{\pm}$ have the type $\{j, -k\}$ $(j, k = 1, 2)$. They are bounded in the space $H_s(\Gamma) \times H_{s-1}(\Gamma)$. It is easy to verify that

$$(P^{\pm})^2 = P^{\pm}\,, \quad P^{+} + P^{-} = I\,, \quad \text{and} \quad P^{+}P^{-} = P^{-}P^{+} = 0\,. \tag{5.67}$$

Thus, P^+ and P^- are mutually complementary projectors in $H_s(\Gamma) \times H_{s-1}(\Gamma)$. These operators are called the Calderón projectors for the Helmholtz equation (5.51).

Let $\varphi = (\varphi_1, \varphi_2)'$ be an arbitrary vector-valued function from $H_s(\Gamma) \times H_{s-1}(\Gamma)$, $s \geq 3/2$. Then the formula

$$u(x) = \pm \int_{\Gamma} \{\Phi(x-y)\varphi_2(y) - (\gamma_{1,y}\Phi)(x-y)\varphi_1(y)\}\, dS_y \qquad (x \notin \Gamma) \tag{5.68}$$

defines a solution of the Helmholtz equation in $G_{\pm}$ that belongs to $H_{s+\frac{1}{2}}$ in G_+ and locally in G_- near Γ, and satisfies the radiation condition at infinity. In the same way as above, we obtain that

$$\gamma^{\pm}u = P^{\pm}\varphi\,. \tag{5.69}$$

As the final result we find that at least for $s \geq 3/2$ the projectors P^+ and P^- provide a decomposition of the space $H_s(\Gamma) \times H_{s-1}(\Gamma)$ into the direct sum of the spaces of the Cauchy data for the Helmholtz equation in G_+ and in G_- with the radiation condition at infinity.

The principal symbols of the pseudodifferential operators P^+ and P^- are

$$\begin{pmatrix} \frac{1}{2} & \pm i\sigma \\ \mp\frac{1}{4}i\sigma^{-1} & \frac{1}{2} \end{pmatrix}, \tag{5.70}$$

where σ is the principal symbol of T_1. These matrices are mutually complementary projectors in $\mathbb{C}^2$. Their rank is equal to 1.

The paper (Seeley 1966) begins with analysis of another example: the *Cauchy–Riemann operator* $\partial/\partial\bar{z} = (\partial_1 + i\partial_2)/2$ in $\mathbb{R}^2$ is considered as defining the decomposition of the space $H_{1/2}(S)$ on the unit circle into the direct sum of the Hardy spaces $H_{\pm}$ of boundary values of the functions holomorphic inside and outside the unit disk with zero at infinity.

b. Now let A be an elliptic differential operator of order $m = 2q$ on M. We first assume that it has an extension that is an elliptic invertible differential operator on the closed manifold M_0. We denote the extension again by A. We need Green's formula (4.12), but at present it is more convenient to write it in the form

$$(u, A^*v)_M - (Au, v)_M = (\mathfrak{A}\gamma^+ u, \gamma^+ v)_\Gamma \,. \tag{5.71}$$

Here and further

$$\gamma^\pm u = (\gamma_0^\pm u, \ldots, \gamma_{m-1}^\pm u)', \tag{5.72}$$

and $\mathfrak{A}$ is a matrix consisting of differential operators on Γ. More precisely,

$$\mathfrak{A} = -i \begin{pmatrix} \mathfrak{A}_{1,1} & \mathfrak{A}_{1,2} & \ldots & \mathfrak{A}_{1,m-1} & \mathfrak{A}_{1,m} \\ \mathfrak{A}_{2,1} & \mathfrak{A}_{2,2} & \ldots & \mathfrak{A}_{2,m-1} & 0 \\ \vdots & \vdots & & \vdots & \vdots \\ \mathfrak{A}_{m,1} & 0 & \ldots & 0 & 0 \end{pmatrix}, \tag{5.73}$$

the order of $\mathfrak{A}_{j,k}$ is equal to $m+1-j-k$, and

$$\mathfrak{A}_{1,m} = \mathfrak{A}_{2,m-1} = \ldots = \mathfrak{A}_{m,1} = A_m \quad \text{on} \quad \Gamma \,, \tag{5.74}$$

where, as in (4.10), A_m is the coefficient of the derivative of highest order in A with respect to $t = x^n$. In view of the ellipticity of A, the function A_m is different from zero everywhere, so the operator $\mathfrak{A}$ has an inverse, which is also a matrix consisting of differential operators.

Since A is invertible on M_0, the formally adjoint operator A^* is also invertible (see e.g. (Agranovich 1990a, Sect. 2.3)). It follows that there exist fundamental solutions $\Phi(x, y)$ and $\Phi^{(*)}(x, y)$ for A and A^*, and since they are the kernels of mutually adjoint operators, we have

$$\Phi^{(*)}(x, y) = \overline{\Phi(y, x)} \tag{5.75}$$

in our local coordinates (see Sect. 0.2).

We now describe formulas for the reconstruction of solutions of the equation $Au = 0$ on M_+ and M_- in terms of their Cauchy data $\gamma^+ u$ and $\gamma^- u$ on Γ: see (5.79) below. We apply formula (5.71) to a function $u(y) \in H_m(M)$ satisfying the equation $Au = 0$ on M_+ and to the function $v(y) = \Phi^{(*)}(y, x)$. We obtain

$$u(x) = (\mathfrak{A}\gamma^+ u(y), \gamma_y^+ \Phi^{(*)}(y, x))_\Gamma \qquad (x \in M). \tag{5.76}$$

(Actually, $v(y)$ has a singularity at $y = x$, and we need to apply Green's formula in $M \setminus V_\varepsilon(x)$, where $V_\varepsilon(x)$ is a neighborhood of x that shrinks to x as $\varepsilon \to 0$; this procedure is well known.) Formula (5.76) is already the representation of u in terms of $\gamma^+ u$, and clearly the right-hand side is a sum of certain surface potentials (or coboundary operators) with densities $\gamma_0^+ u, \ldots, \gamma_{m-1}^+ u$. We now transform this formula. With each smooth function v on M_0 the operator $\gamma = \gamma^+$ associates the set of if its traces on Γ; the adjoint operator γ^* is defined by the relation

$$(\gamma^* g, v)_{M_0} = (g, \gamma v)_\Gamma \,. \tag{5.77}$$

We set

$$\mathcal{H}^{(s)}(\Gamma) = \prod_{j=0}^{m-1} H_{s-j-\frac{1}{2}}(\Gamma) \, . \tag{5.78}$$

Obviously, γ^* is a bounded operator from

$$[\mathcal{H}^{(s)}(\Gamma)]^* = \prod_{j=0}^{m-1} H_{-s+j+\frac{1}{2}}(\Gamma)$$

to $H_s(M)$ for $s > m - (1/2)$. Rewrite formula (5.76) in the form

$$\begin{aligned} u(x) &= (\gamma^* \mathfrak{A} \gamma u(y), \Phi^{(*)}(y,x))_{M_0} \\ &= (\Phi(x,y)\gamma^* \mathfrak{A} \gamma u(y), 1))_{M_0} = (A^{-1}\gamma^* \mathfrak{A}\gamma u)(x) \qquad (x \in M) \, . \end{aligned}$$

An analogous formula, but with the opposite sign, can be written for solutions of the equation $Au = 0$ on M_-. Thus, for solutions $u(x)$ of the equation $Au = 0$ on M_+ and on M_- we obtain the formulas

$$u = W^{\pm}(\gamma^{\pm} u) \quad \text{on} \quad M_{\pm} \, , \quad \text{where} \quad W^{\pm} = \pm \mathcal{R}_{\pm}(A^{-1}(\gamma^{\pm})^* \mathfrak{A} \, \cdot \,) \tag{5.79}$$

and $\mathcal{R}_{\pm}$ is the operation of restriction of functions on M_0 to $M_{\pm}$. These are the key formulas in (Seeley 1966). Now we formulate the main results of this paper (see the definition of Sobolev spaces of negative order in Subsect. 2.1f).

Theorem 5.4.2. *The operators $W^{\pm}$ are bounded from $\mathcal{H}^{(s)}$ to $H_s(M_{\pm})$ for all real s, and their ranges lie in*

$$K_{\pm}(A, s) = \{u \in H_s(M_{\pm}) : \ Au = 0 \ \text{on} \ M_{\pm}\} \, . \tag{5.80}$$

Here the equation $Au = 0$ is understood in the sense of distributions. Since A is elliptic, the elements of these spaces belong to $C^\infty(M_{\pm})$ (we recall that $M_{\pm}$ do not contain Γ). For functions $u \in C^\infty(M_{\pm})$ we set

$$\gamma(t)u(x) = (u, D_t u, \ldots, D_t^{m-1} u)(x', t) \qquad (t = x^n, \ 0 < |t| < 1) \, . \tag{5.81}$$

Theorem 5.4.3. *If $g \in \mathcal{H}^{(s)}(\Gamma)$ $(s \in \mathbb{R})$, then there exist the limits*

$$\lim_{t \to \pm 0} \gamma(t) W^{\pm} g = P^{\pm} g \tag{5.82}$$

in the sense of the norm in $\mathcal{H}^{(s)}(\Gamma)$. The operators $P^{\pm}$ (called Calderón projectors) are matrix pseudodifferential operators, they are bounded operators in $\mathcal{H}^{(s)}$, and they are mutually complementary projectors, i.e. formulas (5.67) *are valid for them.*

Returning to Theorem 5.4.2, let us note that the assertion can be checked directly for $s < 1/2$. Seeley proves it for integers $s = k \geq m$ and then uses interpolation to extend the result to all real s. (We will briefly discuss the interpolation in Subsect. 8.3b.) If $s = k \geq m$ and $g \in \mathcal{H}^{(s)}$, then the function

$u = W^+ g$ belongs at least to $H_0(M)$, but $\lim_{t\to+0} \gamma(t)W^+g$ belongs to $\mathcal{H}^{(s)}$ (cf. Theorem 5.4.3)). Using the fact that $u \in C^\infty$ and $Au = 0$ on M_+, it is possible to continue the function u on M_0 by setting $u = v$ on M_-, where v is a function from $H_k(M_-)$ with Cauchy data such that $Au \in H_{k-m}(M_0)$. Since $u \in H_0(M_0)$, it follows, by virtue of the theorem on improved smoothness on M_0, that $u \in H_k(M_0)$ and hence $W^+g \in H_k(M)$.

Theorem 5.4.4. *The subspaces $P^\pm \mathcal{H}^{(s)}(\Gamma)$ are the closures in $\mathcal{H}^{(s)}(\Gamma)$ of the sets of values $\gamma^\pm u$ for solutions u to the equation $Au = 0$ on $M_\pm$ that belong to $C^\infty(\overline{M}_\pm)$, respectively. These subspaces coincide with the sets of limits $\lim_{t\to\pm 0} \gamma(t)u$ for $u \in K_\pm(A, s)$, and $K_\pm(A, s)$ coincide with $W^\pm P^\pm \mathcal{H}^{(s)}(\Gamma)$.*

In particular, we see that for solutions $u(x) \in H_s(M_\pm)$ of a homogeneous elliptic equation $Au = 0$ of order m, the Cauchy data $\gamma^\pm u$ are meaningful not only for $s > m - (1/2)$, but also for $s \le m - (1/2)$. Cf. Section 7.9 below and also (Lions and Magenes 1968).

Let us write the pseudodifferential operator $P^\pm$ in the form of a matrix $(P^\pm_{j,k})$; then $\operatorname{ord} P^\pm_{j,k} \le j - k$, so that $P^\pm$ has the type $\{j, -k\}$. We now describe the way of calculating their principal symbols. To this end, we reduce the equation $Au = 0$ near Γ to a system of equations of the first order in t, setting $U = (u, D_t u, \dots, D_t^{m-1} u)'$:

$$D_t U + A_m^{-1} \begin{pmatrix} 0 & -A_m & 0 & \dots & 0 & 0 \\ 0 & 0 & -A_m & \dots & 0 & 0 \\ \vdots & \vdots & \vdots & & \vdots & \vdots \\ 0 & 0 & 0 & \dots & 0 & -A_m \\ A_0 & A_1 & A_2 & \dots & A_{m-2} & A_{m-1} \end{pmatrix} U = 0 . \tag{5.83}$$

Replacing all operators A_j for $t = 0$ by their principal symbols $\sigma^{(j)}$, we consider the matrix

$$\mathfrak{S} = -(\sigma^{(m)})^{-1} \begin{pmatrix} 0 & -\sigma^{(m)} & 0 & \dots & 0 & 0 \\ 0 & 0 & -\sigma^{(m)} & \dots & 0 & 0 \\ \vdots & \vdots & \vdots & & \vdots & \vdots \\ 0 & 0 & 0 & \dots & 0 & -\sigma^{(m)} \\ \sigma^{(0)} & \sigma^{(1)} & \sigma^{(2)} & \dots & \sigma^{(m-2)} & \sigma^{(m-1)} \end{pmatrix} . \tag{5.84}$$

Theorem 5.4.5. *The principal symbol $\sigma_0^\pm(x, \xi)$ of the pseudodifferential operator $P^\pm$ coincides with the Riesz projector on the subspace in $\mathbb{C}^m$ that is invariant with respect to the matrix $\mathfrak{S}$ and corresponds to all its eigenvalues lying in the upper (lower) halfplane:*

$$\sigma_0^\pm(x, \xi) = -\frac{1}{2\pi i} \oint_\pm (\mathfrak{S} - \lambda E)^{-1} d\lambda , \tag{5.85}$$

where the contour of integration lies in the upper (lower) halfplane and goes in the positive direction around all eigenvalues of $\mathfrak{S}$ lying in this halfplane.

It remains to add that the operators W^+ and P^+ can be defined without the assumption that A has an invertible elliptic extension on M_0. In this case

$$K_+(A,s) = W^+\mathcal{H}^{(s)}(\Gamma) \dotplus N_0(A) ,$$

where $N_0(A)$ is the subspace of solutions of the Cauchy problem with zero right-hand sides (see Section 5.1); P^+ is a projector, and

$$P^+ = \lim_{t\to+0} \gamma(t)W^+ .$$

Seeley considers operators acting in sections of bundles and uses more general spaces.

§6. Elliptic Boundary Problems for Elliptic Systems

6.1. Definitions.

a. Let $\{l_j\}_1^p$ and $\{m_k\}_1^p$ be two sets of integers. Assume that a matrix differential operator A of type $\{l_j, m_k\}$ is given on M:

$$A = \begin{pmatrix} A_{1,1} & \dots & A_{1,p} \\ \vdots & \ddots & \vdots \\ A_{p,1} & \dots & A_{p,p} \end{pmatrix} , \tag{6.1}$$

where $\operatorname{ord} A_{j,k} \le l_j + m_k$ (and $A_{j,k} = 0$ if $l_j + m_k < 0$). ¿From the outset we assume that the number $\sum(l_j + m_j)$ is even and equal to $2q$, $q \in \mathbb{N}$.

Furthermore, let $\{r_j\}_1^q$ be the third set of integers, and assume that a matrix of boundary operators

$$B = \begin{pmatrix} B_{1,1} & \dots & B_{1,p} \\ \vdots & & \vdots \\ B_{q,1} & \dots & B_{q,p} \end{pmatrix} \tag{6.2}$$

is given on Γ; each of them has the same structure as in (1.3), but $\operatorname{ord} B_{j,k} \le r_j + m_k$ (and $B_{j,k} = 0$ if $r_j + m_k < 0$).

Consider the boundary problem

$$Au = f \quad \text{on} \quad M_+ , \quad Bu = g \quad \text{on} \quad \Gamma . \tag{6.3}$$

Here u and f are columns of height p consisting of functions on M, and g is a column of height q consisting of functions on Γ.

We assume that the operator A is *elliptic in the sense of Douglis–Nirenberg* on M:

$$\det a_0(x,\xi) \ne 0 \quad \text{on} \quad T^*M \setminus 0 , \tag{6.4}$$

where $a_0(x,\xi)$ is the principal symbol of A, i.e. the matrix consisting of the principal symbols of $A_{j,k}$.

We also assume this operator to be *properly elliptic* on Γ: a scalar differential operator with principal symbol $\det a_0(x,\xi)$ is properly elliptic. (From this it follows that the sum $\sum(l_j+m_j)$ has to be even.) For $n>2$, proper ellipticity follows from ellipticity.

Finally, we assume that at each point of the boundary Γ, the Shapiro–Lopatinskij condition holds: in the appropriate local coordinates the boundary problem on the ray

$$a_0(x',0,\xi',D_n)v(t)=0 \quad (t=x_n>0)\,, \quad b_0(x',0,\xi',D_n)v(t)|_{t=0}=h \quad (6.5)$$

has one and only one solution in the space $\mathfrak{M}(\xi')$ of solutions $v(t)$ to the system $a_0(x',0,\xi',D_n)v(t)=0$ with $|v(t)|\to 0$ as $t\to+\infty$, for any $\xi'\neq 0$ and any numerical vector h. Here $b_0(x,\xi)$ is the principal symbol of B, i.e. the matrix consisting of the principal symbols of $B_{j,k}$.

Under these conditions the boundary problem (6.3) is called *elliptic in the Agmon–Douglis–Nirenberg sense*, or elliptic in the general sense.

If $l_1=\ldots=l_p$, then the system $Au=f$ is elliptic in the sense of Petrovskij, and it is then natural to call the boundary problem *elliptic in the sense of Petrovskij.* If $l_1=\ldots=l_p$ and $m_1=\ldots=m_p$, then the system $Au=f$ is elliptic in the usual sense and the boundary problem (6.3) will be called *elliptic in the usual sense* or simply *elliptic.* In the last case it is convenient to assume that $l_1=\ldots=l_p=m$ and $m_1=\ldots=m_p=0$, so that $mp=2q$ and the j-th row in the matrix B consists of boundary operators of order not greater than r_j.

b. Now we present two assertions simplifying the verification of the Shapiro–Lopatinskij condition in the matrix case. Instead of $a_0(x',0,\xi)$ and $b_0(x',0,\xi)$ (where $(x',0)$ are the local coordinates of the boundary point), we will write $a_0(\xi)$ and $b_0(\xi)$. Denote by $a_0^+(\xi',\zeta)$ the polynomial (1.11), where $\zeta_1(\xi'),\ldots,\zeta_q(\xi')$ are all the roots of the polynomial $\det a_0(\xi',\zeta)$ lying in the upper halfplane, and let $a^0=(a^{j,k,0})$ be the matrix of the cofactors of the elements of the matrix $a_0=(a_{j,k,0})$:

$$\sum_k a_{j,k,0}(\xi',\zeta)a^{k,l,0}(\xi',\zeta)=\delta_j^l\det a_0(\xi',\zeta)\,. \quad (6.6)$$

Proposition 6.1.1. (See e.g. (Agmon et al. 1964).) *The Shapiro–Lopatinskij condition (at a fixed boundary point) is equivalent to the following condition: the rows of the matrix*

$$b_0(\xi',\zeta)\,a^0(\xi',\zeta) \quad (6.7)$$

are linearly independent modulo the polynomial $a_0^+(\xi',\zeta)$.

Denote by Γ_+ a closed contour in the upper halfplane surrounding all the roots of the polynomial $a_0^+(\xi',\zeta)$. Using the Laplace transform, it is not hard to verify the following assertion (see e.g. (Volevich 1965)).

Proposition 6.1.2. *The space $\mathfrak{M}(\xi')$ is spanned by the columns of the matrix*

$$\omega(\xi', t) = \oint_{\Gamma_+} e^{i\zeta t} a_0^{-1}(\xi', \zeta)(E, \zeta E, \dots, \zeta^{s-1} E)\, d\zeta\ , \tag{6.8}$$

where E is the $p \times p$ unit matrix and s is the largest of the orders of $A_{j,k}$. Therefore, the Shapiro–Lopatinskij condition is equivalent to the following condition: the rank of the matrix

$$b_0(\xi', D_n)\, \omega(\xi', t)|_{t=0} \tag{6.9}$$

is maximal, i.e. it is equal to q.

c. The ellipticity of the boundary problem

$$Au = f \quad \text{in} \quad \mathbb{R}^n_+\ , \qquad Bu = g \quad \text{on} \quad \mathbb{R}^{n-1} = \partial\mathbb{R}^n_+ \tag{6.10}$$

is defined in an evident way. As in the scalar case, we assume that here the coefficients of the operators $A_{j,k}$ in the matrix A and coefficients of the operators $B_{j,k}$ in the matrix B belong to $B^\infty(\overline{\mathbb{R}^n_+})$. The boundary problem (6.10) is called *uniformly elliptic* if the modulus of the determinant of the principal symbol of A for $x \in \overline{\mathbb{R}^n_+}$, $|\xi| = 1$ and the sum of the moduli of all $q \times q$ minors of matrix (6.9) for $x' \in R^{n-1}$, $|\xi'| = 1$ are bounded from below by positive constants.

6.2. Examples.

a. If $l_j = m_j$ for all j, then we can consider the Dirichlet problem with boundary conditions

$$\gamma_j u_k = g_{j,k} \quad \text{on} \quad \Gamma \quad (j = 0, \dots, m_k - 1;\ k : m_k > 0)\ . \tag{6.11}$$

For a system elliptic in the usual sense, of even order m, conditions (6.11) take the form

$$\gamma_j u = g_j \quad \text{on} \quad \Gamma \quad (j = 0, \dots, m/2)\ . \tag{6.12}$$

Even in this case the Dirichlet problem is not always elliptic. The first example was found by Bitsadze in 1948 (see e.g. (Bitsadze 1966)):

$$A = \begin{pmatrix} \partial_2^2 - \partial_1^2 & 2\partial_1\partial_2 \\ -2\partial_1\partial_2 & \partial_2^2 - \partial_1^2 \end{pmatrix}. \tag{6.13}$$

If we set $z = x^1 + ix^2$ and $w = u_1 + iu_2$, then the system $Au = 0$ can be written in the form $\partial^2 w/\partial\bar{z}^2 = 0$; in the disk $\{z : |z| < \varepsilon\}$, $\varepsilon > 0$, it has the solutions $w = (\varepsilon^2 - |z|^2)\psi(z)$, where $\psi(z)$ is an arbitrary holomorphic function. These solutions vanish for $|z| = \varepsilon$, so that the Dirichlet problem has an infinite-dimensional kernel and is hence not elliptic (see the next section).

For $n > 2$ the Dirichlet problem with two unknown functions is always elliptic: see (Boyarskij 1960). For a matrix differential operator elliptic in the usual sense, of even order m, the Dirichlet problem is elliptic if A is *strongly elliptic* in the sense of Vishik (Vishik 1950, 1951):

$$\operatorname{Re} a_0(x,\xi) = \frac{1}{2}[a_0(x,\xi) + a_0^*(x,\xi)] > 0 \tag{6.14}$$

for $x \in \overline{G}$, $\xi \neq 0$ (i.e. the corresponding quadratic form is positive definite). The definition of strong ellipticity remains meaningful for matrix operators of the type $\{l_j, m_k\}$ with $l_j = m_j$ for all j (see (Nirenberg 1955)), and for these systems the Dirichlet problem remains elliptic (see (Agmon et al. 1964)).

We also formulate the following result, which was obtained for systems elliptic in the usual sense in (Lopatinskij 1956); see also (Agranovich 1965).

Theorem 6.2.1. *Let A be a matrix properly elliptic differential operator of even order m. Then the Dirichlet problem for the system $Au = f$ on M_+ is elliptic if and only if at each boundary point the principal symbol a_0 of A, written in local coordinates of Sect. 0.2, admits the factorization*

$$a_0(\xi',\zeta) = \sigma_-(\xi',\zeta)\sigma_+(\xi',\zeta) \tag{6.15}$$

for $\xi \neq 0$, where $\sigma_+(\xi',\zeta)$ and $\sigma_-(\xi',\zeta)$ are matrix polynomials of degree $m/2$ in ζ; here the coefficient in the principal term of σ_+ is equal to the unit matrix, and the roots of the polynomials $\det \sigma_+(\xi',\zeta)$ and $\det \sigma_-(\xi',\zeta)$ coincide with the roots of the polynomial $\det a_0(\xi',\zeta)$ lying in the upper and lower halfplanes, respectively.

Here it is essential that both multipliers, σ_- and σ_+, have degree $m/2$ with respect to ζ and nondegenerate principal coefficients. More general factorization that is polynomial with respect to ζ and retains separation of the roots indicated above is always possible. For example, the (principal) symbol of the operator (6.13) admits the following factorization (with $\zeta = \xi_2$):

$$\begin{pmatrix} \xi_2^2 - \xi_1^2 & 2\xi_1\xi_2 \\ -2\xi_1\xi_2 & \xi_2^2 - \xi_1^2 \end{pmatrix} = \begin{pmatrix} \xi_2^2 - \xi_1^2 & i \operatorname{sgn} \xi_1 \\ -2\xi_1\xi_2 & -1 \end{pmatrix} \cdot \begin{pmatrix} 1 & i \operatorname{sgn} \xi_1 \\ 0 & (i\xi_2 - |\xi_1|)^2 \end{pmatrix}.$$

Factorization of the type (6.15) can be used for the description of elliptic boundary problems in a halfspace for a given elliptic system (see (Samoĭlenko 1972)).

b. Any system elliptic in the Douglis–Nirenberg sense can be reduced to a first order system elliptic in the Douglis–Nirenberg sense, i.e. with $l_j + m_k \leq 1$, by introduction of new unknown functions. In addition, an elliptic boundary problem for the initial system is then reduced to an elliptic boundary problem for the new system. This is pointed out in (Agmon et al. 1964) (with a reference to a remark due to Atyah and Singer). For example, the Laplace equation $\partial_1^2 u + \partial_2^2 u = 0$ is equivalent to the system

$$\partial_1 u_2 + \partial_2 u_3 = 0\,, \quad \partial_1 u_1 - u_2 = 0\,, \quad \partial_2 u_1 - u_3 = 0 \tag{6.16}$$

with principal symbol

$$\begin{pmatrix} 0 & \xi_1 & \xi_2 \\ \xi_1 & -1 & 0 \\ \xi_2 & 0 & -1 \end{pmatrix};$$

it is of type $\{l_j, m_k\}$ with $l_1 = m_1 = 1$ and $l_2 = m_2 = l_3 = m_3 = 0$ and is elliptic in the Douglis–Nirenberg sense. The Dirichlet problem for the Laplace equation is equivalent to the Dirichlet problem for (6.16), in which only γu_1 is prescribed on the boundary.

c. Elliptic systems are encountered in various problems in mechanics of continuous media. We mention the matrix *Lame equation*

$$Lu \equiv \mu\Delta + (\lambda + \mu) \operatorname{grad} \operatorname{div} u = f\ , \tag{6.17}$$

which describes equilibrium states of elastic homogeneous isotropic media; here $u = (u_1, u_2, u_3)' = \widetilde{z} - z$ is the vector of a displacement (the point z is shifted to the position $\widetilde{z}$); λ and μ are the so-called Lame constants. It is easy to verify that L is elliptic in the usual sense for $\mu \neq 0$, $\lambda + 2\mu \neq 0$: the determinant of the principal symbol of this operator is equal to $-\mu^2(\lambda + 2\mu)|\xi|^6$. There are 6 basic boundary problems for system (6.17) (see (Kupradze 1976, Chapter I, §14)). The first of them is the boundary problem with the displacements prescribed on the boundary of the domain,

$$\gamma_0 u = g \quad \text{on} \quad \Gamma\ , \tag{6.18}$$

i.e. the Dirichlet problem. The second is the problem with given stresses, the boundary condition has the form

$$\gamma_0 \Big[\lambda\nu_j \operatorname{div} u + \mu \sum_{k=1}^{3} (\partial_j u_k + \partial_k u_j)\nu_k\Big] = 0 \quad \text{on} \quad \Gamma \quad (j = 1, 2, 3)\ , \tag{6.19}$$

where $\nu = (\nu_1, \nu_2, \nu_3)$ is the unit inner normal. Let $\mu(\lambda + 2\mu) \neq 0$; the Shapiro–Lopatinskij condition is satisfied in the case of the boundary condition (6.18) for $\lambda + 3\mu \neq 0$, and in the case of the boundary condition (6.19) for $\lambda + \mu \neq 0$. See (Kozhevnikov 1993), where the third and the forth boundary problems are also considered.

The nearest generalizations of these boundary problems are obtained in the following way. First, instead of the equilibrium state, it is possible to consider harmonic oscillations of the medium, with the dependence on time described by the factor $e^{i\omega t}$. Then in L an additional term $\rho\omega^2 u$ appears (ρ is the density, and ω is the frequency of oscillations). Second, it is possible to consider equilibrium state or harmonic oscillations of a transversally-isotropic body, in which the properties of the medium are the same along all directions orthogonal to a given axis (see (Kupradze 1976, Ch. XIV, §2)).

In hydrodynamics the following system is used to describe small oscillations of a viscous compressible barotropic fluid:

$$\mu\left[-\Delta-\left(\beta+\frac{1}{3}\right)\operatorname{grad}\operatorname{div}\right]v+\operatorname{grad}p=\lambda v\,,$$
$$\operatorname{div}v=\lambda p \tag{6.20}$$

in a three-dimensional domain G, with boundary condition

$$\gamma_0 v=0 \quad \text{on} \quad \Gamma\,. \tag{6.21}$$

Here $v=(v_1,v_2,v_3)'$ is the fluid velocity vector, p is the deviation of the pressure from its value in the equilibrium state, $\mu\,(>0)$ and $\beta\mu\,(\geq 0)$ are viscosity coefficients, $\lambda=-i\omega$, and ω is the frequency of oscillations. The system (6.20) is not elliptic in the usual sense, but as a system of the type $\{l_j,m_k\}$ with $l_j=m_j=1$ $(j=1,2,3)$, $l_4=m_4=0$ it is elliptic in the Douglis–Nirenberg sense for $\lambda\neq[(\beta+(4/3))\mu]^{-1}$ ($-\lambda$ enters into the right lower element of the matrix principal symbol). For all other values of λ, except $\lambda=[(\beta+(7/3))\mu]^{-1}$, the Shapiro–Lopatinskij condition is satisfied, so that the boundary problem (6.20), (6.21) is elliptic. Condition (6.21) corresponds to absolutely hard vessel walls; the boundary condition for absolutely soft walls has the form

$$\gamma_0\left\{-p\nu_j+\mu\sum_k\left[-\left(\beta-\frac{2}{3}\right)\delta_{j,k}\operatorname{div}v+(\partial_k v_j+\partial_j v_k)\right]\nu_k\right\}=0 \ \text{ on } \Gamma \tag{6.22}$$

$(j=1,2,3)$. In this case the Shapiro-Lopatinskij condition is satisfied for $\lambda\neq[(\beta+(1/3))\mu]^{-1}$. See (Levitin 1993).

If the fluid is considered as incompressible, then the last equation in (6.20) is replaced by $\operatorname{div}v=0$, β is equal to $-1/3$, and the boundary problem (6.20), (6.21) becomes the Dirichlet problem for the three-dimensional *Stokes system*

$$-\mu\Delta v+\operatorname{grad}p=\lambda v\,, \quad \operatorname{div}v=0\,, \tag{6.23}$$

$\gamma_0 v=0$ on Γ. It is elliptic but can also be considered as "elliptic in a subspace;" see (Métivier 1978) and Section 9.7 below.

d. In the paper (Solomyak 1963) an example is presented of a matrix elliptic operator in $\mathbb{R}^4$ for which no elliptic boundary conditions exist (for topological reasons), even in a halfspace and with pseudodifferential boundary conditions (see Sect.7.8 below):

$$A=\begin{pmatrix}\partial_1+i\partial_2 & -\partial_3-i\partial_4\\ \partial_3-i\partial_4 & \partial_1-i\partial_2\end{pmatrix}. \tag{6.24}$$

6.3. Main Theorems. Consider the boundary problem (6.3). Introduce the spaces

$$H_{\{s+m_k\}}(M)=H_{s+m_1}(M)\times\ldots\times H_{s+m_p}(M)\,, \tag{6.25}$$
$$H_{\{s-l_j\}}(M)=H_{s-l_1}(M)\times\ldots\times H_{s-l_p}(M)\,, \tag{6.26}$$
$$H_s(M,\Gamma)=H_{\{s-l_j\}}(M)\times H_{\{s-r_j-\frac{1}{2}\}}(\Gamma)\,, \tag{6.27}$$

where the space $H_{\{s-r_j-\frac{1}{2}\}}(\Gamma)$ is defined in (2.22). Assume that

$$s \geq -\min m_k\,, \quad s \geq \max l_j\,, \quad s > \max r_k + \frac{1}{2}\,. \tag{6.28}$$

Then a bounded operator

$$\mathcal{A} = (A, B)\colon H_{\{s+m_k\}}(M) \to H_s(M, \Gamma) \tag{6.29}$$

corresponds to our boundary problem (6.3).

The following theorems are generalizations of those in §2 .

Theorem 6.3.1. *Under conditions* (6.28), *ellipticity of the boundary problem* (6.3) *is equivalent to the Fredholm property of the operator* (6.29) *and to the validity of the a priori estimate*

$$\sum_k \|u_k\|_{s+m_k,M} \leq C\Big(\sum_j \|f_j\|_{s-l_j,M} + \sum_j \|g_j\|_{s-r_j-\frac{1}{2},\Gamma} + \sum_k \|u_k\|_{0,M}\Big)\,. \tag{6.30}$$

If uniqueness holds for (6.3), *then the last sum in the right-hand side can be omitted.*

Theorem 6.3.2. *Let the boundary problem* (6.3) *be elliptic, and let* $s' > s$, *where* s *satisfies inequalities* (6.28). *Let*

$$u \in H_{\{s+m_k\}}(M)\,, \quad f \in H_{\{s'-l_j\}}(M)\,, \quad \textit{and} \quad g \in H_{\{s'-r_j-\frac{1}{2}\}}(\Gamma)\,.$$

Then $u \in H_{\{s'+m_k\}}(M)$.

In particular, $\operatorname{Ker}\mathcal{A}$ *consists of infinitely smooth functions and does not depend on* s.

Theorem 6.3.3. *Assume that the boundary problem* (6.3) *is elliptic and that* s *satisfies inequalities* (6.28). *Then the index of the operator* $\mathcal{A}$ *does not depend on* s.

The proofs are similar to those outlined in §2. In particular, for an elliptic boundary problem a rough right parametrix is constructed and used; but it follows that a precise two-sided parametrix exists (cf. Theorem 2.4.5).

Assume that $H_{\{s+m_k\}}(M) \subset H_{\{s-l_j\}}(M)$. Then, as in the scalar case, we can introduce the operator A_B in $H_{\{s-l_j\}}(M)$:

$$A_B u = Au \quad \text{on} \quad D(A_B) = \{u \in H_{\{s+m_k\}}(M) : \; Bu = 0 \; \text{ on } \; \Gamma\}\,.$$

¿From a priori estimate (6.30) it follows that this operator is closed; but it is not always densely defined: its domain is not dense in $H_{\{s-l_j\}}(M)$ if at least one of the boundary conditions retains its sense in this space. However, if A is elliptic in the usual sense, $m_1 = \ldots = m_p = 0$ and $l_1 = \ldots = l_p = m$, and if the orders r_j of the boundary operators are less than m, then we can take $s = m$; in this case A_B acts in $[H_0(M)]^p$ and its domain is dense in this space.

Note that much more general operators A_B can be considered as acting in $H_0(M)$: see Sect. 6.5 below.

6.4. Ellipticity with Parameter. We now formulate generalizations of definitions and assertions of §3. Consider the boundary problem

$$A(\lambda)u = f \quad \text{on} \quad M_+ , \quad B(\lambda)u = g \quad \text{on} \quad \Gamma . \tag{6.31}$$

Here $A(\lambda)$ and $B(\lambda)$ are matrices of the same size as in (6.1) and (6.2), and they depend on λ polynomially:

$$A(\lambda) = \sum \lambda^r A_r , \qquad B(\lambda) = \sum \lambda^r B_r . \tag{6.32}$$

As in Sect. 6.1, we fix the sets $\{l_j\}$, $\{m_k\}$, and $\{r_j\}$ of integers. The parameter λ will have a weight $\tau(\in \mathbb{N})$ with respect to the differentiation. The degree of the element $A_{j,k}(\lambda)$ of the matrix $A(\lambda)$ in λ is not greater than $(l_j + m_k)/\tau$, and the degree of the element $B_{j,k}(\lambda)$ of the matrix $B(\lambda)$ in λ is not greater than $(r_j + m_k)/\tau$. In the matrix A_r the element $A_{r,j,k}$ is a differential operator of order not greater than $l_j + m_k - r\tau$; in the matrix B_r the element $B_{r,j,k}$ is a differential operator of order not greater than $r_j + m_k - r\tau$. The principal symbols $a_{r,0}(x,\xi)$ and $b_{r,0}(x,\xi)$ of the matrix operators A_r and B_r are defined in the obvious way, and we set

$$a_0(x,\xi,\lambda) = \sum \lambda^r a_{r,0}(x,\xi) , \qquad b_0(x,\xi,\lambda) = \sum \lambda^r b_{r,0}(x,\xi) . \tag{6.33}$$

Conditions of *ellipticity with parameter* of the boundary problem (6.31) in an angle $\mathcal{L}$ are formulated essentially in the same way as in Sect. 3.1. First,

$$\det a_0(x,\xi,\lambda) \neq 0 \quad \text{for} \quad (x,\xi) \in T^*M, \ \lambda \in \mathcal{L}, \ (\xi,\lambda) \neq 0 . \tag{6.34}$$

Second, if we write the boundary problem (6.31) in local coordinates near a point of the boundary, then for $(\xi',\lambda) \neq 0$, $\lambda \in \mathcal{L}$ the boundary problem

$$\begin{aligned} a_0(x',0,\xi',D_n,\lambda)\, v(t) &= 0 \qquad (t = x^n > 0) , \\ b_0(x',0,\xi',D_n,\lambda)\, v(t)|_{t=0} &= h \end{aligned} \tag{6.35}$$

is required to have one and only one solution $v(t)$ with $|v(t)| \to 0$ as $t \to +\infty$, for any numerical vector h.

Theorem 6.4.1. *Assume that the boundary problem* (6.31) *is elliptic with parameter in $\mathcal{L}$ and that s satisfies inequalities* (6.28). *Then for $\lambda \in \mathcal{L}$ with sufficiently large modulus, the boundary problem has one and only one solution $u \in H_{\{s+m_k\}}(M)$ for any $(f,g) \in H_s(M,\Gamma)$. In addition, the estimate*

$$\sum_{k=1}^{p} |||u_k|||_{s+m_k,M} \leq C\Big(\sum_{j=1}^{p} |||f_j|||_{s-l_j,M} + \sum_{j=1}^{q} |||g_j|||_{s-r_j-\frac{1}{2},\Gamma} \Big) \tag{6.36}$$

holds with a constant C not depending on λ and u.

Conversely, if such an estimate holds for $\lambda \in \mathcal{L}$ with sufficiently large modulus, then the boundary problem is elliptic with parameter in $\mathcal{L}$.

Let us dwell, in particular, on the boundary problem

$$(A - \lambda I)u = f \quad \text{on} \quad M_+ , \quad Bu = g \quad \text{on} \quad \Gamma , \tag{6.37}$$

where B does not depend on λ. Here A and B have the same structure as in Sect. 6.1, but we additionally assume that

$$l_1 + m_1 = \ldots = l_p + m_p = \tau . \tag{6.38}$$

¿From Theorem 6.4.1 it follows that if the boundary problem (6.37) is elliptic with parameter in $\mathcal{L}$, then the resolvent of the operator A_B corresponding to this problem for $g = 0$ exists for $\lambda \in \mathcal{L}$ with sufficiently large modulus. In addition, the resolvent is compact, so that A_B has a discrete spectrum in $H_{\{s-l_j\}}(M) = H_{\{s-\tau+m_k\}}(M)$ not depending on s. If A is elliptic in the usual sense and we can take $s = 0$ ($l_1 = \ldots = l_p = m = \tau$; $m_1 = \ldots = m_p = 0$; $r_j < m$), then the estimate takes the form

$$|\lambda| \|(A_B - \lambda I)^{-1} f\|_{0,M} \leq C \|f\|_{0,M}, \tag{6.39}$$

so that the norm of the resolvent decreases as $1/|\lambda|$ as $\lambda \to \infty$ in $\mathcal{L}$. In the general case, for $u = (A_B - \lambda I)^{-1} f$ we only obtain

$$\sum_{k=1}^{p} \|u_k\|_{s-l_k,M} \leq C|\lambda|^{\rho} \sum_{k=1,}^{p} \|f_k\|_{s-l_k,M} \tag{6.40}$$

with some ρ, i.e. the norm of the resolvent is estimated by some power of $|\lambda|$.

Examples of boundary problems polynomially depending on a parameter for systems elliptic in the usual sense are encountered in papers on oscillations of isotropic or transversally-isotropic bodies of cylindrical form.

Boundary problems of the form (6.37) for a system elliptic in the Douglis–Nirenberg sense arise naturally in the procedure of linearization with respect to a parameter of a problem depending on it polynomially. Let us consider, for example, a scalar boundary problem (3.1) with $r_j < m$ and (for simplicity of notation) $A_0 = -1$ and $\tau = 1$. Setting

$$u_1 = u, \; u_2 = \lambda u, \; \ldots, \; u_m = \lambda^{m-1} u \tag{6.41}$$

and denoting the vector-valued function $(u_1, \ldots, u_p)'$ again by u, we obtain a boundary problem of the form (6.37) with

$$A = \begin{pmatrix} 0 & I & 0 & \ldots & 0 & 0 \\ \vdots & \vdots & \vdots & & \vdots & \vdots \\ 0 & 0 & 0 & \ldots & 0 & I \\ A_m & A_{m-1} & A_{m-2} & \ldots & A_2 & A_1 \end{pmatrix} \tag{6.42}$$

and the right-hand side $(0,\dots,0,f)'$ in the system on M_+. It is not hard to verify that the boundary problem (6.37) obtained in this way is elliptic with parameter in $\mathcal{L}$ if the original boundary problem was elliptic with parameter in $\mathcal{L}$. A similar remark on the linearization of a matrix boundary problem (6.31) with respect to parameter is also true.

6.5. Reduced Cauchy Data and L^2-realizations. In this section we essentially follow the papers (Grubb 1977a,b, 1979); see also (Grubb and Geymonat 1977, 1979).

a. For simplicity we assume that $l_j = m_j$ for all j (see the notation in Sect. 6.1). This class of systems is especially important for the spectral theory, since it contains formally adjoint systems and strongly elliptic systems. We also assume that the operator A is elliptic (in the Douglis–Nirenberg sense) on the closed manifold M_0. Let $m = m_1 \geq m_2 \geq \dots \geq m_p \geq 0$ and, more precisely,

$$\begin{aligned} m = m_1 = \dots = m_{r_1} > m_{r_1+1} = \dots = m_{r_1+r_2} > \dots \\ > m_{r_1+\dots+r_{s-1}+1} = \dots = m_{r_1+\dots+r_s} \geq 0. \end{aligned} \tag{6.43}$$

We set $\mu_1 = m_{r_1}, \dots, \mu_s = m_{r_1+\dots+r_s}$ and agree to write a vector-valued function $u(x) = \big(u_1(x),\dots,u_p(x)\big)'$ in the form $\big(u^1(x),\dots,u^s(x)\big)'$, where $\dim u^j(x) = r_j$. Furthermore, set

$$\begin{aligned} N = \{1,\dots,s\}, \quad M_0 = \{0,\dots,m-1\}, \\ M_1 = \{m,\dots,2m-1\}, \quad M = M_0 \cup M_1, \end{aligned} \tag{6.44}$$

$$N_j = \{t \in N : \mu_t - m + j \geq 0\},\ \beta_j u = \{\gamma_{\mu_t - m + j} u^t\}_{t \in N_j},\ n_j = \sum_{t \in N_j} r_t. \tag{6.45}$$

The set $\beta u = \{\beta_j u\}_{j \in M}$ is the total trace of u. Generally (more precisely, if we do not have $m_1 = \dots = m_p > 0$), this set of data is in some sense too large (we will comment this at the end of Subsect. b), and it is convenient to replace it by the smaller set of the *reduced Cauchy data*.

For this we need Green's formula. It can be written in the form

$$(Au, v)_M - (u, A^*v)_M = i(I^\times \mathcal{C}\beta u, \beta v)_\Gamma \tag{6.46}$$

for $u,\ v \in C^\infty(M)$. Here $\mathcal{C}$ is a $2m \times 2m$ block matrix $(\mathcal{C}_{j,k})_{j,k\in M}$, where the block $\mathcal{C}_{j,k}$ is an $n_{2m-1-j} \times n_k$ matrix consisting of differential operators of order $j-k$ on Γ, and $I^\times = (I^\times_{j,k})_{j,k\in M}$ is a "skew-unit" block matrix with blocks $I^\times_{j,k}$ equal to the zero matrix for $2m-1-j \neq k$ and to the unit matrix for $2m-1-j = k$. In particular, $\mathcal{C}$ is a triangular matrix. Generally, in distinction to the case $m_1 = \dots = m_p$, $\mathcal{C}$ is not invertible.

Now we set

$$\beta^0 u = \{\beta_j u\}_{j\in M_0},\ \beta^1 u = \{\beta_j u\}_{j \in M_1},\ \mathcal{C}^{\varepsilon\delta} = (\mathcal{C}_{i,j})_{i \in M_\varepsilon, j \in M_\delta}, \tag{6.47}$$

where ε, $\delta \in \{0,1\}$. The reduced Cauchy data for u relative to A and for v relative to A^* are defined by the formulas

$$\begin{aligned} \chi u &= \{\chi_0 u, \chi_1 u\} = \{\beta^0 u, \mathcal{C}^{11}\beta^1 u\}, \\ \chi' v &= \{\chi'_0 v, \chi'_1 v\} = \{\beta^0 v, I^\times \mathcal{C}^{00*} I^\times \beta^1 v\}, \end{aligned} \tag{6.48}$$

where $I^\times$ are $m \times m$ skew–unit block matrices. Here $\chi_0 u$ is the Dirichlet data of u, and $\chi_1 u$, or rather $I^\times \chi_1 u$, is a choice of reduced Neumann data of u.

For example, in the case of the Stokes operator (cf. (6.23))

$$\begin{pmatrix} -\Delta I_3 & -\operatorname{grad} \\ \operatorname{div} & 0 \end{pmatrix}$$

we have $m = m_1 = m_2 = m_3 = \mu_1 = 1$, $m_4 = \mu_2 = 0$, $N = \{0,1\}$, $u^1 = \mu(v_1, v_2, v_3)$, $u^2 = -p$; $\beta^0 u = \gamma_0 u^1$, $\beta^1 u = \{\gamma_1 u^1, \gamma_0 u^2\}$, and it can be verified that the Neumann data are $\partial_\nu u^1 + \gamma_0 u^2 \nu$, where ν is the inner normal.

Now Green's formula (6.46) can be rewritten in the form

$$(Au, v)_M - (u, A^* v)_M = i\left(\begin{pmatrix} I^\times \mathcal{C}^{10} & I^\times \\ I^\times & 0 \end{pmatrix} \chi u, \chi' v\right). \tag{6.49}$$

b. We set

$$\mathcal{H}_\sigma(M) = \prod_{k \in N} \left[H_{\sigma - m + \mu_k}(M)\right]^{r_k} \tag{6.50}$$

and introduce the space

$$D_+^{\sigma,0}(A) = \left\{u \in \mathcal{H}_\sigma(M) : Au \in \left[L^2(M)\right]^p\right\} \tag{6.51}$$

with the graph topology. If $\sigma \geq 2m$, then obviously $C^\infty(M)$ is dense in $D_+^{\sigma,0}(A)$ (recall that $M = \overline{M}$), and $\chi = \{\chi_0, \chi_1\}$ is a continuous map

$$\mathcal{H}_\sigma(M) \to \mathcal{H}_{\sigma,1}(\Gamma) \times \mathcal{H}_{\sigma,2}(\Gamma),$$

where

$$\mathcal{H}_{\sigma,1}(\Gamma) = \prod_{j \in M_0} \left[H_{\sigma - j - \frac{1}{2}}(\Gamma)\right]^{n_j}, \quad \mathcal{H}_{\sigma,2}(\Gamma) = \prod_{j \in M_1} \left[H_{\sigma - j - \frac{1}{2}}(\Gamma)\right]^{n_{2m-1-j}}. \tag{6.52}$$

Theorem 6.5.1. (Grubb.) *The space $\left[C_0^\infty(M)\right]^p$ is dense in $D_+^{\sigma,0}(A)$ for all $\sigma \leq 0$. For all integers $\sigma \leq 0$, χ extends by continuity to $D_+^{\sigma,0}(A)$ and maps this space continuously into $\mathcal{H}_{\sigma,1}(\Gamma) \times \mathcal{H}_{\sigma,2}(\Gamma)$.*

In the proof of the first assertion a parametrix for A on M_0 is used. The second assertion is proved by means of the first assertion, Green's formula, and the duality between the spaces $H_r(\Gamma)$ and $H_{-r}(\Gamma)$.

Green's formula (6.49) extends to $u \in D_+^{m,0}(A)$ and $v \in D_+^{m,0}(A^*)$.

Note that the analog of the second assertion in Theorem 6.5.1 is generally not true for the total trace.

c. Now consider a boundary condition of the form $B\chi u = 0$ on Γ. Here $B = (B_{jk})_{j,k \in M}$, and B_{jk} is a $q_j \times n_k$ matrix for $k \in M_0$ and $q_j \times n_{2m-1-k}$ matrix for $k \in M_1$, with some $q_j \geq 0$, consisting of differential (or pseudodifferential, cf. Sect. 7.8) operators of order $j-k$ on Γ. If we impose the Shapiro–Lopatinskij condition, then $\sum q_j = \sum m_k$. The matrix B is triangular; the boundary condition is normal if the diagonal blocks define surjective morphisms.

Using Theorem 6.5.1, it is possible to define the *L^2-realization* A_B of A determined by B as follows: A_B is the operator mapping u into Au, with domain

$$D(A_B) = \{u \in [L^2(M)]^p : Au \in [L^2(M)]^p,\ B\chi u = 0\}. \tag{6.53}$$

(Note that for us the notation $A_{B\chi}$ would be more precise.)

In view of Theorem 6.5.1, here the boundary conditions are meaningful, and it is not difficult to check that A_B is a closed operator with dense domain. If the Shapiro–Lopatinskij condition holds, then actually $D(A_B) \subset D_+^{m,0}(A)$.

We add that the analogs of Calderón projectors are studied in (Grubb 1977a). Some further results for operators A_B will be mention below in Subsects. 7.1b and 9.1d.

6.6. Elliptic Boundary Problems in Sections of Bundles. Let E and F be complex vector bundles over M having the same dimension p, and let G_j ($j = 1, \dots, q$) be one-dimensional vector bundles over Γ. Our considerations in the previous sections can be generalized to the case in which the operator A acts from $C^\infty(M, E)$ (the space of infinitely smooth sections of E) to $C^\infty(M, F)$ and the boundary operators B_j act from from $C^\infty(M, E)$ to $C^\infty(\Gamma, G_j)$. For simplicity we here have in mind ellipticity in the ordinary sense. Locally, over a neighborhood O of a fixed point $x \in M_+$ or a semi-neighborhood O^+ of a point $x \in \Gamma$ and over $O^+ \cap \Gamma$, these bundles admit trivializations (i.e. they are direct products $O \times \mathbb{C}^p$ or $O^+ \times \mathbb{C}^p$ and $(O^+ \cap \Gamma) \times \mathbb{C}$), and the operators A and B_j can be written as in Sect. 6.1. Hence, it is easy to define the orders (m and r_j) of these operators and to formulate conditions of ellipticity; the formulations can be given in an invariant form. In particular, $q = mp/2$. Using partitions of unity on M and Γ, we can introduce the Sobolev norms $\|\cdot\|_{s,E}$, $\|\cdot\|_{s-m,F}$, and $\|\cdot\|_{s-r_j-\frac{1}{2},G_j}$. The Sobolev spaces $H_s(M,E)$, $H_{s-m}(M,F)$, and $H_{s-r_j-\frac{1}{2}}(\Gamma, G_j)$ of sections of the bundles E, F, and G_j are defined as the completions of $C^\infty(M,E)$, $C^\infty(M,F)$, and $C^\infty(\Gamma, G_j)$, respectively, with respect to the corresponding norms. The main theorem consists in equivalence of the ellipticity of the boundary problem and of the Fredholm property of the corresponding operator in Sobolev spaces. The proofs can be carried out in the spirit of considerations of §2. A theorem on the unique solvability of a boundary problem elliptic with parameter is also valid.

Elliptic boundary problems in sections of bundles arise, in particular, if we consider boundary problems for differential forms.

Details and variants can be found in many books and papers, for example, in (Hörmander 1985a) or (Grubb 1986).

§7. Generalizations and Variants

7.1. Variational Boundary Problems.

a. A boundary problem may be called variational if it appears in the search for a function minimizing a given functional. The elliptic equation is then the Euler equation for this functional.

For example, consider the Dirichlet problem for the Laplace equation:

$$-\Delta u = f \quad \text{in} \quad G\,, \qquad \gamma_0 u = 0 \quad \text{on} \quad \Gamma\,. \tag{7.1}$$

For simplicity we first assume that $u \in H_2(G)$, so that $f \in H_0(G)$. Denote by $\mathcal{H}$ the closure $\mathring{H}_1(G)$ of the linear submanifold $C_0^\infty(G)$ in $H_1(G)$. Introduce the sesquilinear form

$$a[u,v] = \sum_1^n (\partial_k u, \partial_k v)_G\,. \tag{7.2}$$

Integrating by parts, we see that

$$a[u,v] = (f,v)_G \qquad (v \in \mathcal{H})\,. \tag{7.3}$$

If we assume all functions to be real-valued, then there exists a unique function $u \in \mathcal{H}$ on which the functional $a[u,u] - 2(f,u)$ attains its least value. This function coincides with the solution of the boundary problem (7.1) and belongs to $\mathring{H}_1(G) \cap H_2(G)$.

Variational problems arise in many branches of applied mathematics. The variational point of view on the boundary problem has at least two advantages. First, if it is necessary to minimize the assumptions on the smoothness of the coefficients and the boundary, then the weak setting of the boundary problem permits this to a much greater extent. (See Sect. 7.2 below.) Second, as is well known, there are direct methods of solving variational problems.

Instead of (7.1), it is possible to consider the Dirichlet problem for a strongly elliptic equation (or system) of any order, and now we will briefly dwell on this, following e.g. (Agmon 1965), (Nečas 1967), and (Lions and Magenes 1968). Historically, this was the first "break-through" to boundary problems for elliptic equations of higher order, and it was this approach that led to the appearance of strongly elliptic systems, first in (Vishik 1950, 1951), and later in papers of Gårding (see (Gårding 1953)) and other mathematicians.

Let $A = A(x,D)$ be a differential operator of order $m = 2q$. We write it in the "divergent" form

$$Au = \sum_{|\alpha|,|\beta| \le q} D^\alpha (a_{\alpha,\beta}(x) D^\beta u)\,. \tag{7.4}$$

Assume that it is strongly elliptic:

$$\operatorname{Re} \sum_{|\alpha|=|\beta|=q} a_{\alpha,\beta}(x) \xi^{\alpha+\beta} \ge \gamma\, |\xi|^{2q} \qquad (\gamma = \text{Const} > 0)\,. \tag{7.5}$$

Generalizing the notation introduced above, we set $\mathcal{H} = \overset{\circ}{H}_q(G)$ (this is the closure of $C_0^\infty(G)$ in $H_q(G)$) and introduce the form

$$a[u,v] = \sum_{|\alpha|,|\beta|\leq q} (a_{\alpha,\beta}D^\beta u, D^\alpha v)_G \,, \tag{7.6}$$

which coincides with $(Au,v)_G$ on functions $u \in \mathcal{H} \cap H_{2q}(G)$ and $v \in \mathcal{H}$.

Theorem 7.1.1. (Gårding's inequality.) *Under condition* (7.5), *there exist constants* $\varepsilon > 0$ *and* $C_1 \geq 0$ *such that*

$$\varepsilon \,\|u\|_{q,M}^2 \leq \operatorname{Re} a[u,u] + C_1 \|u\|_{0,M}^2 \qquad (u \in \mathcal{H}) \,. \tag{7.7}$$

Conversely, (7.5) *follows from* (7.7).

The proof can be carried out beginning with the case of operator (7.4) in a halfspace, with constant coefficients and without lower order terms (see e.g. (Agmon 1965) or (Nečas 1967)).

Lemma 7.1.2. (Lax–Milgram Lemma.) *Let* $a[u,v]$ *be a sesquilinear form on a Hilbert space* H *with the properties*

$$|a[u,v]| \leq C_2\|u\|\|v\| \quad \text{and} \quad \varepsilon\|u\|^2 \leq |a[u,u]| \qquad (u,v \in H) \,, \tag{7.8}$$

where C_2 *and* ε *are positive constants. If* $F(v)$ *is a bounded semi-linear functional on* H, *then there exists an element* $u \in H$ *such that* $F(u) = a[u,v]$.

The proof is elementary; see e.g. (Nečas 1967).

Now we can assume that $C_1 = 0$ in (7.7) or insert a parameter in the problem. We choose the second possibility and consider the boundary problem

$$Au - \lambda u = f \quad \text{in} \quad G \,, \quad \gamma_j u = 0 \quad \text{on} \quad \Gamma \quad (j = 1, \ldots, q-1) \,. \tag{7.9}$$

Integrating by parts, we can attach a weak form to it:

$$a_\lambda[u,v] = (f,v) \qquad (v \in \mathcal{H}) \,, \tag{7.10}$$

where $a_\lambda[u,v] = a[u,v] - \lambda(u,v)_G$. Using the Gårding inequality and the Lax–Milgram Lemma, we easily obtain

Theorem 7.1.3. *Let the operator* A *be strongly elliptic and* $\operatorname{Re}\lambda \leq -C_1$, *where* C_1 *is the constant in* (7.7). *Then the integral identity* (7.10) *has a unique solution* $u \in \mathcal{H} = \overset{\circ}{H}_q(G)$ *for any* $f \in H_0(G)$.

Here, moreover, instead of $f \in H_0(G)$, we can assume that $f \in \mathcal{H}^* = H_{-q}(G)$ (see Subsect. 2.1f). If $f \in H_0(G)$, then $u \in \overset{\circ}{H}_q(G) \cap H_{2q}(G)$. This follows from the theory on elliptic boundary problems in the complete scale of the spaces $H_s(G)$, $s \in \mathbb{R}$, more precisely, from the theorem on improved smoothness of solutions, proved in (Lions and Magenes 1968) (cf. Sect. 7.9 below).

b. Now we pass to boundary problems with more general boundary conditions:

$$Au - \lambda u = f \quad \text{in} \quad G\,, \quad B_j u = 0 \quad (j = 1, \dots, q) \quad \text{on} \quad \Gamma\,. \tag{7.11}$$

Here $\{B_j\}$ is a normal system of boundary operators of orders $r_j < m = 2q$. Let $r_j < q$ for $j \le r - 1$ and $r_j \ge q$ for $j \ge r$. According to the terminology of the calculus of variations, the first boundary conditions (with $r_j < q$) are called *stable*, and the second ones (with $r_j \ge q$) are called *natural.* Let $\{F_j\}_1^q$ be a Dirichlet system consisting of boundary operators of orders less than q, in which $F_j = B_j$ $(j < r)$. The following variant of Green's formula holds for functions $u \in H_{2q}(M)$, $v \in H_q(M)$:

$$a[u, v] = (Au, v)_G - \sum_1^q (\Phi_j u, F_j v)_\Gamma\,, \tag{7.12}$$

where $\{\Phi_j\}_1^q$ is also a normal system of boundary operators and $\operatorname{ord}\Phi_j + \operatorname{ord}F_j = 2q - 1$. We assume the following: $\{F_j\}$ can be chosen in such a way that $\Phi_j = B_j$ for $j = r, \dots, q$. Of course, this is an essential restriction on the structure of boundary operators in (7.11) and even on the set of orders of the boundary operators. Now let u be a solution of the boundary problem (7.11), and let v be a function from $H_q(M)$ satisfying the stable boundary conditions. Then the sum in (7.12) disappears, and we again come to the weak formulation (7.10) of the boundary problem. However now the space $\mathcal{H}$ of solutions is defined as follows:

$$\mathcal{H} = \{u \in H_q(G) : B_1 u = \ldots = B_{r-1} u = 0 \text{ on } \Gamma\}\,. \tag{7.13}$$

Here the stable boundary conditions are presented explicitly, while the natural boundary conditions are taken into account implicitly in the equation (7.10).

The form $a[u, v]$ is called *coercive* on the space $\mathcal{H}$ if for $u \in \mathcal{H}$ an inequality of the form (7.7) is valid with some positive ε and nonnegative C_1. The term was introduced by Aronszajn. The conditions for coerciveness have been studied by Aronszajn (Aronszajn 1955) and other authors. A very general result was obtained in (Agmon 1958):

Theorem 7.1.4. *The form $a[u, v]$ is coercive on the space $\mathcal{H}$ if the following two conditions hold.* 1) *The operator $A(x, D)$ is strongly elliptic in $\overline{G}$.* 2) *Let x_0 be any point of the boundary, and let $a_0(\xi)$ and $b_{j,0}(\xi)$ be the principal symbols of operators A and B_j at this point in local coordinates of Sect.* 0.2. *Then for $\xi' \ne 0$ and solutions $w(t)$ of the boundary problem*

$$a_0(\xi', D_n) w(t) = 0 \ \ (t > 0)\,, \quad b_{j,0}(\xi', D_n) w(0) = 0 \ \ (j = 1, \dots, r-1) \tag{7.14}$$

with $|w(t)| \to 0$ (exponentially) as $t \to +\infty$, the following inequality holds:

$$\operatorname{Re} \int_0^\infty \sum_{|\alpha|=|\beta|=q} a_{\alpha,\beta}(x_0) (\xi')^{\alpha'+\beta'} D_n^{\beta_n} w(t) \overline{D_n^{\alpha_n} w(t)}\, dt > 0\,. \tag{7.15}$$

Theorem 7.1.3 has a generalization to this situation (see (Lions and Magenes 1968):

Theorem 7.1.5. *Let the form* $a[u, v]$ *be coercive on the space* (7.13). *Then for* $\operatorname{Re}\lambda \leq -C_1$ *(see* (7.7)), *the integral identity* (7.10) *has a unique solution* $u \in \mathcal{H}$ *for any* $f \in \mathcal{H}^*$. *If* $f \in H_0(G)$, *then* u *is the solution of the boundary problem* (7.11) *belonging to* $H_m(G)$.

The considerations of this section have generalizations to the case of matrix variational problems: see (Figueiredo 1963) and (Grubb 1979). The latter paper contains an investigation of the form $a_B(u, v)$ associated with the operator A_B in the sense of Sect. 6.5. Conditions necessary and sufficient for the coerciveness of this form on the corresponding subspace of $\mathcal{H}_m(M)$ are obtained.

Besides the variational boundary problems discussed here, variational boundary problems "with constraints" have been considered. Spectral problems of this type will be discussed below in Sect. 9.7.

7.2. Boundary Problems in Nonsmooth Domains.

a. At first we briefly discuss the variational approach to boundary problems with generalized homogeneous boundary conditions in domains with nonsmooth boundaries. (See the definition of Sobolev spaces in such domains in Subsect. 2.1f.)

Let G be a bounded domain in $\mathbb{R}^n$; at this moment we do not assume anything about its boundary. Let a sesquilinear form (7.6) be given. For simplicity, we assume that it has smooth coefficients. Consider $\mathcal{H} = \overset{\circ}{H}_q(G)$ as a domain of definition of this form, and denote by $a_0(x, \xi)$ its *principal symbol*:

$$a_0(x, \xi) = \sum_{|\alpha|=|\beta|=q} a_{\alpha,\beta}(x)\xi^{\alpha+\beta} . \tag{7.16}$$

Assume that condition (7.5) of strong ellipticity is satisfied. Then Theorem 7.1.1 remains true. Using Lax–Milgram Lemma 7.1.2, we can define a closed operator A in $H = L_2(G)$ with domain $D(A) \subset \mathcal{H}$, such that

$$(Au, v) = a[u, v] \quad \text{for} \quad u \in D(A),\ v \in \mathcal{H} . \tag{7.17}$$

If we include the term $-\lambda u \cdot \overline{v}$ in the form, then we obtain the operator $A - \lambda I$ instead of A. This operator is invertible for $\operatorname{Re}\lambda \leq -C_1$ (see (7.7)). Moreover, we actually have

$$|\arg a_0(x, \xi)| \leq \theta \tag{7.18}$$

with some $\theta < \pi/2$, and if $\theta_1 > \theta$, then the operator $(A - \lambda I)^{-1}$ exists for all λ with $|\arg\lambda| \geq \theta_1$ and sufficiently large $|\lambda|$, and

$$\|(A - \lambda I)^{-1}\| \leq C|\lambda|^{-1} \tag{7.19}$$

for these λ. As to the (unbounded) operator A itself, it is a *Fredholm operator* in the following sense: its kernel is finite-dimensional, its range is closed, and the range has a finite-dimensional complement. See e.g. (Agmon 1965) and especially (Kato 1966, Chapter VI), where the corresponding notions from the abstract theory of forms and operators are discussed in detail. Formula (7.4) for A remains true in the sense of distributions. It is convenient to write $A = A_D$, since here we discuss the generalized Dirichlet problem (with homogeneous boundary conditions).[11]

Now assume that G is a *Lipschitz domain* or, what is the same, its boundary is a closed *Lipschitz surface*. This means that the boundary Γ can locally be represented, after an appropriate rotation of the coordinate system in $\mathbb{R}^n$, as the graph of a function satisfying *Lipschitz condition*: $x^n = f(x') = f(x^1, \dots, x^{n-1})$, where

$$|f(x') - f(y')| \leq C|x' - y'|. \tag{7.20}$$

(An equivalent assumption is: Γ satisfies the so-called uniform cone property introduced by Agmon, see (Agmon 1965); the equivalence is proved in (Grisvard 1985).) Then we can consider the *generalized Neumann problem.* It corresponds to the choice of $\mathcal{H} = H_q(G)$ as the domain of definition of the form (7.6). The condition of strong ellipticity is insufficient to obtain an analog of Gårding's inequality for $u \in H_q(G)$. We introduce the *generalized principal symbol*

$$a_0(x, \zeta) = \sum_{|\alpha|-|\beta|-q} a_{\alpha,\beta}(x)\zeta_\alpha\overline{\zeta_\beta} \tag{7.21}$$

with $\zeta_\alpha \in \mathbb{C}^n$, and we subject it to the condition of *very strong ellipticity*:

$$\operatorname{Re} a_0(x, \zeta) \geq \gamma \sum_{|\alpha|=q} |\zeta_\alpha|^2 \qquad (x \in \overline{G},\ \zeta_\alpha \in \mathbb{C}^n,\ |\alpha| = q)\ , \tag{7.22}$$

where γ is a positive constant. It is easy to prove that under this condition the analog of Gårding's inequality (7.7) is true for functions $u \in \mathcal{H} = H_q(G)$. See e.g. (Nečas 1967). Using the Lax–Milgram Lemma, we define a closed operator $A = A_N$ such that formula (7.17) is valid, $D(A) \subset \mathcal{H}$, and $A_N - \lambda I$ is invertible for sufficiently large $\operatorname{Re}\lambda$. Moreover, we have $|\arg a_0(x, \zeta)| \leq \theta$ with some $\theta < \pi/2$, and if $\theta_1 > \theta$, then the resolvent $(A - \lambda I)^{-1}$ exists for λ with $|\arg \lambda| \geq \theta_1$ and sufficiently large $|\lambda|$ and satisfies estimate (7.19) for these λ. In addition, A_N is a Fredholm operator.

The difference in assumptions concerning Γ is connected with differences in properties of the spaces $\mathcal{H} = \overset{\circ}{H}_q(G)$ and $\mathcal{H} = H_q(G)$. In particular, we have to insure the existence of a bounded operator of extension of functions $u \in \mathcal{H}$

[11] In the terminology of Kato's book, the form $a[u, v]$ with domain $\mathcal{H}$ is closed and sectorial, and the operator A defined by this form is m-sectorial. The two relations $u \in D(A)$ and (7.10) together are equivalent to the equation $(A - \lambda I)u = f$. Only if the boundary Γ is sufficiently smooth, we have $D(A) = \mathcal{H} \cap H_{2q}(G)$; in general, $D(A)$ can be larger.

to functions from $H_q(\mathbb{R}^n)$ and the validity of an interpolation inequality of the form (2.6) for $u \in \mathcal{H}$.

Instead of $\mathcal{H} = \overset{\circ}{H}_q(G)$ or $\mathcal{H} = H_q(G)$, we can choose $\mathcal{H} = V$ as a domain of definition of the form, where V is any closed subspace in $H_q(G)$ containing $\overset{\circ}{H}_q(G)$:

$$\overset{\circ}{H}_q(G) \subset V \subset H_q(G) . \tag{7.23}$$

As in the case of a smooth Γ, the form (7.6) is called *coercive* on V if inequality (7.7) is valid for $u \in V$. The condition of very strong ellipticity in a Lipschitz domain G is sufficient for this, and then the corresponding closed operator $A = A_V$ can be defined. The operator $A_V - \lambda I$ is invertible for sufficiently large $\operatorname{Re}\lambda$ and even for λ with $|\arg\lambda| \geq \theta_1$ and large $|\lambda|$. The operator A_V itself is a Fredholm operator. See e.g. the book (Nečas 1967).

Note that some integrals along Γ can be included in the form $a[u, v]$ if the boundary is piecewise–smooth or at least Lipschitz. See e.g. the same book (Nečas 1967).

b. There is a deep theory of elliptic boundary problems with boundaries smooth outside some singularities, such as corners, conical points, edges, etc. In particular, the investigation of the asymptotic behavior of a solution near such singularities is possible, and the structure of $D(A_D)$, $D(A_N)$, or $D(A_V)$ can be investigated using the information about the geometry of singularities. This theory, originated in (Kondrat'ev 1967), is the subject of the survey (Plamenevskij 1996) in the present volume.

c. There is a quite different approach to elliptic boundary problems in nonsmooth domains, based on the theory of potentials on nonsmooth boundaries. The first purpose in this theory is to solve the Dirichlet and Neumann boundary problems for homogeneous second order elliptic equations or systems

$$A(x, D)u(x) = 0 \quad \text{in} \quad G \quad \text{or} \quad \text{in} \quad \mathbb{R}^n \setminus \overline{G} \tag{7.24}$$

with inhomogeneous boundary conditions

$$\gamma_0 u = f \quad \text{or} \quad \gamma_1 u = g \quad \text{on} \quad \Gamma \tag{7.25}$$

by means of the single layer potential

$$u(x) = \int_\Gamma \Phi(x, y)\phi(y)\, dS_y \tag{7.26}$$

and the double layer potential

$$v(x) = \int_\Gamma \frac{\partial \Phi(x, y)}{\partial \nu_y} \psi(y)\, dS_y . \tag{7.27}$$

Here $\Phi(x, y)$ is a fundamental solution of equation (7.24) with singularity at y, and dS is the surface element on Γ; at ∞, an appropriate behavior of solutions is assumed. If G is Lipschitz, then Γ has a tangent hyperplane almost everywhere, so the Neumann condition and the double layer potential are meaningful. In this case it is possible to generalize usual integral formulas, including Green's formula, integral representation of solutions, and the formulas describing the behavior of the potentials as $x \to \Gamma$. However, in this case the investigation of the usual integral equations on Γ is much more difficult than in the case of a smooth Γ. In particular, the direct value of the double layer potential and of the normal derivative of the single layer potential are now singular integral operators, they are bounded operators in $L^2(\Gamma)$ but generally lose the compactness. Here we do not go into details, since the potential theory is the subject of the survey (Maz'ya 1988) in vol. 27 of EMS. We only indicate some new papers devoted to this theory for Lipschitz domains: (Costabel 1988), (Dahlberg et al. 1988), (Fabes et al. 1988), (Verchota 1990), (Gao 1991), (Jerison and Kenig 1995), (Pipher and Verchota 1995), and the monograph (Kenig 1994). Further references can be found in these papers and especially in (Kenig 1994).

In Section 9.8 we will touch some spectral problems for elliptic equations in nonsmooth domains.

7.3. Transmission Elliptic Problems. Let M be a compact manifold with boundary Γ. Assume that it is divided into two parts M_1 and M_2 by an $(n-1)$-dimensional closed C^∞ submanifold S having, for simplicity, no points in common with Γ. More precisely, we assume that $M_1 \cap M_2 = S$ and S is the boundary of M_1, while the boundary of M_2 consists of S and Γ. Consider the following boundary problem:

$$\begin{aligned} A^{(1)}u^{(1)} &= f^{(1)} \quad \text{on} \quad M_1 \setminus S\,, \\ A^{(2)}u^{(2)} &= f^{(2)} \quad \text{on} \quad M_2 \setminus (S \cup \Gamma)\,, \\ B_j u^{(2)} &= g_j \quad (j = 1, \dots, q) \quad \text{on} \quad \Gamma\,, \\ B_j^{(1)}u^{(1)} + B_j^{(2)}u^{(2)} &= h_j \quad (j = 1, \dots, m) \quad \text{on} \quad S\,. \end{aligned} \tag{7.28}$$

Here $A^{(1)}$ and $A^{(2)}$ are differential operators of the same (for simplicity) order $m = 2q$ with C^∞ coefficients on M_1 and M_2, respectively; $u^{(k)}$ and $f^{(k)}$ are functions on M_k ($k = 1, 2$). The boundary operators B_j on Γ of orders r_j are of the usual form (1.3). The boundary operators $B_j^{(k)}$ on S ($j = 1, 2$) of orders ρ_j have a similar structure. For simplicity we assume that $B_j^{(1)}$ and $B_j^{(2)}$ have the same order.

Such a boundary problem is called the *transmission problem.* For simplicity we restrict ourselves to the scalar case. Transmission problems have been considered by many authors. See e.g. (Schechter 1960a) and (Sheftel' 1965, 1966), where further references can be found. Transmission problems arise in applied fields when processes in piecewise-smooth media are considered.

The results discussed in the previous sections are easily generalized to transmission problems. Let us formulate the ellipticity conditions for the transmission problem (7.28). Of course, they include the conditions of ellipticity of the operators $A^{(1)}$ and $A^{(2)}$ on M_1 and M_2, their proper ellipticity, and the Shapiro–Lopatinskij condition for the system $\{B_j\}$ on Γ. It remains to formulate the analog of the Shapiro–Lopatinskij condition on S.

Let x be any point on S, and let local coordinates be chosen in a neighborhood of this point such that S is defined by the equation $x^n = 0$, so that $x = (x', 0)$, and x^n grows, for definiteness, in the direction of M_1. Let $a_0^{(k)}(x', 0, \xi)$ and $b_{j,0}^{(k)}(x', 0, \xi)$ $(k = 1, 2)$ be the principal symbols of the operators $A^{(k)}$ and $B_j^{(k)}$ at this point. Consider the following boundary problem on the line $\mathbb{R}^1$ $(t = x^n,\ \xi' \neq 0)$:

$$a_0^{(1)}(x', 0, \xi', D_n)v^{(1)}(t) = 0 \quad \text{on} \quad \mathbb{R}_+ \,, \tag{7.29}$$

$$a_0^{(2)}(x', 0, \xi', D_n)v^{(2)}(t) = 0 \quad \text{on} \quad \mathbb{R}_- \,, \tag{7.30}$$

$$\begin{aligned} &b_{j,0}^{(1)}(x', 0, \xi', D_n)v^{(1)}(t)|_{t=+0} \\ &\quad + b_{j,0}^{(2)}(x', 0, \xi', D_n)v^{(2)}(t)|_{t=-0} = h_j \quad (j = 1, \ldots, 2q) \,. \end{aligned} \tag{7.31}$$

Denote by $\mathfrak{M}^{(1)}(x', \xi')$ the subspace of solutions $v^{(1)}(t)$ of equation (7.29) on $\mathbb{R}_+$ with $|v^{(1)}(t)| \to 0$ as $t \to +\infty$ and by $\mathfrak{M}^{(2)}(x', \xi')$ the subspace of solutions $v^{(2)}(t)$ of equation (7.30) with $|v^{(2)}(t)| \to 0$ as $t \to -\infty$. In view of proper ellipticity, these subspaces are of dimension q. For any $\xi \neq 0$ and any numbers h_j, the problem (7.29)–(7.31) is required to have one and only one solution with $v^{(k)} \in \mathfrak{M}^{(k)}(x', \xi')$ $(k = 1, 2)$. This is the desired analog of the Shapiro–Lopatinskij condition on S.

If all these conditions are satisfied, then the transmission problem (7.28) is called elliptic.

The ellipticity of this problem is equivalent to the Fredholm property of the corresponding operator

$$(u^{(1)}, u^{(2)}) \mapsto (f^{(1)}, f^{(2)}, g_1, \ldots, g_q, h_1, \ldots, h_m) \tag{7.32}$$

acting in the corresponding spaces:

$$u^{(k)} \in H_s(M_k),\ f^{(k)} \in H_{s-m}(M_k),\ g_j \in H_{s-r_j-\frac{1}{2}}(\Gamma),\ h_j \in H_{s-\rho_j-\frac{1}{2}}(S) \,. \tag{7.33}$$

Here it suffices to subject s to the conditions

$$s \geq m, \quad s > r_j + \frac{1}{2} \quad (j = 1, \ldots, q) \,, \quad s > \rho_j + \frac{1}{2} \quad (j = 1, \ldots, m) \,. \tag{7.34}$$

The proof follows the scheme outlined in §2.

A particular transmission problem arises if we require that $u^{(1)}$ and $u^{(2)}$ coincide on S along with their normal derivatives of orders $1, \ldots, m-1$. Actually, we then have a boundary problem for an elliptic equation with coefficients having, in general, a jump at S. For this reason, the transmission problems are sometimes called *elliptic problems for elliptic equations with discontinuous coefficients* (see e.g. (Sheftel' 1965)). The third possible term is *conjugation problems.*

For elliptic problems of the form (7.28), a local theorem on smoothness of solutions is also valid. It is also possible to include a parameter in the equations (7.28), and to define conditions of ellipticity with parameter λ in an angle $\mathcal{L}$ on the complex plane; under these conditions the problem has one and only one solution for $\lambda \in \mathcal{L}$ with sufficiently large modulus, with corresponding a priori estimate. It is possible to deduce an analog of Green's formula, to generalize the results to elliptic systems, etc. There are spectral transmission problems; some of them are mentioned in Sect. 9.4 below.

7.4. Exterior Elliptic Boundary Problems. Many papers are devoted to this subject, but we will mention only few of them. In its simplest setting, a solution of an elliptic equation in $\mathbb{R}^n$ in a complement of a closed bounded domain is to be found; the domain has an $(n-1)$-dimensional C^∞ boundary Γ, and boundary conditions of the usual form are prescribed on Γ. Besides the condition of proper ellipticity and the Shapiro–Lopatinskij condition on Γ, the conditions on the coefficients at infinity and the type of ellipticity condition must be specified; sometimes the last can be uniform ellipticity. In theorems on the Fredholm property or unique solvability, the choice of spaces for solutions and for the right-hand sides is of key importance. For an equation of the second order coinciding with the Helmholtz equation in a neighborhood of infinity, the problem with the radiation condition at infinity is well known; we mentioned it in Section 5.4. Generalizations of this condition are investigated in papers of Vainberg; see (Vainberg 1982), (Egorov and Shubin 1988a, §7), and references therein. Elliptic equations in classes of functions with growth at infinity were studied in (Bagirov and Kondrat'ev 1975).

7.5. Nonlocal Elliptic Boundary Problems. The term "nonlocal elliptic boundary problem" is used in the literature in quite different situations. In this section we have in mind boundary problems whose formulation is suggested by the paper (Bitsadze and Samarskij 1969). In this paper, in particular, the following boundary problem is considered; it arises in the physics of plasma. A solution of an elliptic second order equation $Au = f$ is to be found in a bounded domain $G \subset \mathbb{R}^n$, under the conditions

$$u(x) = u(\omega(x)) \quad \text{on} \quad \Gamma_1\,, \quad u(x) = g(x) \quad \text{on} \quad \Gamma_2\,,$$

where Γ_1 is a subset of the boundary $\Gamma = \partial G$, $\Gamma_2 = \Gamma \backslash \Gamma_1$, $y = \omega(x)$ is a diffeomorphic mapping of a neighborhood Ω of Γ_1 onto a set $\omega(\Omega)$, and $\omega(\Gamma_1) \subset \overline{G}$. Roĭtberg and Sheftel' have considered the following general modification of this boundary problem in the simple case $\Gamma_1 = \Gamma$, $\Gamma_2 = \varnothing$, $\widetilde{\Gamma} = \omega(\Gamma) \subset G$ (see (Roĭtberg and Sheftel' 1972, 1973)). The boundary problem has the following form:

$$A(x,D)u(x) = f(x) \quad \text{in} \quad G\,, \tag{7.35}$$

$$B_j^{(1)}(x,D)u(x) + B_j^{(2)}(y,D_y)u(y)|_{y=\omega(x)} = g_j(x) \quad (j=1,\dots,q) \quad \text{on} \quad \Gamma\,, \tag{7.36}$$

where $A(x,D)$ is a properly elliptic differential operator of order $m = 2q$, while $B_j^{(1)}$ and $B_j^{(2)}$ are boundary operators of orders r_j and of the usual form.

It turns out that the analog of the Shapiro–Lopatinskij condition for this boundary problem coincides with the Shapiro–Lopatinskij condition for the boundary problem

$$Au = f \quad \text{on} \quad G\,, \quad B_j^{(1)}u = g_j \quad (j=1,\dots,q) \quad \text{on} \quad \Gamma \tag{7.37}$$

obtained from (7.35)–(7.36) by omitting the terms with $B_j^{(2)}$. Thus, the boundary problem (7.35)–(7.36) can be called elliptic if the boundary problem (7.37) is elliptic. In this case the usual theorems are valid for the boundary problem (7.35)–(7.36): the corresponding operator is Fredholm in Sobolev spaces, and a local assertion on improved smoothness of solutions is true. It is also possible to include a parameter λ in equations (7.35)–(7.36) and to impose the condition of ellipticity with parameter in an angle $\mathcal{L}$ on the boundary problem; then the theorem on unique solvability is true for $\lambda \in \mathcal{L}$ with sufficiently large modulus.

In (Roĭtberg and Sheftel' 1972, 1973) one can find other generalizations of the results mentioned here. More complicated situations are considered by Skubachevskij. The equations may contain some abstract operators that do not affect the formulation of the Shapiro–Lopatinskij condition; Γ_2 can be nonempty; more than one map $x \mapsto \omega(x)$ can be present. See (Skubachevskij 1983, 1985, 1986, 1991) and references therein.

7.6. Elliptic Boundary Conditions on Submanifolds of Various Dimensions. A boundary problem for the polyharmonic equation $\Delta^m u = 0$ with boundary conditions prescribed on submanifolds of codimensions greater than 1 was considered by Sobolev (see (Sobolev 1950)). General boundary problems of such type were considered in (Sternin 1966). Below we essentially follow this paper.

Let M be a closed manifold[12] (i.e. a compact manifold without boundary), and let Y be a submanifold of codimension k ($1 \le k \le n-1$). (It is possible to consider more general case, when several pairwise disjoint submanifolds Y_j of various dimensions are given; however, for simplicity we consider the case of only one given submanifold Y.) Let $A = A(x,D)$ be an elliptic differential operator of order m on M, and let $B_j(x,D)$ $(j = 1,\dots,r)$ be boundary operators of orders r_j having the form[13]

[12]Only in this section.

[13]For complete analogy with notation in other sections, we should write $y' = x'$ instead of y and $t'' = x''$ instead of t; similarly, below we should write ξ' instead of ξ and η'' instead of η.

$$B_j = \sum_{|\alpha|+|\beta|\le r_j} b_{\alpha,\beta}(y,t) D_y^\alpha D_t^\beta \tag{7.38}$$

near a point on Y. Here $y = (y^1, \dots, y^{n-k})$ are local coordinates along Y and $t = (t^{n-k+1}, \dots, t^n)$ are coordinates along directions transversal to Y, so that $(y,t) = (y^1, \dots, t^n)$ are local coordinates on M. Denote by ρ_j the order of B_j with respect to t. Consider the following boundary problem:

$$Au = f \quad \text{on} \quad M \setminus Y \,, \tag{7.39}$$

$$B_j u|_Y = g_j \qquad (j = 1, \dots, r) \,. \tag{7.40}$$

With it we associate the operator

$$\mathcal{A}: u \to (f, g) = (f, g_1, \dots, g_r) \,. \tag{7.41}$$

We assume that

$$s > \max \left(\rho_j + \frac{k}{2} \right) ; \tag{7.42}$$

then $\mathcal{A}$ is a bounded operator in the following Sobolev spaces:

$$\mathcal{A}: H_s(M) \to [H_{s-m}(M)/\Delta_{s-m}(M,Y)] \times \prod_j H_{s-r_j-\frac{k}{2}}(Y) \,, \tag{7.43}$$

where $\Delta_{s-m}(M,Y)$ is the subspace in $H_{s-m}(M)$ consisting of all distributions supported on Y. The number r of boundary conditions depends on s in the following way. Set

$$l_k(s) = \begin{cases} [m - s - (k/2)] & \text{if } s + \frac{k}{2} \notin \mathbb{Z} \,, \\ m - s - (k/2) - 1 & \text{otherwise} \end{cases} \tag{7.44}$$

(where $[h]$ is the integral part of h). Then r must be equal to the number of pairwise different monomials $\eta_{n-k+1}^{\beta_1} \cdots \eta_n^{\beta_k}$ with $|\beta| \le l_k(s)$.

This condition can be explained as follows. Let $P(D_t)$ be a differential operator of order m in $\mathbb{R}^k$ with constant coefficients, such that $P(\eta) \ne 0$ for all real η, including $\eta = 0$. Consider the equation

$$P(D_t)v(t) = 0 \quad \text{for} \quad t \ne 0 \,. \tag{7.45}$$

If $v(t) \in H_s(\mathbb{R}^k)$, then $P(D_t)v(t) \in H_{s-m}(\mathbb{R}^k)$, and this is a distribution supported at $t = 0$, i.e. a linear combination of the delta-function $\delta(t)$ and its derivatives. From this it can be deduced that $v(t)$ is a linear combination of the fundamental solution $F_{\eta \mapsto t}^{-1}[1/P(\eta)]$ and its derivatives. Such a linear combination belongs to $H_s(\mathbb{R}^k)$ if and only if the maximal order l of these derivatives satisfy the condition

$$2s - 2m + 2l < -k \,. \tag{7.46}$$

In (7.44) l_k is the maximal integer satisfying this inequality. Thus, $r = r_k(s)$ is the dimension of the space of solutions of the equation (7.45) in $H_s(\mathbb{R}^k)$. Below in this section it will be clear why r must be equal to this dimension.

It follows that if $l_k(s) < 0$ (in particular, if $s > m - (k/2)$), then no boundary condition may be added to the equation (7.39). We assume that $l_k(s) \geq 0$.

The conditions of ellipticity of the boundary problem (7.39)–(7.40) are formulated as follows. First, the operator A must be elliptic on M. Second, if $a_0(y,t,\xi,\eta)$ and $b_{j,0}(y,t,\xi,\eta)$ are the principal symbols of A and B_j written in the local coordinates near a point on Y, then for any $\xi \neq 0$ the boundary problem

$$a_0(y,0,\xi,D_t)v(t) = 0 \qquad (t \neq 0)\,, \tag{7.47}$$

$$b_{j,0}(y,0,\xi,D_t)v(t)|_{t=0} = 0 \qquad (j = 1,\dots,r) \tag{7.48}$$

must have only trivial solution in $H_s(\mathbb{R}^k)$. The last condition relates to all points on Y. From it and from what was said above it is clear why r must be chosen in the manner described above.

The main result is the equivalence of the Fredholm property of the operator (7.41) to the ellipticity conditions; the proof is similar to that outlined in §2, and it is carried out by constructing a parametrix.

In addition, in (Sternin 1966) there are a theorem on improved smoothness of solutions and a theorem about the asymptotics of solutions near Y.

In (Sternin 1976) the main result is generalized to pseudodifferential operators A and B_j, the ellipticity conditions are formulated in an invariant form, and, moreover, the index of $\mathcal{A}$ is calculated.

7.7. General Realizations of a Differential Elliptic Operator. Let $A = A(x,D)$ be a scalar properly elliptic differential operator of order $m = 2q$ in a domain $G \subset \mathbb{R}^n$. As above, we assume that G is bounded and has a smooth boundary, and that the coefficients of $A(x,D)$ belong to $C^\infty(\overline{G})$; for simplicity, we will not consider more general situations. The *minimal operator* $A_{\min}$ and the *maximal operator* $A_{\max}$ corresponding to $A(x,D)$ are operators in $H_0(G)$ defined by the relations

$$A_{\min}u = A(x,D)u(x)\,, \qquad u \in D(A_{\min}) = \overset{\circ}{H}_m(G)\,, \tag{7.49}$$

where $\overset{\circ}{H}_m(G)$ is the closure of $C_0^\infty(G)$ in $H_m(G)$, and

$$\begin{gathered} A_{\max}u = A(x,D)u(x)\,, \qquad u \in D(A_{\max})\,, \\ D(A_{\max}) = \{u \in H_0(G) \cap H_{m,\mathrm{loc}}(G) : A(x,D)u \in H_0(G)\}\,. \end{gathered} \tag{7.50}$$

Consider closed operators $\mathbb{A}$ in $H_0(G)$ satisfying the conditions

$$A_{\min} \subseteq \mathbb{A} \subseteq A_{\max}\,. \tag{7.51}$$

Such $\mathbb{A}$ are called *realizations* of the differential expression $A(x, D)$. Obviously, operators A_B corresponding to elliptic boundary problems with homogeneous differential boundary conditions (see Subsect. 2.4d) or pseudodifferential boundary conditions (see Sect. 7.8 below) are special cases of realizations. Each realization $\mathbb{A}$ is defined by its domain $D(\mathbb{A})$ described by "boundary conditions" that can have very general form. The investigation of general realizations goes back to the well-known paper (Vishik 1952) ($m = 2$); see also (Grubb 1968) and references in these papers.

For simplicity we assume that the differential operator $A(x, D)$ is formally selfadjoint. Then the operator $A_{\min}$ is symmetric and semibounded, and $A_{\max}$ is its adjoint: $A_{\max} = A^*_{\min}$. For definiteness, let the principal symbol $a_0(x, \xi)$ be positive; then $A_{\min}$ is bounded from below.

As we pointed out in Subsect. 4.2b, the Dirichlet problem for $A(x, D)$ is formally selfadjoint. Let $\{S_j\}$ be a normal system of boundary operators on Γ with $\operatorname{ord} S_j = m - j - 1$ such that Green's formula

$$(Au, v)_G - (u, Av)_G = \sum_{j=0}^{q-1} \{(S_j u, \gamma_j v)_\Gamma - (\gamma_j u, S_j v)_\Gamma\} \tag{7.52}$$

is true. We assume that the Dirichlet problem for $A(x, D)$ has a zero kernel; then the corresponding operator A_D has an inverse. Introduce the following notation: $\Lambda_\Gamma = (I - \Delta_\Gamma)^{1/2}$, where Δ_Γ is the Beltrami–Laplace operator on Γ, $H = [H_0(\Gamma)]^q$,

$$\widetilde{u} = A_D^{-1} A(x, D) u \, , \qquad u \in H_m(G) \, ; \tag{7.53}$$

$$B_1 u = (\Lambda_\Gamma^{1/2} S_0 \widetilde{u}, \Lambda_\Gamma^{3/2} S_1 \widetilde{u}, \ldots, \Lambda_\Gamma^{q-(1/2)} S_{q-1} \widetilde{u}) \, , \tag{7.54}$$

$$B_2 u = (\Lambda_\Gamma^{-1/2} S_0 \widetilde{u}, \Lambda_\Gamma^{-3/2} S_1 \widetilde{u}, \ldots, \Lambda_\Gamma^{-q+(1/2)} S_{q-1} \widetilde{u}) \, . \tag{7.55}$$

In (Mikhailets 1990) the following results were obtained.

1. Each realization coincides with the restriction of the maximal operator to functions satisfying a boundary condition of the form

$$P B_2 u - C B_1 u = 0 \, , \tag{7.56}$$

where P is an orthoprojector and C is a closed linear operator in H; their domains can be nondense. Conversely, all such restrictions are realizations.

2. The "correctly solvable realizations," i.e. $\mathbb{A}$ such that there exists a bounded inverse operator $\mathbb{A}^{-1}$, are described by boundary conditions of the form

$$B_2 u - C B_1 u = 0 \, , \tag{7.57}$$

where C is any bounded operator in H. We denote the corresponding realization by $\mathbb{A}_C$.

3. $(\mathbb{A}_C)^* = \mathbb{A}_{C^*}$; in particular, $\mathbb{A}_C = (\mathbb{A}_C)^*$ if and only if $C = C^*$.

4. Let $A_{\min} > 0$. Then $\mathbb{A}_C = (\mathbb{A}_C)^* > 0$ if and only if $C = C^* > 0$.

5. If $A_{\min} > 0$, then the operator $\mathbb{A}_C$ is maximally sectorial relative to

$$\mathcal{L}_{\alpha,\beta} = \{\lambda : \arg\lambda \in [\alpha, \beta]\} ,$$

where $-\pi \leq \alpha \leq 0 \leq \beta \leq \pi$, $0 < \beta - \alpha \leq \pi$, if and only if C is maximally sectorial in H relative to $\mathcal{L}_{-\beta,-\alpha}$. An operator is called *maximally sectorial relative to* $\mathcal{L}$ if its numerical range is contained in $\mathcal{L}$ and this operator has no extensions with this property.

6. $\mathbb{A}_{C_1}^{-1} - \mathbb{A}_{C_2}^{-1} \in \mathfrak{S}_p$ in $H_0(G)$ if and only if $C_1 - C_2 \in \mathfrak{S}_p$ in H. Here $\mathfrak{S}_p$ is the Neumann–Shatten class of compact operators if $0 < p < \infty$ and the set of all compact operators if $p = \infty$. A similar assertion is true for some other known two-sided ideals in the ring of all bounded operators.

7. Let $t \geq 0$ and $s \in [0, m]$. Then for any $f \in H_t(G)$ the solution of the equation $\mathbb{A}_C u = f$ belongs to $H_{s+t}(G)$ if and only if $C[H_s(\Gamma)]^q \subseteq [H_{s+t}(\Gamma)]^q$. In particular, $D(\mathbb{A}_C) \subset H_t(G)$ if and only if $R(C) \subseteq [H_t(\Gamma)]^q$.

These results admit generalizations to the case of matrix elliptic differential operators $A(x, D)$, and the case of L_p-spaces instead of L_2-spaces. Moreover, conditions for the Fredholm property and other types of solvability of realizations are also described.

In Sect. 9.6 we will discuss some spectral properties of these realizations.

7.8. Pseudodifferential Elliptic Boundary Problems.

a. Using the notation from Section 0.2, we introduce pseudodifferential boundary operators of the form

$$B_j u(x) = \sum_{k=0}^{r_j} B_{j,k} \gamma_k u \qquad (j = 1, \ldots, q) , \tag{7.58}$$

where $B_{j,k}$ belong to $\Psi_{\text{ph}}^{r_j - k}(\Gamma)$, i.e. they are polyhomogeneous pseudodifferential operators of orders $r_j - k$ on Γ (see (Agranovich 1990a, §2)), and B_{j,r_j} are operators of multiplication by a function. We define the principal symbol of B_j by the equality

$$b_{j,0}(x', \xi) = \sum_{k=0}^{r_j} b_{j,k}(x', \xi') \xi_n^k , \tag{7.59}$$

where $b_{j,k}(x', \xi')$ are the principal symbols of $B_{j,k}$. The ellipticity conditions for the boundary problem

$$Au = f \quad \text{on} \quad M_+ , \quad B_j u = g_j \quad (j = 1, \ldots, q) \quad \text{on} \quad \Gamma \tag{7.60}$$

are formulated, actually, as in the case of differential B_j, and the results of §2 remain true with minor modifications in the proofs.

Of course, elliptic boundary problems for a differential operator $A(x, D)$ with homogeneous pseudodifferential boundary conditions define realizations in the sense of Sect. 7.7. Now we present a situation in which a pseudodifferential boundary condition naturally arises.

Example 7.8.1. (Agranovich 1977.) Using the same notation as in Example 5.4.1, consider the equation

$$\Delta u(x) + \ldots = f(x) \tag{7.61}$$

in $\mathbb{R}^3$, where the dots denote lower order terms, and assume that this equation coincides with the Helmholtz equation $\Delta u(x) + k^2 u(x) = 0$ in G_-. At infinity, we subject the solution to the radiation condition (5.52). This boundary problem is equivalent to the following transmission problem: the function $u(x)$ satisfies equation (7.61) in G_+, the Helmholtz equation in G_-, the radiation condition at infinity, and the conditions

$$u^+ = u^-, \quad \partial_\nu u^+ = \partial_\nu u^- \quad \text{on} \quad \Gamma = \partial G_\pm \,. \tag{7.62}$$

In G_- we have the integral representation (5.54″) for $u(x)$. Passing to the limit as $x \to \Gamma$, we obtain relation (5.55″). Replacing in this relation u^- and $\partial_\nu u^-$ by u^+ and $\partial_\nu u^+$, respectively, we obtain the boundary problem for equation (7.61) in G_+ with a pseudodifferential boundary condition

$$u^+ = \left(\frac{1}{2} + T_2\right) u^+ - T_1 \partial_\nu u^+ \,. \tag{7.63}$$

It is easy to verify that this boundary problem is elliptic. Of course, the same is true for other dimensions n. We also mention that other boundary problems, with more general transmission conditions instead of (7.62), can also be considered.

Elliptic boundary problems with pseudodifferential boundary conditions were first considered by Dynin in 1961; see (Dynin 1961). The pseudodifferential operators (of nonnegative order) were defined there as linear combinations of differential operators with singular integral operators as coefficients. The term "singular differential operator" was used; but it has been ousted in the literature by the term "pseudodifferential operator." This generalization of the theory of elliptic boundary problems is natural; besides, it was stimulated by the necessity to obtain a freedom of homotopies to simplify the calculation of the index of the operator corresponding to an elliptic boundary problem. Indeed, unlike the case of differential operators, the symbol of a pseudodifferential operator need not be a polynomial. As we mentioned in (Agranovich 1990a), the problem of calculating the index of a general elliptic operator was formulated in (Gel'fand 1960), and it had a strong influence on the development of the theory of elliptic equations. In the author's papers (Agranovich 1964, 1965) the elliptic operator A in G is also assumed to be a pseudodifferential operator, but with the condition that it must be differential at the boundary Γ in the normal direction. It is already possible in this framework to obtain relative index theorems (see below). In (Grubb 1974) the notion of a normal system of boundary operators was extended to pseudodifferential $\{B_j\}$, Green's formula was established, and a boundary problem formally adjoint to the given boundary problem for a differential operator A with pseudodifferential $\{B_j\}$ was constructed.

The final step of generalizations in this direction was the construction of the general theory of elliptic boundary problems for pseudodifferential operators $A = A(x, D)$ on a manifold with boundary in the papers of Vishik and Eskin (see in particular (Vishik and Eskin 1964, 1967) and (Eskin 1973)), and Boutet de Monvel (see (Boutet de Monvel 1966, 1971)). The paper of Brenner and Shargorodsky in this volume is devoted to these generalizations. We only note that in the context of Boutet de Monvel's calculus of operators corresponding to boundary problems (see (Boutet de Monvel 1966, 1971) and the detailed and elaborated presentation in (Rempel and Schulze 1982)), the proof of the theorem "ellipticity implies the Fredholm property" becomes similar to the proof of the analogous theorem for pseudodifferential operators on a closed manifold and therefore does not need a localization: a parametrix is constructed by means of the calculus. In particular, the structure of the parametrix for a usual elliptic differential boundary problem becomes especially clear in terms of this calculus. However, the construction of the calculus requires considerable work. Only in the framework of this theory it is possible to justify all homotopies that are needed for the calculation of the index of a general differential elliptic boundary problem (see (Boutet de Monvel 1971) and (Rempel and Schulze 1982)). In the book (Grubb 1986) a calculus of pseudodifferential boundary problems with parameter is constructed; it leads to theorems on unique solvability of boundary problems elliptic with parameter that are much more general than those formulated in §3 above.

b. Now we formulate two *relative index theorems*.

Consider two elliptic boundary problems

$$Au = f \quad \text{on} \quad M_+ , \quad B^{(1)}u = g^{(1)} \quad \text{on} \quad \Gamma \tag{7.64}$$

and

$$Au = f \quad \text{on} \quad M_+ , \quad B^{(2)}u = g^{(2)} \quad \text{on} \quad \Gamma . \tag{7.65}$$

Here A is a $p \times p$ matrix differential operator elliptic in the usual sense, and $B^{(j)}$ are $q \times p$ matrix pseudodifferential boundary operators. Using an invertible scalar elliptic pseudodifferential operator Λ of order 1 on Γ, we can equalize the main orders of scalar operators in the matrices $B^{(1)}$ and $B^{(2)}$. Therefore, we assume that they have the same order r. Let $\mathcal{A}_1$ and $\mathcal{A}_2$ be the operators in Sobolev spaces corresponding to the boundary problems (7.64) and (7.65) (see (6.29)).

Theorem 7.8.2. *The difference $\kappa(\mathcal{A}_2) - \kappa(\mathcal{A}_1)$ of the indices of the operators $\mathcal{A}_2$ and $\mathcal{A}_1$ is equal to the index of a $q \times q$ matrix elliptic pseudodifferential* *operator of order zero on Γ. The principal symbol of this operator is equal to $L_2L_1^{-1}$, where L_1 and L_2 are the Lopatinskij matrices of the boundary problems* (7.64) *and* (7.65).

This result easily follows from what has been said in Subsect. 5.2c. We only need to take into account that the index of the product of two Fredholm operators is equal to the sum of their indices. Actually, considerations on the level of principal symbols are sufficient, and they are easier than in Section 5.2. The result is obtained in (Agranovich and Dynin 1962); a particular case is contained in (Dynin 1961). See also (Agranovich 1965) and (Rempel and Schulze 1982).

Now consider elliptic boundary problems

$$A^{(1)}u = f^{(1)} \quad \text{on} \quad M_+ , \quad Bu = g \quad \text{on} \quad \Gamma , \tag{7.66}$$

$$A^{(2)}u = f^{(2)} \quad \text{on} \quad M_+ , \quad Bu = g \quad \text{on} \quad \Gamma , \tag{7.67}$$

with the same boundary condition for $p \times p$ matrix elliptic (in the usual sense) operators $A^{(1)}$ and $A^{(2)}$ of the same order. Let $a_0^{(1)}(x,\xi)$ and $a^{(2)}(x,\xi)$ be their principal symbols. Assume that they coincide for $x \in \Gamma$. Denote by $\mathcal{A}^{(1)}$ and $\mathcal{A}^{(2)}$ the operators in Sobolev spaces corresponding to the boundary problems (7.76) and (7.67).

Theorem 7.8.3. *Under these conditions the difference $\kappa(\mathcal{A}^{(2)}) - \kappa(\mathcal{A}^{(1)})$ of the indices of the operators $\mathcal{A}^{(2)}$ and $\mathcal{A}^{(1)}$ is equal to the index of a $p \times p$ matrix elliptic operator of order* 0 *on M_0 with principal symbol equal to*

$$a_0^{(2)}(x,\xi)[a_0^{(1)}(x,\xi)]^{-1}$$

for $x \in M$, and to the unit matrix for $x \in M_-$.

A theorem close to this was obtained in (Agranovich 1962, 1965).

Theorems 7.7.2 and 7.7.3 have some useful corollaries. First, consider scalar elliptic boundary problems. We know that in this case the index of the Dirichlet problem is equal to zero. Therefore, calculation of the index of a general elliptic boundary problem reduces to calculation of the index of some pseudodifferential operator on a closed manifold, Γ, with known principal symbol. Another corollary: for any matrix elliptic boundary problem in a domain $G \subset \mathbb{R}^n$, the index does not depend on boundary operators if $q < n-1$. (This follows from the fact that the index of a $q \times q$ matrix elliptic operator on a hypersurface in $\mathbb{R}^n$ is equal to zero if $q < n-1$.) Using Theorem 7.8.3 and Theorem 6.2.1, one can verify that the index of the matrix Dirichlet problem, if it is elliptic, in a domain $G \subset \mathbb{R}^n$ is equal to the index of some $p \times p$ matrix elliptic operator on a sphere S^n; from this it follows that this index is equal to zero for $p < n$ even if $p > 1$.

These and some other corollaries from Theorems 7.8.2 and 7.8.3 are discussed in (Agranovich 1965). As to the complete solution of the problem of calculating the index for any elliptic boundary problem, we refer the reader to (Atiyah and Bott 1964) and (Boutet de Monvel 1971); see also (Rempel and Schulze 1982), (Hörmander 1985a), and (Fedosov 1990).

7.9. Elliptic Boundary Problems in Complete Scales of Banach Spaces. This section is devoted to generalizations of the main theorems on elliptic boundary problems in two directions: here we consider the Sobolev spaces H_s with arbitrary real s, and the spaces $H_{s,p}$ with $p > 1$ instead of $p = 2$.

General results in these directions were obtained in the 60's by Lions and Magenes, Berezanskij, S. Kreĭn, Roĭtberg, and other mathematicians; see (Lions and Magenes 1968), (Berezanskij et al. 1963), (Berezanskij 1965), (Roĭtberg 1991), and references therein. The theorems presented in the book (Lions and Magenes 1968) are well known, therefore we consider another approach.

a. We begin with the notion of a generalized solution. Consider the boundary problem (1.1), assuming for simplicity that $M = G$ is a bounded domain in $\mathbb{R}^n$ with a smooth boundary Γ. Let u be a function from $C^\infty(\overline{G})$. Set

$$\begin{aligned} u_0 &= u\,; \quad u_j = \gamma_{j-1}u \quad (j = 1, \ldots, r)\,; \\ f_0 &= f\,; \quad f_j = \gamma_{j-1}f \quad (j = 1, \ldots, r-m)\,, \end{aligned} \tag{7.68}$$

where

$$m = 2q \quad \text{and} \quad r = \max\{m, r_1 + 1, \ldots, r_q + 1\}\,. \tag{7.69}$$

From (4.12) and (4.1) it follows that

$$(u_0, A^*v) + \sum_{l=0}^{m-1} (u_{l+1}, L_{m-l}v) = (f_0, v) \qquad (v \in C^\infty(\overline{G}))\,, \tag{7.70}$$

and

$$\sum_{k=0}^{r_j} B_{jk}u_{k+1} = g_j \qquad (j = 1, \ldots, q)\,. \tag{7.71}$$

In addition, if $r > m$, then the consistency conditions

$$\gamma_{j-1}A(x, D)u = f_j \qquad (j = 1, \ldots, r-m) \tag{7.72}$$

have to be satisfied; if $r = m$, then they are superfluous.

Let u_0 be a function defined in G, and let $u_1, \ldots, u_r$ be functions defined on Γ. We call the tuple $U = (u_0, u_1, \ldots, u_r)$ a *generalized solution* of the boundary problem (1.1) if the relations (7.70)–(7.72) are satisfied. Here, instead of f, we consider the tuple $\Phi = (f_0, \ldots, f_{r-m})$. If u is a classical solution of (1.1), then, in terms of it, a generalized solution is constructed in the indicated manner. However the generalized solution can exist in a more general situation. Consider the following mapping defined by the boundary problem (1.1) (or, what is the same, by the relations (7.70)–(7.72)):

$$\mathcal{A}: U = (u_0, \ldots, u_r) \to \Psi = (f; g) = (f_0, \ldots, f_{r-m}; g_1, \ldots, g_q) . \tag{7.73}$$

b. To investigate this mapping, we need to introduce some appropriate spaces of functions.

For any $s \in \mathbb{R}$ and $p > 1$, denote by $H_{s,p} = H_{s,p}(\mathbb{R}^n)$ the Banach space of distributions f in $\mathbb{R}^n$ with finite norm

$$\|f, \mathbb{R}^n\|_{s,p} = \|F^{-1}(1+|\xi|^2)^{s/2} F f\|_{L_p(\mathbb{R}^n)} . \tag{7.74}$$

It is easy to verify that $C_0^\infty(\mathbb{R}^n)$ is dense in $H_{s,p}$ and that $H_{s,p}$ and $H_{-s,p'}$, where

$$\frac{1}{p} + \frac{1}{p'} = 1 , \tag{7.75}$$

are mutually adjoint relative to the extension of the scalar product $(u, v)_{\mathbb{R}^n}$.

For a domain $G \subset \mathbb{R}^n$ we set

$$H_{s,p;0}(G) = \{u \in H_{s,p} : \operatorname{supp} u \subset \overline{G}\} . \tag{7.76}$$

This is a subspace in $H_{s,p}$, and $C_0^\infty(G)$ is dense in $H_{s,p;0}(G)$.

By $H_{s,p}(G)$ $(s \geq 0,\ p > 1)$ we denote the space of restrictions to G of elements u from the space $H_{s,p} = H_{s,p}(\mathbb{R}^n)$, with norm

$$\|u, G\|_{s,p} = \inf \|v, \mathbb{R}^n\|_{s,p} \qquad (s \geq 0,\ p > 1) , \tag{7.77}$$

where inf is taken over all $v \in H_{s,p}$ equal to u in G (almost everywhere). It is easy to see that $H_{s,p}(G)$ is isometrically isomorphic to the factor-space $H_{s,p}/H_{s,p;0}(\mathbb{R}^n \setminus \overline{G})$. For $p = 2$ it coincides with $H_s(G)$.

By $H_{-s,p}(G)$ $(s > 0,\ p > 1)$ we denote the space dual to $H_{s,p'}(G)$ (where p and p' are connected by relation (7.75)) with respect to the extension of the scalar product in $L_2(G)$. The norm in $H_{-s,p}(G)$ is defined by the equality

$$\|u, G\|_{-s,p} = \sup\{|(u, v)_G| / \|v, G\|_{s,p'} : v \in H_{s,p'}(G)\} . \tag{7.78}$$

The space $H_{-s,p}(G)$ is isometrically isomorphic to the subspace $H_{-s,p;0}(G)$ of the space $H_{-s,p}(\mathbb{R}^n)$. We note that the delta-function $\delta(x-y)$ supported at the point $y \in \overline{G}$ obviously belongs to $H_{-s,p}(G)$ for $s > n/p'$ (the last inequality guarantees the continuity of the embedding of $H_{s,p'}(G)$ in $C(\overline{G})$). From this it is clear that $H_{-s,p}(G)$ $(s > 0)$ is generally not a space of distributions in G: if $s > 1/p'$, then $H_{-s,p}(G)$ contains elements supported on Γ.[14]

[14] We emphasize once again that according to our definitions $H_{s,p}(G) = H_{s,p;0}(G)$ $(s < 0)$. In (Lions and Magenes 1968), $H_{s,p}(G)$ $(s < 0,\ p > 1)$ denotes the space $(\overset{\circ}{H}_{-s,p'}(G))^*$ adjoint, with respect to $(\ ,\ )_G$, to the closure $\overset{\circ}{H}_{-s,p'}(G)$ of the linear manifold $C_0^\infty(G)$ in $H_{-s,p'}(G)$. One more variant of the definition of the space $H_{s,p}(G)$ $(s < 0)$ is used in (Hörmander 1983a), (Triebel 1978), and (Grubb 1990). Namely, there $H_{s,p}(G)$, $s \leq 0$, as in the case $s > 0$, denotes the factor-space $H_{s,p}(\mathbb{R}^n)/H_{s,p;0}(\mathbb{R}^n \setminus G)$. Note that $(\overset{\circ}{H}_{-s,p'}(G))^*$ $(s < 0)$ coincides with this factor-space if $-s - (1/p')$ is not an integer; if $-s - (1/p')$ is an integer, then these spaces are generally different.

By $B_{s,p}(\Gamma)$ ($s \in \mathbb{R}$, $p > 1$) we denote the Besov space with norm $\langle\!\langle\ \ \rangle\!\rangle_{s,p}$ (see e.g. (Triebel 1978)). If $s > 0$, then $B_{s,p}(\Gamma)$ is the space of the traces on Γ of elements of $H_{s+1/p,p}(G)$, with norm

$$\langle\!\langle g, \Gamma \rangle\!\rangle_{s,p} = \inf\{\|u, G\|_{s+1/p,p} : u \in H_{s+1/p,p}(G),\ u|_\Gamma = g\}\ . \tag{7.79}$$

If $s < 0$, then $B_{s,p}(\Gamma)$ is the space adjoint to $B_{-s,p'}(\Gamma)$ with respect to the extension of the scalar product in $L_2(\Gamma)$, and

$$\langle\!\langle g, \Gamma \rangle\!\rangle_{s,p} = \sup\{|(g, v)_\Gamma| / \langle\!\langle v, \Gamma \rangle\!\rangle_{-s,p} : v \in B_{-s,p'}(\Gamma)\}\ . \tag{7.80}$$

Finally, the space $B_{0,p}(\Gamma)$ is defined by means of complex interpolation (see Subsect. 8.3b below). If $p = 2$, then we have $B_{s,p}(\Gamma) = H_s(\Gamma)$ for all s.

Now we introduce the space $\widetilde{H}^{s,p,(r)}(G)$ (Roĭtberg 1964, 1971); it plays a basic role in the approach discussed in this section. Here $s \in \mathbb{R}$, $p > 1$, and $r \in \mathbb{N}$. For $s \neq k+1/p$ ($k = 0, \ldots, r-1$) this space is defined as the completion of $C^\infty(\overline{G})$ with respect to the norm

$$\|u\|_{s,p,(r)} = \left(\|u, G\|^p_{s,p} + \sum_{j=1}^{r} \langle\!\langle \gamma_{j-1} u, \Gamma \rangle\!\rangle^p_{s-j+1-1/p,p} \right)^{1/p} . \tag{7.81}$$

If $s > r - 1 + 1/p$, then the norm (7.81) is equivalent to the norm $\|u, G\|_{s,p}$, and $\widetilde{H}^{s,p,(r)}(G) = H_{s,p}(G)$. For $s < r - 1 + 1/p$ these norms are not equivalent. If $s = k + 1/p$ ($k = 0, \ldots, r-1$), then we define the space $\widetilde{H}^{s,p,(r)}(G)$ and the norm $\|\ \ \|_{s,p,(r)}$ by complex interpolation. Finally, for $r = 0$ we set $\widetilde{H}^{s,p,(0)}(G) = H_{s,p}(G)$ and $\|u\|_{s,p,(0)} = \|u, G\|_{s,p}$.

The mapping

$$u \mapsto (u, \gamma_0 u, \ldots, \gamma_{r-1} u) \tag{7.82}$$

can obviously be extended by continuity to a (continuous) isomorphism between $\widetilde{H}^{s,p,(r)}(G)$ and some subspace of the space

$$\mathcal{H}^{s,p,(r)} = H_{s,p}(G) \times \prod_{j=1}^{r} B_{s-j+1-1/p,p}(\Gamma)\ . \tag{7.83}$$

This permits us to identify $u \in \widetilde{H}^{s,p,(r)}(G)$ with the corresponding tuple

$$U = (u_0, u_1, \ldots, u_r) \in \mathcal{H}^{s,p,(r)}\ ,$$

and also to identify $f \in \widetilde{H}^{s-m,p,(r-m)}(G)$ with the corresponding tuple

$$\Phi = (f_0, f_1, \ldots, f_{r-m}) \in \mathcal{H}^{s-m,p,(r-m)}$$

if $r > m$.

The mapping (7.73) corresponding to the boundary problem (1.1) is extended to the continuous mapping

$$\mathcal{A}_{s,p} \colon \widetilde{H}^{s,p,(r)}(G) \to K_{s,p}\ , \tag{7.84}$$

where

$$K_{s,p} = \widetilde{H}^{s-m,p,(r-m)}(G) \times \prod_{j=1}^{q} B_{s-r_j-1/p,p}(\Gamma) . \tag{7.85}$$

Until now, ellipticity was not essential.

Theorem 7.9.1. *Let the boundary problem* (1.1) *be elliptic. Then* (7.73) *is a Fredholm operator for any $s \in \mathbb{R}$ and $p > 1$.*

The proof in the case $s \geq r$ is carried out by following the scheme of §2. If, in addition, $r = m$ and the boundary problem is normal, then the same result is true for the formally adjoint boundary problem, and this yields the desired result for $s \leq 0$. The result is extended to $s \in (0, m)$ by means of interpolation (see (Roĭtberg 1964) and (Berezanskij 1965, Chapter 3, §6)). These "methods of transposition and interpolation" are similar to those used in (Lions and Magenes 1968).

In (Roĭtberg 1970) Theorem 7.9.1 is extended to boundary problems with $r = m$ not being normal: for such boundary problems Green's formula is obtained, and the formally adjoint boundary problem is investigated (in general it is pseudodifferential). As a consequence, the methods of transposition and interpolation are extended to this case.

The methods of transposition and interpolation are not applicable in the case of general boundary problems for elliptic systems. Moreover, these methods are not profitable if we need to minimize assumptions on smoothness of the coefficients and the boundary. In (Roĭtberg 1975, 1991) a direct proof is presented for Theorem 7.9.1 by means of the left and right parametrix for $\mathcal{A}_{s,p}$.

c. In (Roĭtberg 1971, 1991) the following theorem is proved.

Theorem 7.9.2. *Let $A(x, D)$ be a properly elliptic operator of order m. Then for any $s \in \mathbb{R}$ and $p > 1$, the norm $\|u\|_{s,p,(m)}$* (see (7.81)) *is equivalent to the norm*

$$\|u\|_{H_A^{s,p}(G)} = \|u, G\|_{s,p} + \|A(x, D)u, G\|_{s-m,p} \tag{7.86}$$

on functions in $C^\infty(\overline{G})$. Therefore, the space $\widetilde{H}^{s,p,(m)}(G)$ coincides with the completion $H_A^{s,p}(G)$ of $C^\infty(\overline{G})$ with respect to this norm.

As a corollary we find that for $u \in H_A^{s,p}(G)$ the traces

$$\gamma_k u \in B_{s-k-1.p,p}(\Gamma)$$

are well defined.

In addition we see that in the case $r = m$ we may replace $H^{s,p,(r)}(G)$ by $H_A^{s,p}(G)$ in the formulation of Theorem 7.9.1.

Other theorems on isomorphisms can be obtained by using the following simple statements (see (Roĭtberg 1968)).

Proposition 7.9.3. ("The graph method.") *Let B_1, B_2, and Q_2 be Banach spaces, and let Q_2 be linearly and continuously embedded in B_2. Let T be a linear (continuous) isomorphism $B_1 \to B_2$. Then the linear submanifold $Q_1 = T^{-1}Q_2$ in B_1 is a Banach space with respect to the graph norm $\|x\|_{B_1} + \|Tx\|_{Q_2}$; denote it by Q_1^T. The restriction of the operator T to this submanifold defines an isomorphism $Q_1^T \to Q_2$.*

Proposition 7.9.4. ("The matching method.") *Let B_1 and B_2 be Banach spaces, and let T be a continuous isomorphism $B_1 \to B_2$. Let E_1 be a subspace in B_1, and $E_2 = TE_1$. Then T naturally defines an isomorphism of factor-spaces $B_1/E_1 \to B_2/E_2$.*

Below in this section we assume that $r = m$, i.e. $r_j < m$ for all j. We present two examples of using Propositions 7.9.3 and 7.9.4. To simplify the formulations, we assume that $\mathcal{A}_{s,p}$ has no defect, i.e. no kernel or cokernel.

As a first example, we derive some results due to Lions and Magenes from those stated above. Let $s \in [0, m]$, and let $X = X_{s-m,p}$ be a Banach space of distributions in G containing $C^\infty(\overline{G})$ and embedded in $H_{s-m,p}(G)$ (for example, $X = L_p(G)$). We set

$$Q_2 = X_{s-m,p} \times \prod_{j=1}^{q} B_{s-r_j-1/p,p}(\Gamma) \, .$$

Using the isomorphism $H_A^{s,p}(G) \to K_{s,p}$ and applying Proposition 7.9.3, we obtain the isomorphism

$$D_A^{s,p}(G) \to X_{s-m,p}(G) \times \prod_{j=1}^{q} B_{s-r_j-1/p,p}(\Gamma) \qquad (0 \le s \le m) \, , \tag{7.87}$$

where

$$D_A^{s,p}(G) = \{u : \; u \in H_{s,p}(G), \; Au \in X_{s-m,p}\} \tag{7.88}$$

and it is assumed that the graph norm

$$\|u\|_{D_A^{s,p}(G)} = \|u, G\|_{s,p} + \|Au\|_{X_{s-m,p}} \tag{7.89}$$

is introduced in this space. A theorem on isomorphism of the type (7.87) was obtained by Lions and Magenes by means of transposition and interpolation for normal elliptic boundary problems.

The second example is a theorem on isomorphism for boundary problems with homogeneous boundary conditions (see (Berezanskij et al. 1963), where the methods of transposition and interpolation were used):

$$H^{s,p}(\mathrm{bnd}) \to (H^{m-s,p'}(\mathrm{bnd})^+)^* \qquad (s \in \mathbb{R}, \; p > 1) \, . \tag{7.90}$$

Here the space $H^{s,p}(\mathrm{bnd})$ is defined as follows: for $s \ge m$ it is a subspace in $H_{s,p}(G)$ defined by homogeneous boundary conditions $B_j u = 0$ ($j = 1, \ldots, q$), and for $s < m$ it is the closure of the space $H^{m,p}(\text{bnd})$ in $H_{s,p}(G)$. Replacing B_j by C_j (see (4.15)), we obtain the definition of the space $H^{s,p}(\text{bnd})^+$. Finally, $(\quad)^*$ is the space dual to the space inside the parentheses with respect to the extension of the scalar product in $L_2(G)$.

The theorem on isomorphism (7.90) can be obtained by using the isomorphism

$$\widetilde{H}^{s,p,(m)}(\text{bnd}) = \{u \in \widetilde{H}^{s,p,(m)}(G) : \ B_j u = 0 \ (j = 1, \dots, q)\} \to H_{s-m,p}(G)$$

defined by the restriction of the operator $\mathcal{A}_{s,p}$ to the left-hand side. Here, if $s < m$, a matching is necessary, since different elements in $\widetilde{H}^{s,p,(m)}(\text{bnd})$ can have coinciding first components in this case.

d. Now we formulate a local assertion on improved smoothness of solutions. Let Γ_1 be an open subset of the boundary Γ, and let G_1 be a subdomain in G adjacent to Γ_1 ($\partial\overline{G}_1 \cap \Gamma = \Gamma_1$). We agree to write $u \in \widetilde{H}^{s_1,p_1,(r)}_{\text{loc}}(G_1, \Gamma_1)$ if for any function $\chi \in C^\infty(\overline{G})$ equal to zero in some neighborhood in $\overline{G}$ of the set $\overline{G} \setminus (G_1 \cup \Gamma_1)$ we have $\chi u \in \widetilde{H}^{s_1,p_1,(r)}(G)$. The relation $F \in K_{s_1,p_1,\text{loc}}(G_1, \Gamma_1)$ is defined similarly.

Theorem 7.9.5. (See (Roĭtberg 1991).) *Let* $u \in \widetilde{H}^{s,p,(r)}(G)$ ($s \in \mathbb{R}$, $p > 1$) *be a generalized solution of the boundary problem* (1.1) *with* $F = (f, g) \in K_{s,p}$. *If* $F \in K_{s_1,p_1,\text{loc}}(G_1, \Gamma_1)$ ($s_1 \geq s$, $p_1 \geq p$), *then* $u \in \widetilde{H}^{s_1,p_1,(r)}_{\text{loc}}(G_1, \Gamma_1)$.

From this theorem it is easy to obtain local assertions on improved smoothness of weak solutions of the boundary problem (1.1).

e. These theorems can naturally be extended to boundary problems elliptic in the Agmon–Douglis–Nirenberg sense. See (Grubb 1990), (Roĭtberg 1991) and references therein.

f. Theorems on complete sets of isomorphisms find many applications. Here we mention two of them.

1. Local properties of Green's function. For simplicity, assume that there are no kernel and cokernel and that $r_j < m$ for all j. The vector-valued function

$$R(x, y) = (R_0(x, y), R_1(x, y), \dots, R_q(x, y)) \qquad (x \neq y) \tag{7.91}$$

is called *Green's function* of the boundary problem (1.1) if, for sufficiently smooth f and g, the function

$$u(x) = \int_G R_0(x, y) f(y)\, dy + \sum_{j=1}^{q} \int_\Gamma R_j(x, y) g_j(y)\, dy \tag{7.92}$$

is a solution of the boundary problem. Here, in particular, the function R_0 can be defined as the solution of the boundary problem

$$A(x, D_x)R_0(x,y) = \delta(x-y)\ , \quad B_j(x, D_x)R_0(x,y)|_{\partial G} = 0 \quad (j = 1, \dots, q)\ .$$

Since $\delta(x-y)$ is a continuous function of $y \in \overline{G}$ with values in the space $H_{-s,p}(G)$ $(s > n/p')$, $R_0(x,y)$ is a continuous function with values in the space $\widetilde{H}^{m-s,p,(m)}(G)$. It is C^∞ with respect to the collection of all variables outside the diagonal $x = y$. In addition, if $\omega_\alpha(x,y)$ is a C^∞ function on $\overline{G}\times\overline{G}$ that is different from zero only in a neighborhood of the diagonal and coincides with $(x-y)^\alpha$ in a smaller neighborhood (where $\alpha = (\alpha_1, \dots, \alpha_n)$, $\alpha_j \in \mathbb{Z}_+$), then

$$\omega_\alpha(x,y)R_0(x,y) \in \widetilde{H}^{m-s+|\alpha|,p,(m)}(G)\ .$$

Another way of investigating Green's function is analysis of formulas for the parametrix.

2. Investigation of elliptic boundary problems with power singularities in the right-hand sides near some submanifolds of different dimensions. It is possible to prove the existence of a solution and to investigate its behavior near these submanifolds. Additional conditions are indicated to select a unique solution (such conditions are needed even in the absence of a defect). See (Roĭtberg 1991).

Such boundary problems are related to boundary problems with a strong degeneracy:

$$\begin{aligned} \rho_0(x)A(x,D)u(x) &= f(x) \quad \text{in } G\ , \\ \rho_j(x)B_j(x,D)u(x) &= g_j \quad (j = 1, \dots, q) \quad \text{on } \Gamma\ , \end{aligned} \tag{7.93}$$

where the functions $\rho_0, \dots, \rho_q$ have a power degeneracy near some submanifolds. Dividing these equations by $\rho_0(x)$ and $\rho_j(x)$, we obtain boundary problems with power singularities in the right-hand sides.

Generalizations and other applications of the results mentioned here can be found in (Berezanskij 1965), (Lions and Magenes 1968), and (Roĭtberg 1991).

§8. Some Functions of the operator A_B

In Sects. 8.1–8.4 we consider the operator A_B corresponding to the scalar normal elliptic boundary problem

$$Au = f \quad \text{on} \quad M\ , \quad B_j u = 0 \quad (j = 1, \dots, q) \quad \text{on} \quad \Gamma\ , \tag{8.1}$$

assuming that after replacing A by $A(\lambda) = A - \lambda I$, we obtain a boundary problem elliptic with parameter in an angle $\mathcal{L}$ with bisectrix R_-. In essence, we follow (Seeley 1969a,b) in Sects. 8.1 and 8.3 and (Seeley 1969b) and (Greiner 1971) in Sect. 8.4. In Sect. 8.5 we indicate some generalizations.

8.1. Parametrix for $A_B - \lambda I$.

a. First we consider the boundary problem in $\mathbb{R}^n_+$

$$(A - \lambda I)u = f \quad \text{in} \quad \mathbb{R}^n_+ , \quad B_j u = 0 \quad (j = 1, \dots, q) \quad \text{for} \quad x^n = 0 \qquad (8.2)$$

that is elliptic with parameter in $\mathcal{L}$, assuming for simplicity that the coefficients in A do not depend on x for sufficiently large $|x|$ and that the coefficients in B_j do not depend on x' for sufficiently large $|x'|$. Let $\theta(t)$ be a function from $C^\infty(\overline{\mathbb{R}}_+)$ that is equal to zero for $t < 1/2$ and 1 for $t > 1$. Set

$$\theta_1 = \theta_1(\xi, \lambda) = \theta(|\xi|^2 + |\lambda|^{1/q}) , \quad \theta_2 = \theta_2(\xi', \lambda) = \theta(|\xi'|^2 + |\lambda|^{1/q}) . \qquad (8.3)$$

We define the right parametrix of order N for the boundary problem (8.1) by the formula

$$P^{(N)}(\lambda) = \sum_{l=0}^{N} P_{1,l}(\lambda) - \sum_{l=0}^{N} P_{2,l}(\lambda) , \qquad (8.4)$$

where

$$(P_{1,l}(\lambda)f)(x) = (P_{1,l}f)(x) = \frac{1}{(2\pi)^n} \int_{\mathbb{R}_n} e^{ix\cdot\xi}\theta_1 c_l(x, \xi, \lambda)(Ff)(\xi)\, d\xi , \qquad (8.5')$$

$$(P_{2,l}(\lambda)f)(x) = (P_{2,l})f(x) = \frac{1}{(2\pi)^n} \int_{\mathbb{R}_n} e^{ix'\cdot\xi'}\theta_2 d_l(x, \xi, \lambda)(Ff)(\xi)\, d\xi, \qquad (8.5'')$$

for $\lambda \in \mathcal{L}$, $|\lambda| \geq \text{const} > 0$ on functions $f \in C_0^\infty(\mathbb{R}^n_+)$; in the right-hand side we set $f(x) = 0$ for $x^n \leq 0$.

The functions $c_l(x, \xi, \lambda)$ are positive homogeneous in $(\xi, \lambda^{1/m})$ $(m = 2q)$ of degree $-m - l$[15] and are constructed like the components of the symbol of the parametrix for $A - \lambda I$ in $\mathbb{R}^n$ (see (Agranovich 1990a, Sect. 4.2), where they are denoted by b_l). Namely, we set

$$[a(x, \xi) - \lambda] \circ \sum_{l \geq 0} c_l(x, \xi, \lambda) = 1 , \qquad (8.6)$$

where $a(x, \xi)$ is the complete symbol of the differential operator A and $\circ$ is the composition of symbols depending on (x, ξ); λ is considered as a parameter. Separating the homogeneous terms of degree $0, 1, \dots$ in (8.6), we obtain equations for consecutive definition of $c_0, c_1, \dots$. In particular,

$$c_0(x, \xi, \lambda) = [a_0(x, \xi) - \lambda]^{-1} . \qquad (8.7)$$

All the $c_l(x, \xi, \lambda)$ are rational functions in (ξ, λ) whose denominators are powers of the difference $a_0(x, \xi) - \lambda$; therefore, these functions remain homogeneous and holomorphic in (ξ_n, λ) for (ξ, λ) such that $a_0(x, \xi) - \lambda \neq 0$. There exist positive constants C_1 and C_2 such that if $\zeta = \xi_n$ does not belong to the set

$$\{\zeta : \; C_1(|\xi'| + |\lambda|^{1/m}) \leq |\operatorname{Im} \zeta|, \; |\zeta| \leq C_2(|\xi'| + |\lambda|^{1/m})\} , \qquad (8.8)$$

[15] In the following sense: $c_l(x, t\xi, t^m\lambda) = t^{-m-l} c_l(x, \xi, \lambda)$ for $t > 0$, $(\xi, \lambda) \neq 0$, $\lambda \in \mathcal{L}$.

then the functions $c_l(x, \xi', \zeta, \lambda)$ are holomorphic in (ζ, λ) and C^∞ with respect to the collection of all variables and remain positively homogeneous in $(\xi', \zeta, \lambda^{1/m})$ of degree $-m-l$ for $x \in \overline{\mathbb{R}^n_+}$, $\xi' \in \mathbb{R}_{n-1}$, $\lambda \in \mathcal{L}$.

The first sum in (8.4) provides an approximation of the solution of the equation $(A - \lambda I)u = f$. After subtracting the second sum we must obtain an approximation of the solution of the boundary problem (8.1). We assume the functions $d_l(x, \xi, \lambda)$ to be positively homogeneous in $((x^n)^{-1}, \xi, \lambda^{1/m})$ of degree $-m-l$. To determine them, we replace the coefficients $a_\alpha(x)$ in the symbol $a(x, \xi)$ by their Taylor expansions in powers of x^n and denote by $a^{(p)}(x, \xi, \lambda)$ the sum of all terms in $a(x, \xi) - \lambda$ positive homogeneous in $((x^n)^{-1}, \xi, \lambda^{1/m})$ of degree $m - p$ $(p = 0, 1, \ldots)$. In particular,

$$a^{(0)}(x, \xi, \lambda) = a_0(x', 0, \xi) - \lambda \, .$$

We write the symbol of the boundary operator B_j in the form of the sum of terms $b_{j,r}(x', \xi) = b_{j,r}(x', 0, \xi)$ homogeneous in ξ of degree $r_j - r$ $(r = 0, \ldots, r_j)$. Now we set (cf. Subsect. 5.2b)

$$\sum_{p \geq 0} a^{(p)}(x, \xi', D_n, \lambda) \circ \sum_{l \geq 0} d_l(x, \xi, \lambda) = 0 \qquad (x^n > 0) \, , \tag{8.9}$$

$$\sum_{0 \leq r \leq r_j} b_{j,r}(x', \xi', D_n) \circ \sum_{l \geq 0} d_l(x, \xi, \lambda) \tag{8.10}$$

$$= \sum_{0 \leq r \leq r_j} b_{j,r}(x', \xi) \circ \sum_{l \geq 0} c_l(x, \xi, \lambda) \quad \text{for} \quad x^n = 0 \quad (j = 1, \ldots, q) \, ,$$

and in addition we assume that for all l

$$d_l(x, \xi, \lambda) \to 0 \quad \text{as} \quad x^n \to +\infty \, . \tag{8.11}$$

The symbol $\circ$ in the right-hand side of (8.10) has the same meaning as in (8.6). In the left-hand side it has the sense of a composition of symbols depending on (x', ξ'), while the variables x^n and λ are considered as parameters. Actually, this follows from the structure of the integrals in (8.5′) and (8.5″). More precisely, relations (8.9) and (8.10) can be written in the form

$$\sum_{\alpha'; p; l} \frac{1}{\alpha'!} [\partial_{\xi'}^{\alpha'} a^{(p)}](x, \xi', D_n, \lambda) D_{x'}^{\alpha'} d_l(x, \xi, \lambda) = 0 \qquad (x^n > 0) \, , \tag{8.12}$$

$$\sum_{\alpha'; r; l} \frac{1}{\alpha'!} [\partial_{\xi'}^{\alpha'} b_{j,r}](x, \xi', D_n) D_{x'}^{\alpha'} d_l(x, \xi, \lambda) \tag{8.13}$$

$$= \sum_{\beta; t; k} \frac{1}{\beta!} [\partial_\xi^\beta b_{j,t}](x', \xi) D_x^\beta c_k(x, \xi, \lambda) \quad \text{for} \quad x^n = 0 \quad (j = 1, \ldots, q) \, .$$

Separating the terms in (8.12) with $|\alpha'| + p + l = s$ and the terms in (8.13) with $|\alpha'| + r + l = s = |\beta| + t + k$ $(s = 0, 1, \ldots)$, we obtain a sequence of boundary problems for consecutive definition of $d_0, d_1, \ldots$. The first of these boundary problems has the form

$$[a_0(x', 0, \xi', D_n) - \lambda]\, d_0(x, \xi, \lambda) = 0 \quad (x_n > 0), \tag{8.14}$$

$$\begin{gathered} b_{j,0}(x', \xi', D_n) d_0(x, \xi, \lambda) = b_{j,0}(x', \xi) c_0(x', 0, \xi, \lambda) \\ \text{for} \quad x^n = 0 \quad (j = 1, \ldots, q)\,, \end{gathered} \tag{8.15}$$

$$d_0(x, \xi, \lambda) \to 0 \quad (x^n \to +\infty)\,. \tag{8.16}$$

In the subsequent boundary problems, d_0 is replaced by d_l ($l = 1, 2, \ldots$), and the right-hand sides in (8.15) and (8.16) are replaced by some known functions that are expressed in terms of $d_0, \ldots, d_{l-1}$. In view of our assumption on ellipticity with parameter, the boundary problem (8.14)–(8.16) has one and only one solution for $(\xi', \lambda) \neq 0$, $\lambda \in \mathcal{L}$; moreover, this solution can be written explicitly and investigated:

$$d_0(x, \xi, \lambda) = \sum_{j,k=1}^{q} L^{k,j}(x', \xi', \lambda) b_{j,0}(x', \xi) c_0(x', 0, \xi, \lambda) v_k(x', \xi', \lambda, x^n) \tag{8.17}$$

(cf. formula (3.22)). Automatically, this function is homogeneous of the necessary degree. It is no longer rational in (ξ, λ); however, since $L^{k,j}$ and v_k do not depend on $\zeta = \xi_n$, then on ζ-plane, the function $d_0(x, \xi', \zeta, \lambda)$ is defined just where $c_0(x', 0, \xi', \zeta)$ exists, i.e. outside the set of ζ such that $a_0(x', 0, \xi', \zeta) = \lambda$. In addition, for ζ not belonging to the set (8.8), the function $d_0(x, \xi', \zeta, \lambda)$ is holomorphic in (ζ, λ) and C^∞ with respect to the collection of all variables, and remains homogeneous in $((x_n)^{-1}, \xi, \zeta, \lambda)$.

The same is true for the subsequent d_l.

Now we transform the right-hand sides in (8.5″) as follows: we substitute

$$(Ff)(\xi) = \int e^{-iy^n \xi_n} (F'f)(\xi', y^n)\, dy^n$$

and change the order of integration with respect to y_n and ξ_n. We obtain

$$\begin{aligned} &(P_{2,l} f)(x) \\ &\quad = (2\pi)^{1-n} \int_{\mathbb{R}_{n-1}} e^{ix'\cdot\xi'}\, d\xi' \int_0^\infty \theta_2(\xi', \lambda) \widetilde{d}_l(x, \xi', y^n, \lambda) (F'f)(\xi', y^n)\, dy^n\,, \end{aligned} \tag{8.18}$$

where

$$\begin{aligned} \widetilde{d}_l(x, \xi', y_n, \lambda) &= \frac{1}{2\pi} \int e^{-iy^n \xi_n} d_l(x, \xi, \lambda)\, d\xi_n \\ &= \frac{1}{2\pi} \int_{L_-} e^{-iy^n \zeta} d_l(x, \xi', \zeta, \lambda)\, d\zeta \end{aligned} \tag{8.19}$$

and L_- is a closed contour in the lower halfplane surrounding the lower half of the set (8.8). The function $\widetilde{d}_l(x, \xi', y_n, \lambda)$ is C^∞ for $x \in \mathbb{R}^n_+$, $\xi' \in \mathbb{R}_{n-1}$, $y^n \in \mathbb{R}_+$, and $\lambda \in \mathcal{L}$ ($(\xi', \lambda) \neq 0$) and is positively homogeneous in

$((x^n)^{-1}, \xi', (y^n)^{-1}, \lambda^{1/m})$ of degree $1 - m - l$. This can be seen from the first of equalities (8.19), while from the second one we obtain the estimate

$$|\widetilde{d}_l(x, \xi, y^n, \lambda)| \leq C_3 \exp[-C_4(x^n + y^n)(|\xi'| + |\lambda|^{1/m})](|\xi'| + |\lambda|^{1/m})^{1-m-l}, \tag{8.20}$$

where C_3 and C_4 are positive constants. Similar estimates for derivatives of $\widetilde{d}_l$ can also be written.

We have finished the description of the construction of the operators $P_{1,l}$ and $P_{2,l}$. One can verify that they are integral operators; their kernels are C^∞ outside the diagonal in $\overline{\mathbb{R}^n_+} \times \overline{\mathbb{R}^n_+}$ and have, in general, a weak singularity at the diagonal. If $m + l > n$, then these kernels are continuous and dominated by $\mathrm{Const}(1 + |\lambda|^{1/m})^{n-m-l+1}$.

b. The right parametrix $P^{(N)}$ for the operator $A_B - \lambda I$ on M is constructed in the form

$$P^{(N)}(\lambda)f = \sum_{k=1}^{K} \psi_k P_k^{(N)}(\lambda)(\varphi_k f) ; \tag{8.21}$$

here $\{\varphi_k\}$ and $\{\psi_k\}$ are the systems of functions used in (2.35), $P_k^{(N)}$ is a parametrix of order N for $A - \lambda I$ if $k \leq K'$, and an operator of the form (8.4) in local coordinates if $k > K'$.

Theorem 8.1.1. *For $N + m > n$ and $\lambda \in \mathcal{L}$ with sufficiently large modulus, the operator $(A_B - \lambda I)^{-1} - P^{(N)}(\lambda)$ has a continuous kernel with estimate $O(|\lambda|^{\frac{n-N}{m}-1})$.*

The proof can be found in (Seeley 1969). We only dwell on the following detail of the proof. The range of $P^{(N)}(\lambda)$ is not contained in the domain of A_B: the function $u = P^{(N)}(\lambda)f$ does not satisfy the homogeneous boundary conditions $B_j u = 0$ on Γ. Therefore, it is necessary to construct a correction for $P^{(N)}(\lambda)$. For this purpose, write B_j in the form

$$B_j = \sum_{k=1}^{m} B_{j,k} D_n^{m-k}$$

near Γ (the derivatives in x^n of order not less than m can be excluded from the boundary conditions by using the equation $Au = f$ and equations obtained from it by differentiation). Denote by $\mathcal{B}$ the matrix $(B_{j,k})$. The order of $B_{j,k}$ does not exceed $r_j - m + k$. It can be proved that there exists a matrix pseudodifferential operator $\Psi = (\Psi_{j,k})$ with $\operatorname{ord} \Psi_{j,k} \leq m - j - r_k$ such that

$$\mathcal{B}\Psi\mathcal{B} = \mathcal{B} . \tag{8.22}$$

Let Φ be a linear continuous operator from $\prod_1^m [C^{(k+\sigma)}(\Gamma)]^q$ into $[C^{(m+\sigma)}(M)]^q$ (with sufficiently large σ) such that

$$D_n^{m-k}\Phi(g_1, \ldots, g_m)|_\Gamma = g_k \qquad (k = 1, \ldots, m) \tag{8.23}$$

(cf. (2.12)). Set

$$R^{(N)}(\lambda) = P^{(N)}(\lambda) - \Phi\Psi B P^{(N)}(\lambda) . \tag{8.24}$$

Then, for sufficiently large N, the last term in (8.24) has a continuous kernel, and $R^{(N)}(\lambda)H_0(M)$ lies in the domain of A_B. Moreover, the kernel of the last term in (8.24) is $O(|\lambda|^{-\nu})$, where $\nu \to \infty$ as $N \to \infty$.

8.2. Kernel and Trace of the Resolvent. In this section we assume that $m > n$. In this case, in view of Theorem 2.1.1, the resolvent $R_{A_B}(\lambda) = (A_B - \lambda I)^{-1}$ is an integral operator,

$$R_{A_B}(\lambda)f(x) = \int_M K(x, y, \lambda)f(y)\,dy , \tag{8.25}$$

with a kernel $K(x, y, \lambda)$ continuous in $M \times M$. We want to find its asymptotics for $x = y$, $\lambda \to \infty$ in $\mathcal{L}$, and also the asymptotics of the trace

$$\operatorname{tr} R_{A_B}(\lambda) = \int_M K(x, x, \lambda)\,dx \tag{8.26}$$

(see Theorem 2.1.3).

a. For $m > n$ the kernel $K^{(N)}(x, y, \lambda)$ of the parametrix $P^{(N)}(\lambda)$ is also continuous and, according to Theorem 8.1.1, serves as a uniform approximation to $K(x, y, \lambda)$ up to $O(|\lambda|^{\frac{n-N}{m}-1})$. Let V be a closed subset in M lying 1) in a coordinate neighborhood O_k (inside M) or 2) in a coordinate semi-neighborhood O_k^+ (adjacent to Γ). At first we assume that on V the corresponding function φ_k is equal to 1, while other φ_j are equal to 0.

In the case 1), we obtain an expansion in powers of $(-\lambda)$ for $K(x, x, -\lambda)$ uniform in $x \in V$; it is similar to that given in (Agranovich 1990a, Sect. 5.7):

$$K(x, x, \lambda) \sim \sum_{l=0}^{\infty} (-\lambda)^{\frac{n-l}{m}-1} (2\pi)^{-n} \int_{\mathbb{R}_n} c_l(x, \eta, -1)\,d\eta . \tag{8.27}$$

In the case 2),

$$K(x, x, \lambda) = \sum_{l=0}^{N} (-\lambda)^{\frac{n-l}{m}-1} (2\pi)^{-n} \int_{\mathbb{R}_n} c_l(x, \eta, -1)\,d\eta \tag{8.28}$$
$$+ \sum_{l=0}^{N} (2\pi)^{1-n} \int_{\mathbb{R}_{n-1}} \tilde{d}_l(x, \xi', x^n, \lambda)\,d\xi' + O(|\lambda|^{\frac{n-N}{m}-1}) .$$

Here in view of (8.20)

$$\left| \int_{R_{n-1}} \tilde{d}_l d\xi' \right| \le C_5 \exp(-C_5|\lambda|^{1/m}) \quad \text{for} \quad x^n \ge C_6 > 0,\ |\lambda| \ge C_7 > 0 . \tag{8.29}$$

We see that the terms of the second sum in (8.28) decrease exponentially as $\lambda \to \infty$ in $\mathcal{L}$; however, the last estimate is not uniform relative to x^n (C_5 depends on C_6). It is possible to derive an asymptotic expansion in powers of $(-\lambda)$ for the integral of $K(x, x, \lambda)$ with respect to x^n. Instead, we pass to the trace (8.25) and find its asymptotics.

b. To this end, in view of Theorem 8.1.1 we need to calculate the traces of the terms in (8.21). In turn, for this we need to calculate the traces of operators

$$(2\pi)^{-n} \int_{\mathbb{R}^n} \psi_k(x) e^{ix\cdot\xi} c_l^{(k)}(x, \xi, \lambda) F(\varphi_k f)(\xi)\, d\xi \tag{8.30}$$

$(k = 1, \dots, K')$ and

$$(2\pi)^{1-n} \int_{\mathbb{R}_{n-1}} e^{ix'\cdot\xi'} d\xi' \int_0^\infty \psi_k(x) \widetilde{d}_l(x, \xi', y^n, \lambda) F'(\varphi_k f)(\xi', y^n)\, dy^n \tag{8.31}$$

$(k = K' + 1, \dots, K)$ in local coordinates. Obviously, the trace of operator (8.30) is equal to

$$(2\pi)^{-n} \int \varphi_k(x)\, dx \int_{\mathbb{R}_n} c_l^{(k)}(x, \xi, \lambda)\, d\xi\,.$$

By using first the substitution $\xi = |\lambda|^{1/m}\eta$ for $\lambda < 0$ and then the analytic continuation with respect to λ, we extract the factor $(-\lambda)^{\frac{n-l}{m}-1}$ from the integrand (cf. (Agranovich 1990a, Sect. 5.7)). To calculate the trace of operator (8.31), we use the Taylor expansion of the function $\varphi_k(x)$ in powers of x^n:

$$\varphi_k(x) = \sum_{s=0}^{N} \frac{(x^n)^s}{s!} (\partial_{x^n}^s \varphi_k)(x', 0) + O((x^n)^{N+1}) \tag{8.32}$$

uniformly with respect to x'. The function $(x^n)^s \widetilde{d}_l^{(k)}(x', x^n, \xi', x^n, \lambda)$ is positively homogeneous of degree $1 - m - (s + l)$ in $((x^n)^{-1}, \xi', \lambda^{1/m})$, and the integral

$$\widetilde{d}_{s,l}^{(k)}(x', \xi', \lambda) = \int_0^\infty (x^n)^s \widetilde{d}_l^{(k)}(x', x^n, \xi', x^n, \lambda)\, dx^n \qquad (s, l \in \mathbb{Z}_+) \tag{8.33}$$

converges absolutely (see (8.20)) and is positively homogeneous in $(\xi', \lambda^{1/m})$ of degree $-m - (s + l)$. From the corresponding part of the trace of (8.31) we extract the factor $(-\lambda)^{\frac{n-1-s-l}{m}-1}$ by using first the substitution $\xi' = |\lambda|^{1/m}\eta'$ and then the analytic continuation with respect to λ.

As the final result, we obtain

Theorem 8.2.1. *We have*

$$\operatorname{tr} R_{A_B}(\lambda) \sim \sum_{l=0}^{\infty} (c_l - d_{l-1})(-\lambda)^{\frac{n-l}{m}-1} \tag{8.34}$$

as $\lambda \to \infty$ *in* $\mathcal{L}$ *; here*

$$c_l = (2\pi)^{-n} \sum_{k=1}^{K} \int \varphi_k(x)\, dx \int_{\mathbb{R}_n} c_l^{(k)}(x, \xi, -1)\, d\xi \tag{8.35}$$

and

$$d_l = (2\pi)^{1-n} \sum_{k=K'+1}^{K} \sum_{s_1+s_2=l} \frac{1}{s_1!} \int (\partial_{x_n}^{s_1} \varphi_k)(x', 0)\, dx' \int_{\mathbb{R}_{n-1}} \widehat{d}_{s_1,s_2}^{(k)}(x', \xi', -1)\, d\xi' \tag{8.36}$$

for $l \geq 0$, *while* $d_{-1} = 0$; *each term in the sums with respect to* k *is calculated in its own local coordinates.*

The coefficients c_l are determined in terms of the local complete symbol of A in essentially the same way as in the case of a closed manifold. In particular, the formula for c_0 can be written in the form

$$c_0 = (2\pi)^{-n} \iint_{T^*M} [a_0(x, \xi) + 1]^{-1} d\xi dx\,. \tag{8.37}$$

Unlike c_l, the coefficients d_j depend also on boundary conditions. They do not affect the main term of the asymptotics of the resolvent. More precisely,

$$\operatorname{tr} R_A(\lambda) = c_0(-\lambda)^{\frac{n}{m}-1} + O(|\lambda|^{\frac{n-1}{m}-1}) \qquad (\lambda \to \infty \text{ in } \mathcal{L})\,. \tag{8.38}$$

8.3. Powers of A_B. In this section we do not assume that $m > n$. However, in addition to the assumptions made at the beginning of §8, we assume that there are no eigenvalues of the operator A_B in $\mathcal{L}$. In particular, we assume that A_B is invertible; it follows that it has no eigenvalues in some disk $O_\varepsilon = \{\lambda : |\lambda| < \varepsilon\}$ $(\varepsilon > 0)$. This always can be achieved by passing from A to $A + cI$ with a sufficiently large positive constant c.

a. In accordance with the general definition of a power of an operator of positive type (see e.g. (Agranovich 1990a, Sect. 5.2)), for $\operatorname{Re} z < 0$ the operator A_B^z is defined by the formula

$$A_B^z = \frac{1}{2\pi i} \int_{\Gamma_\delta} \lambda^z R_{A_B}(\lambda)\, d\lambda\,. \tag{8.39}$$

Here Γ_δ, $\delta > 0$, is a contour consisting of the lower and upper sides of the ray $(-\infty, \delta]$ on the halfline $\mathbb{R}_-$, and the arc of the circle $\{\lambda : |\lambda| = \delta\}$ joining the ends of the sides of the ray, with counterclockwise direction on the arc. For $\operatorname{Re} z < -n/m$ this operator has a continuous kernel $K_z(x, y)$. As in the case of an elliptic operator on a closed manifold, the properties of the analytic continuation in z of the kernel and the trace $\operatorname{tr} A_B^z$ can be investigated. For convenience, we will discuss this question after replacing z by $-z$.

From Theorem 8.1.1 we see that the difference $A_B^{-z} - (A_B^{-z})^{(N)}$, where

$$(A_B^{-z})^{(N)} = \frac{1}{2\pi i} \int_{\Gamma\delta} \lambda^{-z} P^{(N)}(\lambda)\, d\lambda \tag{8.40}$$

and $N + m > n$, is an integral operator with a kernel continuous in (x, y, z) and holomorphic in z if $\operatorname{Re} z > (n - N)/m$. Since we can choose N arbitrarily large, we consider the kernel of the operator (8.40). The operator $P^{(N)}(\lambda)$ is defined by (8.21), where, in turn, the operators $P_k^{(N)}(\lambda)$ are defined by a formula of the form (8.4) with all terms depending on k. Let $\varphi(x)$ be a function from $C^\infty(M)$. If its support lies inside M, then the product $\varphi(x)K_z(x, y)$ can be investigated in the same way as in the case of a closed manifold (see (Agranovich 1990a, Sect. 5.5)). In particular, for $x = y$ this product has a meromorphic continuation in z. Otherwise, if the support of $\varphi(x)$ is adjacent to the boundary and lies on a collar of the manifold, then only the integral in x^n of $\varphi(x)K_z(x, x)$ admits a meromorphic continuation. Omitting the details (see (Seeley 1969b)), we only formulate a corollary following from these considerations for the *zeta-function of* A_B:

$$\zeta(z, A_B) = \operatorname{tr} A_B^{-z} = \int_M K_{-z}(x, x)\, dx\,. \tag{8.41}$$

We set

$$z_l = (n - l)/m \qquad (l = 0, 1, \ldots)\,. \tag{8.42}$$

Theorem 8.3.1. *For* $\operatorname{Re} z > n/m$ *the function* (8.41) *is holomorphic in* z. *It admits a meromorphic continuation on the complex plane with possible poles only at the points* z_l *not belonging to* $\mathbb{Z}_- = \{0, -1, \ldots\}$. *These poles are simple, and the residue* ρ_0 *at the first pole* z_0 *is given by the formula*

$$\rho_0 = \frac{-1}{(2\pi)^n m} \sum_{k=1}^{K} \int \varphi_k(x)\, dx \int_{|\omega|=1} [a_0^{(k)}(x, \omega)]^{-n/m} dS\,. \tag{8.43}$$

Formula (8.43) is essentially the same as in the case of a pseudodifferential operator on a closed manifold (see (Agranovich 1990a, Sect. 5.5)). The residues at the other poles and the values of the function (8.41) at the points $z_l \in \mathbb{Z}_-$ are calculated in terms of the functions $c_l^{(k)}$ and $\widehat{d}_{s_1,s_2}^{(k)}$. The corresponding formulas can be found in (Seeley 1969b).

b. Now consider the operator A_B^α for $0 < \alpha < 1$. We discuss some properties of its domain $D(A_B^\alpha)$. First assume that the operator A_B is selfadjoint.

Proposition 8.3.2. *We have* $D(A_B^\alpha) \subset H_{m\alpha}(M)$, *and*

$$\|u\|_{m\alpha,M} \le C_1 \|A_B^\alpha u\|_{0,M} \le C_2 \|u\|_{m\alpha,M} \quad \text{for} \quad u \in D(A_B^\alpha)\,, \tag{8.44}$$

where C_j *do not depend on* u.

We briefly outline the proof. For this we need to mention the method of complex interpolation of Banach spaces; see e.g. (S. Kreĭn et al. 1970), (Lions and Magenes 1968), or (Triebel 1978) and references therein. Let $\mathfrak{B}_0$ and $\mathfrak{B}_1$ be two Banach spaces with norms $\|\cdot\|_0$ and $\|\cdot\|_1$. Assume that $\mathfrak{B}_1$ is continuously embedded in $\mathfrak{B}_0$ and dense in $\mathfrak{B}_0$. Denote by S the vertical strip $\{s = \sigma + i\tau : 0 \le \sigma \le 1\}$ on the complex plane and by S^0 the interior of S. Introduce the space $\mathfrak{H}(\mathfrak{B}_0, \mathfrak{B}_1)$ of continuous bounded functions $F(s)$ on S with values in $\mathfrak{B}_0$ that are holomorphic in S^0 and have a finite norm

$$\|F\| = \max\left\{\sup_\tau \|F(i\tau)\|_0, \ \sup_\tau \|F(1+i\tau)\|_1\right\}.$$

Let $\mathfrak{B}_\alpha = [\mathfrak{B}_0, \mathfrak{B}_1]_\alpha$, $0 < \alpha < 1$, be the space of values $f = F(\alpha)$ of all the functions $F \in \mathfrak{H}(\mathfrak{B}_0, \mathfrak{B}_1)$, with norm $\|f\|_\alpha = \inf\{\|F\| : F(\alpha) = f\}$. For $0 \le \alpha_1 < \alpha_2 \le 1$ the space $\mathfrak{B}_{\alpha_2}$ is continuously embedded in $\mathfrak{B}_{\alpha_1}$.

The following *Interpolation Theorem* is true. Let $\mathfrak{C}_0$ and $\mathfrak{C}_1$ be a second pair of Banach spaces having the same properties as the pair $\mathfrak{B}_0$ and $\mathfrak{B}_1$, and let $\|\cdot\|'_0$ and $\|\cdot\|'_1$ be the norms in $\mathfrak{C}_0$ and $\mathfrak{C}_1$. Let $\mathfrak{A}$ be a linear operator from $\mathfrak{B}_0$ into $\mathfrak{C}_0$ such that $\|\mathfrak{A}f\|'_j \le C_j\|f\|_j$ for $f \in \mathfrak{B}_j$ $(j = 0, 1)$. Then $\|\mathfrak{A}f\|'_\alpha \le C_\alpha\|f\|_\alpha$ for $f \in \mathfrak{B}_\alpha$, $0 < \alpha < 1$, where $C_\alpha = C_0^{1-\alpha}C_1^\alpha$.

On the other hand, it is known that

$$[H_{t_1}(M), H_{t_2}(M)]_\alpha = H_{\alpha t_2 + (1-\alpha)t_1}(M)$$

for $0 \le t_1 < t_2$, $0 < \alpha < 1$. In particular,

$$H_{m\alpha}(M) = [H_0(M), H_m(M)]_\alpha \,. \tag{8.45}$$

Furthermore, if A_B is a selfadjoint positive operator, then

$$D(A_B^\alpha) = [H_0(M), D(A_B)]_\alpha \tag{8.46}$$

(where $D(A_B)$ is considered as a subspace in $H_m(M)$, see (2.41) with $s = 0$) with norm equivalent to $\|A_B^\alpha \cdot\|_{0,M}$.

Proposition 8.3.2 follows from this and from the Interpolation Theorem if we take $\mathfrak{B}_0 = \mathfrak{C}_0 = H_0(M)$, $\mathfrak{B}_1 = D(A_B)$, $\mathfrak{C}_1 = H_m(M)$, and $\mathfrak{A} = I$ (the identity operator in $H_0(M)$).

Relation (8.46) is verified in (Seeley 1972) even for nonselfadjoint A_B (with the ray $\overline{\mathbb{R}}_-$ of ellipticity with parameter and without eigenvalues on $\overline{\mathbb{R}}_-$), so that Proposition 8.3.2 extends to cover this case. In (Grisvard 1967) the following description of the spaces (8.46) is obtained; see also (Seeley 1972), where more general spaces are considered.

Theorem 8.3.3. *Let the boundary problem* (8.1) *with* $A - \lambda I$ *instead of* A *be elliptic with parameter along* $\overline{\mathbb{R}}_-$, *and let* $\{B_j\}_1^q$ *be a normal system of boundary operators of orders* $r_j < m$. *Let* $0 < \alpha < 1$ *and* $m\alpha - \frac{1}{2} \ne r_j$ *for all* j. *Then* (8.46) *is a (closed) subspace of functions* $u(x)$ *in* $H_{m\alpha}(M)$ *satisfying the boundary conditions* $B_j u = 0$ *on* Γ *for all* j *such that* $r_j < m\alpha - \frac{1}{2}$.

If $r_{j,0} = m\alpha - \frac{1}{2}$ for some j_0, then the condition $B_{j_0}u = 0$ on Γ must be added, but in some weak sense; in this case (8.46) is a nonclosed linear submanifold in $H_{m\alpha}(M)$.

8.4. Kernel and Trace of $e^{-\tau A_B}$. Let the boundary problem (8.1) with $A - \lambda I$ instead of A be elliptic with parameter in the closed left halfplane. Then it is elliptic with parameter in a somewhat larger angle $\mathcal{L} = \{\lambda : |\arg\lambda| \geq \varphi\}$, where $0 < \varphi < \pi/2$. We want to determine the behavior of the kernel and the trace of the operator $e^{-\tau A_B}$ as $\tau \to +0$. There are two possibilities: we can 1) express $e^{-\tau A_B}$ in terms of A_B^{-z} (as in the case of pseudodifferential operators on a closed manifold, see (Agranovich 1990a, Sect. 5.2)), or 2) express $e^{-\tau A_B}$ in terms of $R_{A_B}(\lambda)$ by the formula

$$e^{-\tau A_B} = \frac{1}{2\pi i}\int_{\Gamma_{\delta,\psi}} e^{-\tau\lambda} R_{A_B}(\lambda)\, d\lambda\,. \tag{8.47}$$

Here the contour $\Gamma(\delta,\psi)$ consists of the rays $\{\lambda : \arg(\lambda - \delta) = \pm\psi\}$, $\varphi < \psi < \pi/2$, oriented upwards, with vertex at $\delta \in \mathbb{R}_-$, where $|\delta|$ is so large that all eigenvalues of A_B lie to the right of $\Gamma(\delta,\psi)$. The second possibility is preferable: in this case we do not need to assume that A_B has no eigenvalues in $\mathcal{L}$ and can deduce all that we want directly from the results of Sect. 8.1.

Using the same arguments as in (Agranovich 1990a, Sect. 5.6), we can verify that $e^{-\tau A_B}$, $\tau > 0$, is an integral operator,

$$e^{-\tau A_B} f(x) = \int \Theta(x, y, \tau) f(y)\, dy\,, \tag{8.48}$$

with infinitely smooth kernel $\Theta(x,y,\tau)$ holomorphic in τ. Let $\varphi(x)$ be a function from $C^\infty(M)$; if its support lies inside M, then for $\varphi(x)\Theta(x,x,\tau)$ we have an asymptotic expansion in powers of τ as $\tau \to +0$ (cf. (Agranovich 1990a, Sect. 5.6)). If this support is adjacent to the boundary and lies on a collar of the manifold, then a complete asymptotics in powers of τ can be obtained only for the integral of $\varphi(x)\Theta(x,x,\tau)$ with respect to x_n. We only formulate a corollary for the trace

$$\theta(\tau) = \operatorname{tr} e^{-\tau A_B} = \int_M \Theta(x,x,\tau)\, dx\,. \tag{8.49}$$

Theorem 8.4.1. *For $\tau \to +0$, the trace of $e^{-\tau A_B}$ has the asymptotic expansion of the form*

$$\operatorname{tr} e^{-\tau A_B} \sim \sum_{l=0}^{\infty} \eta_l \tau^{(l-n)/m}\,, \tag{8.50}$$

where

$$\eta_0 = \frac{\Gamma(n/m)}{(2\pi)^n m} \sum_{k=1}^{K} \int \varphi_k(x)\, dx \int_{|\omega|=1} [a_0^{(k)}(x,\omega)]^{-n/m} dS\,. \tag{8.51}$$

The last formula is almost the same as in the case of pseudodifferential operators on a closed manifold (see (Agranovich 1990a, Sect. 5.6). Other coefficients are calculated in terms of the functions $c_l^{(k)}$ and $\widehat{d}^{(k)}_{s_1,s_2}$. See (Seeley 1969b) or (Greiner 1971).

8.5. Some Generalizations.

In Sect. 9.3 we will use the main term of the asymptotics of the trace of the operator-valued function

$$R_{\alpha,\sigma}(\lambda) = [R_{A_B^\alpha}(\lambda)]^\sigma \,, \tag{8.52}$$

where $0 < \alpha < 1$, $\sigma \in \mathbb{N}$, and

$$m\alpha\sigma > n \,. \tag{8.53}$$

(In the case of a closed manifold (see (Agranovich 1990a)) we can pass from A to A^α without difficulty, but in the case of a manifold with boundary it is essential that we can consider precisely the operator (8.52).) It is easy to verify that the operator $R_{\alpha,\sigma}(\lambda)$ and the adjoint operator $[R_{\alpha,\sigma}(\lambda)]^*$ are bounded operators from $H_0(M)$ to $H_{m\alpha\sigma}(M)$. Hence, under condition (8.53), by virtue of Theorem 2.1.1, $R_{\alpha,\sigma}(\lambda)$ is an integral operator with continuous kernel; we denote it by $K_{\alpha,\sigma}(x, y, \lambda)$.

For the investigation of this kernel and of the trace of $R_{\alpha,\sigma}(\lambda)$ we again have two possibilities: this operator can be expressed in terms of A_B^{-z} or in terms of $R_{A_B}(\lambda)$. We use the second possibility:

$$R_{\alpha,\sigma}(\lambda) = \frac{1}{2\pi i} \int_{\Gamma_{\delta'}} (\mu^\alpha - \lambda)^{-\sigma} R_{A_B}(\mu)\, d\mu \,. \tag{8.54}$$

Let $\mathcal{L} = \{\lambda : |\arg \lambda| \geq \varphi\} \cup \{0\}$; in order to consider (8.54) for $\lambda \in \mathcal{L}$, we assume that the contour Γ'_δ consists of the rays $\{\mu : \arg \mu = \pm\psi,\ |\mu| \geq \delta\}$ and the arc $\{\mu : |\mu| = \delta,\ |\arg \mu| \leq \psi\}$ joining their endpoints. Here $0 < \delta < \varepsilon$, while ψ is somewhat less than φ and so close to φ that for $|\arg \mu| \geq \psi$ the ellipticity with parameter holds and A_B has no eigenvalues. The power μ^α is defined by the usual formula $\mu^\alpha = |\mu|^\alpha e^{i \arg \mu}$ ($|\arg \mu| \leq \pi$), and the function μ^α is holomorphic on Γ'_δ.

Replacing $R_{A_B}(\mu)$ by $P^{(N)}(\mu)$, we can construct the complete asymptotics for the kernel on the diagonal $x = y$ inside M, for the integral of this kernel near Γ, and for the trace of $R_{\alpha,\sigma}(\lambda)$, as $\lambda \to \infty$. We do not intend to do all this completely and only find the main term in the asymptotics of the trace. The main term in the asymptotics of the kernel $K_{\alpha,\sigma}(x, x, \lambda)$ for $x \in M$ has the form

$$\frac{1}{(2\pi)^n} \int_{T_x^* M} d\xi \, \frac{1}{2\pi i} \int_{\Gamma'_{\delta(\xi)}} (\mu^\alpha - \lambda)^{-\sigma} c_0(x, \xi, \mu)\, d\mu \tag{8.55}$$

(where $\delta(\xi)$ is so small that $a_0(x, \xi) \neq \mu$ on $\Gamma'_{\delta(\xi)}$). The inner integral is calculated in terms of the residue of the function $c_0 = (a_0 - \mu)^{-1}$ at the point

$\mu = a_0(x,\xi)$. After this, the substitution $\xi = |\lambda|^{1/m\alpha}\eta$ for $\lambda < 0$ transforms (8.55) into

$$|\lambda|^{-\sigma+\frac{n}{m\alpha}} c_{0,\alpha,\sigma}(x), \quad \text{where } c_{0,\alpha,\sigma}(x) = \frac{1}{(2\pi)^n}\int_{T_x^*M} \{[a_0(x,\eta)]^\alpha + 1\}^{-\sigma} d\eta \,. \tag{8.56}$$

To transform this formula into a more convenient form, we set $\eta = r\omega$, where $r > 0$ and $|\omega| = 1$, and then, assuming that $a_0(x,\omega) > 0$, we set

$$[a_0(x,\omega)]^\alpha r^{m\alpha} = \rho \,. \tag{8.57}$$

We obtain (in particular, in the case $\alpha = \sigma = 1$; cf. (8.37))

$$c_{0,\alpha,\sigma}(x) = \frac{b_{n/m\alpha,\sigma-n/m\alpha}}{(2\pi)^n n}\int_{|\omega|=1} [a_0(x,\omega)]^{-n/m} dS \,, \tag{8.58}$$

where $b_{\delta,\sigma} = \delta B(\delta, \sigma - \delta)$. The restrictions $\lambda < 0$ and $a_0(x,\omega) > 0$ are removed by holomorphic continuation. Integrating with respect to x, we obtain the main term of the asymptotics of the function $R_{\alpha,\sigma}(\lambda)$. The order of the remainder is defined by replacing n by $n-1$. The contribution of the functions $\hat{d}^{(k)}_{s_1,s_2}$ is contained in the remainder. Thus, the following theorem holds:

Theorem 8.5.1. *Let $0 < \alpha < 1$, $\sigma \in \mathbb{N}$, and $m\alpha\sigma > n$. Then for $\lambda \to \infty$ in $\mathcal{L}$ we have*

$$\operatorname{tr}[R_{A_B^\alpha}(\lambda)]^\sigma = (-\lambda)^{\frac{n}{m\alpha}-\sigma}\int_M c_{0,\alpha,\sigma}(x)\,dx + O(|\lambda|^{\frac{n-1}{m\alpha}-\sigma}) \,. \tag{8.59}$$

The papers (Seeley 1969a,b) are devoted to matrix boundary problems elliptic with parameter in the usual sense. A parametrix for the resolvent of A_B is constructed in this generality, and the operators A_B^z and $e^{-\tau A_B}$ are investigated. We note that at least with respect to the resolvent and a formula of the type (8.59) it is possible to take one step more and to generalize the results to elliptic with parameter boundary problems of the form (6.37). This is useful for some purposes in spectral theory (see Sect. 9.3 below).

§9. Spectral Properties of Operators Corresponding to Elliptic Boundary Problems

9.1. Selfadjoint Elliptic Boundary Problems with Homogeneous Boundary Conditions.

a. In this subsection we consider the operator A_B in $H_0(M)$ corresponding to the boundary problem (8.1). We assume that it is formally selfadjoint, so that A_B is a selfadjoint operator. In addition, in Subsects. 9.1a and b we assume that the conditions of ellipticity with parameter for $A - \lambda I$ instead of A hold along $\mathbb{R}_-$; in particular, the principal symbol of A is positive:

$$a_0(x,\xi) > 0 \quad \text{on} \quad T^*M \setminus 0 \,. \tag{9.1}$$

Then A_B is semibounded from below; replacing, if necessary, A by $A + cI$ with sufficiently large c, we assume that A_B is a positive operator. Let $\{e_j\}_1^\infty$ be an orthonormal basis in $H_0(M)$ consisting of eigenfunctions of A_B, and let $\{\lambda_j\}_1^\infty$ be the sequence of the corresponding eigenvalues: $Ae_j = \lambda_j e_j$. It is convenient to enumerate the eigenvalues in the nondecreasing order, with multiplicities taken into account. The *spectral function* of A is defined by the formula

$$e(x,y,\lambda) = \sum_{\lambda_j \le \lambda} e_j(x)\overline{e_j(y)} \,. \tag{9.2}$$

The *eigenvalue distribution function*

$$N(\lambda) = \text{card}\,\{j : \lambda_j \le \lambda\} \tag{9.3}$$

is the integral of $e(x,x,\lambda)$ over M.

Theorem 9.1. *Under the conditions indicated above,*

$$N(\lambda) = d_0\lambda^{n/m} + o(\lambda^{n/m}) \qquad (\lambda \to +\infty) \,, \tag{9.4}$$

where[16]

$$d_0 = \frac{1}{(2\pi)^n} \iint_{a_0(x,\xi)<1} dx\, d\xi \,. \tag{9.5}$$

This is a rough asymptotics, and it can be easily deduced, for example, from the asymptotics of the trace of the resolvent $R_{A_B}(\lambda)$ of A_B (for $m > n$) or of a sufficiently large power $[R_{A_B}(\lambda)]^\sigma$ of the resolvent ($m\sigma > n$, $\sigma \in \mathbb{N}$) along $\mathbb{R}_-$. Only the main terms of these asymptotics are needed, i.e. it suffices to use formulas (8.38) and (8.59) with $\alpha = 1$. Indeed, having these formulas, we can use the Hardy–Littlewood Tauberian theorem (see e.g. (Agranovich 1990a, Sect. 6.1)). Instead of $[R_{A_B}(\lambda)]^\sigma$, we can consider $R_{A_B^\sigma}(\lambda)$; the operator A_B^σ corresponds to a formally selfadjoint boundary problem (cf. Remark 4.3.7). Here the proof is carried out by the resolvent method.

Formula (9.4) is proved in (Agmon 1965a,b), where similar results for positive and negative eigenvalues of some not semibounded problems are also obtained.

Formula (9.4) is analogous to formula (6.7) in (Agranovich 1990a) for elliptic operators on a closed manifold. Much more deep is the analog of formula (6.17) in that survey, i.e. of Hörmander's theorem for pseudodifferential operators on a closed manifold with an exact remainder estimate

Theorem 9.1.2. *Under the same assumptions,*

$$N(\lambda) = d_0\lambda^{\frac{n}{m}} + O(\lambda^{\frac{n-1}{m}}) \quad (\lambda \to +\infty) \,. \tag{9.6}$$

[16] This coefficient should not be confused with d_0 in Theorem 8.2.1.

This formula is a result of investigations of many mathematicians. A detailed history, which goes back to papers of Courant and H. Weyl on the Laplace operator, can be found in the surveys (Clark 1967), (Birman and Solomyak 1977), and (Rozenblum et al. 1989) that contain an extensive bibliography. We see the step-by-step strengthening of the remainder estimate, the progress in generality, and the competition of the variational method, the resolvent method, and the method of hyperbolic equation; as we mentioned in (Agranovich 1990a), the latter was proposed by Avakumović and Levitan. In (Agmon and Kannai 1967) and (Agmon 1968) the resolvent method was used to obtain the estimate $O(\lambda^{\frac{n-\theta}{m}})$ for elliptic boundary problems in a bounded domain with $\theta < 1/2$ in the general case, and with $\theta < 1$ in the case of constant coefficients in the principal part of A. Such results were also obtained in (Hörmander 1966). In addition, in these papers the boundary conditions may be very general: it is only required that the domain of the operator A_B^σ lie in $H_{m\sigma}(G)$ for some $\sigma \in \mathbb{N}$ with $m\sigma > n$ (cf. Sect. 9.6). Assume for simplicity that $m > n$. In this case, it is possible to derive the complete asymptotics for the kernel $K(x, x, \lambda)$ of the resolvent on the diagonal not only outside an angular neighborhood of the ray $\mathbb{R}_+$ but also outside a narrow "parabolic" neighborhood

$$\{\lambda :\ \operatorname{Re}\lambda > 0,\ \ |\operatorname{Im}\lambda| \leq (\operatorname{Re}\lambda)^{\frac{m-\theta}{m}}\} \tag{9.7}$$

of this ray. The remainder estimate in this asymptotics contains $|\operatorname{Im}\lambda|$ in the denominator and the distance $\delta(x)$ of the point x under consideration from the boundary. It follows that

$$e(x, x, \lambda) = d_0(x)\lambda^{\frac{n}{m}} + O(\lambda^{\frac{n-\theta}{m}}\delta^{-\theta}(x)) , \tag{9.8}$$

where

$$d_0(x) = \frac{1}{(2\pi)^n}\int_{a_0(x,\xi)<1} d\xi . \tag{9.9}$$

Integrating with respect to x, we obtain the above-mentioned result for $N(\lambda)$. In (Brüning 1974) the estimate (9.8) with $\theta = 1$ was obtained in full generality under the same abstract assumptions as in (Agmon 1968); this yields a result close to (9.6) but with the estimate $O(\lambda^{(n-1)/m}\log\lambda)$ of the remainder.

The exact estimate (9.6) was obtained first in (Seeley 1978, 1980) for the Laplace and Beltrami–Laplace equations under the Dirichlet or Neumann boundary conditions. The method in these papers is that of the hyperbolic equation, i.e. the mixed problem for the operator $\partial_t^2 - A$ in a cylindrical domain is considered. For general scalar elliptic boundary problems the result was obtained by Vassiliev, see (Vassiliev 1984, 1986a) and references therein. In essence, Vassiliev also used the method of the hyperbolic equation. However, for $q > 1$ the equation $\partial_t^{2q}u - Au = 0$ is nonhyperbolic, and the Cauchy problem for it is not well posed. Avoiding the passage from A_B to $A_B^{1/q}$ or $A_B^{1/2q}$, which involves consideration of inherent difficulties in the case of boundary problems, Vassiliev considered this equation with one initial condition

$u(0,x) = u_0(x)$ in the appropriate class of functions (in particular, they are bounded for $t \in \mathbb{R}$), and constructs an approximation of the inverse operator for this (well-posed) boundary problem. The manifold is divided into three zones, inner, intermediate, and adjacent to the boundary, and only the construction in the inner zone is similar to that in the case of a closed manifold.

In (Métivier 1983) formula (9.6) was obtained by the resolvent method with the use of Fourier integral operators. However, an additional "simple reflection condition" is imposed on the principal symbol $a_0(x,\xi)$. We explain this condition in the next subsection.

b. In the general case the estimate (9.6) is the best possible. This can be seen from some examples or from a more profound result: under some additional conditions, the second term of the asymptotics can be singled out. Namely,

$$N(\lambda) = d_0\lambda^{\frac{n}{m}} + d_{-1}\lambda^{\frac{n-1}{m}} + o(\lambda^{\frac{n-1}{m}}) \qquad (\lambda \to +\infty) . \tag{9.10}$$

The hypothesis that such a formula can be true for the Laplace equation was conjectured by H. Weyl in 1913. At first such a result for the Beltrami–Laplace equation appeared in (Ivrii 1980). The results of general character for scalar elliptic boundary problems were obtained in (Vassiliev 1984, 1986a), see also (Safarov and Vassiliev 1992, 1996) and references therein. We briefly dwell on conditions sufficient for the validity of formula (9.10) following (Safarov and Vassiliev 1992).

As before, we assume that the operator A_B corresponding to a scalar elliptic boundary problem (8.1) is selfadjoint and positive. Set $h = (a_0)^{1/2m}$ and consider the Hamiltonian system

$$\dot{x} = \partial_\xi h(x,\xi) \, , \quad \dot{\xi} = -\partial_x h(x,\xi) \tag{9.11}$$

in T^*M. Along trajectories $x = x(\tau)$, $\xi = \xi(\tau)$ of this system, i.e. along the bicharacteristics of the function $h(x,\xi)$, this function is constant. We consider bicharacteristics lying on the $(2n-1)$-dimensional manifold

$$S^*M = \{(x,\xi) \in T^*M : h(x,\xi) = 1\} . \tag{9.12}$$

Assume that a bicharacteristic which issues from an interior point of the manifold at time τ_0 first reaches the boundary at time τ_1: $x(\tau_1) \in \Gamma$. In a neighborhood of the corresponding point $(x(\tau_1), \xi(\tau_1))$ we use the local coordinates described in Sect. 0.2. If $\dot{x}^n(\tau_1 - 0) \neq 0$, then we say that the bicharacteristic approaches the boundary transversally; clearly then

$$(\tau_1 - \tau_0)\dot{x}^n(\tau_1 - 0) < 0 . \tag{9.13}$$

In this case we continue the functions $x(\tau)$, $\xi(\tau)$ beyond τ_1 so that the motion goes further along another bicharacteristic going off transversally from the boundary. Namely, if, for definiteness, $\tau_1 > \tau_0$, then we define the initial values $x(\tau_1+0)$ and $\xi(\tau_1+0)$ for the new solution of the system (9.11) by the conditions

$$x(\tau_1+0)=x(\tau_1-0)\,,\quad \xi'(\tau_1+0)=\xi'(\tau_1-0)\,, \tag{9.14}$$

$$h(x(\tau_1+0),\,\xi(\tau_1+0))=1\,,\quad \dot{x}^n(\tau_1+0)>0\,. \tag{9.15}$$

Thus the functions $x(\tau)$ and $\xi'(\tau)$ remain continuous when τ passes trough τ_1, and the Hamiltonian $h(x(\tau),\xi(\tau))$ remains equal to unity, while, in general, the value of $\xi_n(\tau)$ has a jump: $\xi_n(\tau+0)$ is defined as any real root ξ_n^* of the equation

$$a_0(x(\tau_1),\xi'(\tau_1),\xi_n^*)=1 \tag{9.16}$$

such that $\partial_{\xi_n}a_0$ has the sign $+$ (the sign $-$ if $\tau_1<\tau_0$; see (9.11)). Such a root always exists, and the number of these roots is not greater than q. Note that the projection of the bicharacteristic on M remains continuous. If the root ξ_n^* is always unique, then we say that the *simple reflection condition* holds. Clearly it holds if $m=1$ or if $a_0=[\widetilde{a}_0]^q$, where $\widetilde{a}_0$ is the symbol of an elliptic second order differential operator.

This passage from one bicharacteristic to another is called the *transversal reflection of the bicharacteristic from the boundary.* We now consider *bicharacteristic broken lines* or *billiard trajectories.* Each of them consists of smooth parts lying on bicharacteristics, and it is required that the transition from one part to another be a transversal reflection from the boundary. We also assume that the functions $x(\tau)$ and $\xi(\tau)$ are defined on a maximal interval of values of τ.

Thus, on S^*M a dynamical system arises that prescribes the motion of the point (x,ξ) along the billiard trajectories passing through it. (More precisely, this dynamical system is defined on $S^*M\setminus O_1$, see the definition of the set O_1 below.) This dynamical system is called a *Hamiltonian billiards,* more precisely, a *branching Hamiltonian billiards* if the simple reflection condition does not hold. If $m=1$, $M=G$ is a domain in $\mathbb{R}^n$, and $a_0(x,\xi)=|\xi|^2$, then we have no branching, the projections on G of the smooth parts of a billiard trajectory are segments of straight lines with ends at points on Γ (since a_0 does not depend on x), and conditions (9.14), (9.15) mean that the angle of incidence equals the angle of reflection. More general geodesic billiards arises if a Riemannian metric is introduced in $\overline{G}$ and $-a_0(x,\xi)$ is equal to the principal symbol of the corresponding Beltrami–Laplace operator. The projections of the billiard trajectories on G are then geodesics, and the above angle condition holds in the sense of the Riemann metric.

We now introduce three subsets in S^*M:

1) The set O_1 of "tangential" points. A point (x,ξ) is called tangential if the billiard trajectory issuing from this point at a time $\tau=\tau_0$ approaches the boundary at some time τ_1 tangentially, i.e. non-transversally.

2) The set O_2 of "dead-end" points. A point (x,ξ) is called dead-end if the billiard trajectory issuing from it undergoes an infinite number of transversal reflections in a finite time.

We note that if a point $(x, \xi) \in S^*M$ does not belong to $O_1 \cup O_2$, then the billiard trajectory passing through it is defined for all $\tau \in \mathbb{R}$.

3) The set O_3 of "periodic" points. A point (x, ξ) is called periodic if the billiard trajectory issuing from it at a time $\tau = \tau_0$ returns to this point at some $\tau = \tau_1$.

On S^*M there exists a natural density $dx\, d\xi'$ defined by the relation

$$dx\, d\xi = m^{-1} dx\, d\xi' dh(x, \xi) , \tag{9.17}$$

and hence a Lebesgue measure is defined. It is known that the set O_1 is always of measure zero.

Theorem 9.1.3. *Let* $\operatorname{mes} O_2 = 0$ *and* $\operatorname{mes} O_3 = 0$. *Then a formula of the form* (9.10) *is true.*

Instead of $\operatorname{mes} O_3 = 0$ it is possible to assume that $\operatorname{mes} O_3' = 0$, where O_3' is the set of "absolutely periodic" points. However, in (Safarov and Vassiliev 1988) it is proved that $\operatorname{mes} O_3 \setminus O_3' = 0$ if $\operatorname{mes} O_2 = 0$. The formula for d_{-1} can be found in (Vassiliev 1984, 1986a) and in (Safarov and Vassiliev 1992, 1996). This quantity depends only on the principal symbols $b_{j,0}$ of boundary operators and on the principal symbol $a_0(x, \xi)$ for $x \in \Gamma$.

Some geometrical conditions sufficient for O_2 and O_3 to be of zero measure can be found in these papers. If the simple reflection condition holds, then $\operatorname{mes} O_2 = 0$. If the problem is analytic and M is "Hamilton convex", then $\operatorname{mes} O_3 = 0$. The condition of Hamilton convexity means that

$$\{\partial h / \partial \xi_n, h\}_{(x, \xi', \xi_n^*)} \geq 0$$

over Γ and is not equal to zero identically. Here $\{g, h\}$ is the Poisson bracket $g'_\xi \cdot h'_x - h'_\xi \cdot g'_x$. If $M = G$ is a domain in $\mathbb{R}^n$ and $a_0 = |\xi|^m$, then the Hamilton convexity coincides with the usual convexity.

In particular, the two-term asymptotics (9.10) holds if $M = G$ is a convex domain in $\mathbb{R}^n$ with real-analytic boundary and $a_0 = |\xi|^m$.

Here we do not dwell on the results of the investigation of $N(\lambda)$ in more general situation, when there is no two-term asymptotics. See (Safarov 1988a,b) and the papers of Safarov and Vassiliev.

c. The book (Ivrii 1984) contains an investigation on selfadjoint operators A_B corresponding to general elliptic first and second order systems without restrictions on multiplicities of the eigevalues of the principal symbol. The operator can be not semibounded, therefore it is natural to introduce the distribution functions for positive and negative eigenvalues of A_B:

$$N_\pm(\lambda) = \operatorname{card}\{j : 0 \leq \pm\lambda_j \leq \lambda\} . \tag{9.18}$$

If the principal symbol $a_0(x, \xi)$ of A has positive and negative eigenvalues, then

$$N_\pm(\lambda) = d_0^\pm \lambda^{\frac{n}{m}} + O(\lambda^{\frac{n-1}{m}}) \qquad (\lambda \to +\infty) , \tag{9.19}$$

where

$$d_0^{\pm} = \frac{1}{(2\pi)^n} \iint p^{\pm}(x,\xi)\,dx\,d\xi\ , \qquad (9.20)$$

and $p^{\pm}(x,\xi)$ is the spectral selfadjoint projector for $a_0(x,\xi)$ corresponding to its eigenvalues that lie between 0 and ± 1. If all the eigenvalues of the principal symbol are positive, then the formula (9.19) for $N_+(\lambda)$ remains true, while $N_-(\lambda)$ has an asymptotics of the form

$$N_-(\lambda) = d_{-1}^{-}\lambda^{\frac{n-1}{m}} + O(\lambda^{\frac{n-2}{m}}) \qquad (\lambda \to +\infty)\ , \qquad (9.21)$$

where $O(\lambda^0)$ has to be replaced by $O(\log\lambda)$. The two-term asymptotics formulas for $N_+(\lambda)$ and $N_-(\lambda)$ are valid if the simple reflection condition and the condition $\operatorname{mes} O_3 = 0$ are fulfilled. Ivrii also considers a scalar spectral problem for the equation $(-\Delta)^p u + \ldots = \lambda(-\Delta)^q u + \ldots$, where $0 \le q < p$. Some results were announced earlier, see the references in this book. The generalizations to systems of higher order are formulated in (Ivrii 1987).

d. Spectral properties of L^2–realizations A_B in the sense of Sect. 6.5 were investigated under conditions of strong ellipticity and some additional conditions that ensure the selfadjointness and positiveness of this operator. In (Grubb 1977b) the eigenvalue asymptotics is found in the case when all m_j are positive. Let

$$\begin{pmatrix} A^{11} & A^{12} \\ A^{21} & A^{22} \end{pmatrix},$$

where A^{22} is a $\mu_s \times \mu_s$ matrix. Then $N_{A_B} \sim d_0 t^{n/2\mu_s}$, where the constant d_0 is defined by the trace of the principal symbol of the operator $A^{22} - A^{21}(A^{11})^{-1}A^{12}$ on T^*M. (Note that if $A^{12} = A^{21} = 0$, then this is the asymptotics of the "slowest" series of eigenvalues.) Some remainder estimates are obtained in (Levendorskij 1989, 1990). In (Grubb and Geymonat 1977, 1979) it is assumed that $\mu_{s-1} > \mu_s = 0$. In this case the operator A_B has an essential spectrum: it consists of such λ that the boundary problem $(A_B - \lambda)u = f$ on M, $B\chi u = 0$ on Γ is not elliptic. The discrete part of the spectrum has the asymptotics $N_{A_B}(t) \sim d_0 t^{n/2\mu_{s-1}}$, where d_0 is defined essentially as above by the matrix obtained from A by rejection of the last μ_s rows and μ_s columns.

9.2. Boundary Problems Close to Selfadjoint Ones. Assume that (8.1) is a formally selfadjoint boundary problem, and consider the following boundary problem:

$$A^{(1)}u = f \quad \text{in} \quad M_+\ , \quad B_1^{(1)}u = \ldots = B_q^{(1)}u = 0 \quad \text{on} \quad \Gamma\ . \qquad (9.22)$$

Here

$$A^{(1)} = A + A^{(2)}\ , \quad B_j^{(1)} = B_j + B_j^{(2)} \quad (j = 1, \ldots, q)\ , \qquad (9.23)$$

where $A^{(2)}$ and $B_j^{(2)}$ are differential operators of orders not greater than $2q-r$ and r_j-r ($j=1,\dots,q$), respectively, with $r \geq 1$. Denote by $A^{(1)}_{B^{(1)}}$ the operator corresponding to the boundary problem (9.22). The questions arising here are similar to those considered in (Agranovich 1990a, Sect. 6.2). They relate to the asymptotics of the eigenvalues and to the basic properties of the system of root functions. Assume, for simplicity, that the boundary problem (8.1) with $A-\lambda I$ instead of A is elliptic with parameter along $\mathbb{R}_-$; then the same is true for the boundary problem (9.22). Without loss of generality, we then assume that $A_B > 0$. If $B_j^{(1)}$ coincides with B_j for all j, then $A^{(1)}_{B^{(1)}} = A^{(1)}_B$ is a weak perturbation of a selfadjoint operator A_B, and then the abstract theorems presented in (Agranovich 1990, Sect. 6.2) can directly be applied to $A^{(1)}_B$. In the general case the following proposition is true:

Proposition 9.2.1. *The operator $A^{(1)}_{B^{(1)}}$ is similar to the operator*

$$A_B + TA_B^{\frac{m-r}{m}}, \tag{9.24}$$

where T is a bounded operator in $H_0(M)$.

The term *similarity* is understood in the following sense. Let A_1 and A_2 be unbounded operators in a Hilbert space H. They are called similar if there exists a bounded invertible operator L in H such that

$$D(A_2) = LD(A_1) \quad \text{and} \quad A_1 f = L^{-1}A_2Lf \quad \text{for} \quad f \in D(A_1)\,. \tag{9.25}$$

Similar operators have the same spectrum, and the systems of their root vectors are simultaneously Abel or Riesz bases with brackets.

Proposition 9.2.1 is proved in (Markus and Matsaev 1982) for $r=1$, but the proof can be carried out for $r>1$ analogously. It is based on the construction and analysis of an invertible operator in $H_0(M)$ that transforms

$$H_m^{B^{(1)}}(\Gamma) = \{u \in H_m(M) : B_j^{(1)}u = 0 \;\; (1,\dots,q) \;\text{ on }\; \Gamma\}$$

into H_m^B.

The eigenvalues of the operator $A^{(1)}_{B^{(1)}}$ lie in a domain of the form

$$\{\lambda : \operatorname{Re}\lambda > 0,\ |\operatorname{Im}\lambda| \leq C(\operatorname{Re}\lambda)^{\frac{m-r}{m}}\} \tag{9.26}$$

except, possibly, for a finite number of them (see Theorem 6.2.2 (Agranovich 1990a)). Denote by $N^{(1)}(\lambda)$ the distribution function for the moduli of the eigenvalues of $A^{(1)}_{B^{(1)}}$.

Theorem 9.2.2. *For any $r \geq 1$ the following formula is true:*

$$N^{(1)}(\lambda) = d_0\lambda^{\frac{n}{m}} + O(\lambda^{\frac{n-1}{m}})\,. \tag{9.27}$$

Moreover, if $r \geq 2$ *and formula* (9.10) *holds, then*

$$N^{(1)}(\lambda) = d_0\lambda^{\frac{n}{m}} + d_{-1}\lambda^{\frac{n-1}{m}} + o(\lambda^{\frac{n-1}{m}}), \tag{9.28}$$

where d_0 *and* d_{-1} *are the same as in* (9.10).

This theorem is similar to Theorem 6.2.8 in (Agranovich 1990a) about pseudodifferential operators on a closed manifold and follows from an abstract theorem on operators in a Hilbert space due to Markus and Matsaev (see (Markus and Matsaev 1982) or (Markus 1986, §9)).

In the following theorem we indicate conditions for the existence of an Abel, Riesz, or Bari basis in $H_0(M)$ consisting of finite-dimensional subspaces $\mathcal{L}_l$ invariant with respect to $A^{(1)}_{B^{(1)}}$ $(l = 0, 1, \ldots)$. The corresponding definitions can be found in (Agranovich 1990a, Sect. 6.2), along with abstract theorems from which Theorem 9.2.3 follows and a description of the subspaces $\mathcal{L}_l$.

Theorem 9.2.3. *If* $1 \leq r < n$, *then the system* $\{\mathcal{L}_l\}_1^\infty$ *is an Abel basis of order* γ *of subspaces in* $H_0(M)$, *where* $(n-r)/m < \gamma < (n-r)/m + \varepsilon$ *with* $\varepsilon > 0$ *sufficiently small. If* $r = n$, *then it is a Riesz basis of subspaces in* $H_0(M)$. *Finally, if* $r > n$ *and* $B_j = B_j^{(1)}$ *for all* j, *i.e. the boundary conditions are undisturbed, then* $\{\mathcal{L}_l\}_1^\infty$ *is a Bari basis of subspaces in* $H_0(M)$.

These results can easily be extended to matrix boundary problems elliptic in the usual sense and close to selfadjoint ones. If the operator A_B corresponding to the undisturbed selfadjoint boundary problem is not semibounded, then the eigenvalues of the operator $A^{(1)}_{B^{(1)}}$ lie near the rays $\mathbb{R}_-$ and $\mathbb{R}_+$. If, in addition, we know the asymptotics of the distribution functions $N_-(\lambda)$ and $N_+(\lambda)$ for A_B, then the same asymptotics holds for the moduli of the eigenvalues of $A^{(1)}_{B^{(1)}}$ close to $\mathbb{R}_-$ and $\mathbb{R}_+$. This is true for the one-term and two-term asymptotic formulas if $r \geq 1$ and $r \geq 2$, respectively. Theorem 9.2.3 can also be generalized to this case.

9.3. Boundary Problems Far from Selfadjoint. In this section, in contrast to the two previous ones, we only assume that the boundary problem under consideration is elliptic with parameter in an angle (with sufficiently large opening) or angles on the complex plane. We present results concerning the completeness of the root functions and the rough asymptotics of the moduli of eigenvalues.

a. First we consider the scalar boundary problem (8.1). We assume that the orders r_j of boundary operators are less than $m = 2q$ and that the boundary problem for $A - \lambda I$ instead of A is elliptic with parameter in an angle $\mathcal{L}$ with vertex at the origin. As we know, in this case the spectrum of the operator A_B is discrete, and $\mathcal{L}$ does not contain eigenvalues with large moduli.

Theorem 9.3.1. *Let the opening of the complement of* $\mathcal{L}$ *be not greater than* $m\pi/n$. *Then the system of root vectors of* A_B *is complete in* $H_0(M)$. *Moreover, there exists a system of finite-dimensional subspaces in* $H_0(M)$ *that are* *invariant with respect to A_B and form an Abel basis of order γ in this space, where $n/m < \gamma < (n/m) + \varepsilon$ with ε sufficiently small.*

The condition imposed on the opening of the complement of $\mathcal{L}$ can be replaced by the following condition: the boundary problem is elliptic with parameter along some rays $\Gamma_1, \dots, \Gamma_r$ issuing from the origin with angles between the adjacent rays not exceeding $m\pi/n$.

This theorem is similar to the result formulated in Subsect. 6.2c of (Agranovich 1990a) for elliptic pseudodifferential operators on a closed manifold, and it easily follows from the abstract Theorems 6.4.1–6.4.2 formulated there. It suffices to note the following: 1) the rays Γ_j have angular neighborhoods where the conditions of ellipticity with parameter hold; 2) in such an angle

$$\|(A_B - \lambda I)^{-1}\| \leq C|\lambda|^{-1} \tag{9.29}$$

(see Sect. 3.2); 3) the s-numbers of the operator $(A_B)^{-1}$, i.e. the eigenvalues of a selfadjoint positive operator $[(A_B)^* A_B]^{-1/2}$, have a regular asymptotics $s_j \sim cj^{-m/n}$, $c > 0$. The last assertion follows from the fact that the operator $(A_B)^* A_B$ corresponds to a selfadjoint elliptic boundary problem (see Remark 4.3.7), and its eigenvalues have a regular asymptotics according to Subsect. 9.1a (see also the first formula in (9.35) below).

Corollary 9.3.2. *Under the same assumptions, the system of root functions is complete in the subspace $D(A_B)$ of $H_m(M)$.*

Indeed, if λ_0 does not belong to the spectrum of A_B, then there is a continuous isomorphism

$$(A_B - \lambda_0 I)^{-1} \colon H_0(M) \to D(A_B)$$

that transforms linear combinations of root functions into linear combinations of root functions.

Now, for convenience of notation, we assume that ellipticity with parameter holds in the angle

$$\mathcal{L} = \{\lambda : |\arg \lambda| \geq \varphi\} \cup \{0\}\,, \tag{9.30}$$

where $0 < \varphi < \pi$. In particular, this means that

$$|\arg a_0(x, \xi)| < \varphi \quad \text{on} \quad T^*M \setminus 0\,. \tag{9.31}$$

Assume, additionally, that the angle $\mathcal{L}$ does not contain eigenvalues of A_B; in particular, A_B is invertible. Introduce the quantities

$$d_0 = \frac{1}{(2\pi)^n n} \sum \int_M \varphi_k(x)\,dx \int_{|\omega|=1} [a_0^{(k)}(x, \omega)]^{-n/m} dS \tag{9.32}$$

and

$$d_1 = \frac{1}{(2\pi)^n n} \sum \int_M \varphi_k(x)\,dx \int_{|\omega|=1} |a_0^{(k)}(x, \omega)|^{-n/m} dS\,, \tag{9.33}$$

where $\{\varphi_k(x)\}$ is the partition of unity from Subsect. 2.1c and $a_0^{(k)}$ is the principal symbol of A in local coordinates in O_k or O_k^+. If $a_0 > 0$, then (9.32) coincides with (9.5). In the general case a_0^α is to be understood as $|a_0|^\alpha e^{i\alpha \arg a_0}$. The formula for d_1 can be rewritten in the form

$$d_1 = \frac{1}{(2\pi)^n} \iint_{|a_0(x,\xi)|\leq 1} dx\, d\xi \,. \tag{9.34}$$

It is easy to check that $|d_0| \leq d_1$ always (see (Agranovich and Markus 1989)).

Denote by $N(\lambda)$ the distribution function for the moduli of eigenvalues of A_B, and by $N_1(\lambda)$ the distribution function for $[s_j(A_B^{-1})]^{-1}$. We have

$$\lim_{\lambda\to\infty} N_1(\lambda)\lambda^{-n/m} = d_1 \quad \text{and} \quad \limsup_{\lambda\to\infty} N(\lambda)\lambda^{-n/m} < \infty \,; \tag{9.35}$$

here the latter relation follows from the former. Both relations are true without the assumption of ellipticity with parameter. It suffices to assume that A_B corresponds to an elliptic boundary problem and has a discrete spectrum.

Theorem 9.3.3. *The following relations hold:*

$$\liminf_{\lambda\to\infty} \lambda^{-n/m} N(\lambda) > 0 \quad \text{if } d_0 \neq 0\,; \quad \liminf_{\lambda\to\infty} \lambda^{-n/m} N(\lambda) \leq d_1 \,; \tag{9.36}$$

$$\limsup_{\lambda\to\infty} \lambda^{-n/m} N(\lambda) \geq |d_0| \,, \quad \limsup_{\lambda\to\infty} \lambda^{-n/m} N(\lambda) \leq d_1 e \,. \tag{9.37}$$

These results are similar to Theorems 6.4.5 and 6.4.6 in (Agranovich 1990a) and are obtained in the same manner (see (Agranovich and Markus 1989)). In particular, formula (8.59) is utilized. The second inequalities in (9.36) and (9.37) are true without the assumption of ellipticity with parameter.

In particular, if the case $d_0 \neq 0$,

$$N(\lambda) \asymp \lambda^{n/m} \,, \tag{9.38}$$

i.e. the quotient of the left and the right sides is bounded from below and from above by positive constants. If the limit $\lim_{\lambda\to\infty} \lambda^{-n/m} N(\lambda)$ exists, then it necessarily belongs to the segment $[|d_0|, d_1]$. The quantity d_0 is certainly nonzero if $|\arg a_0| \leq q\pi/n$.

Remark 9.3.4. In the just considered case of a scalar boundary problem on a connected manifold, the set of all directions containing the values of the principal symbol $a_0(x,\xi)$ is connected, but the set of directions of ellipticity with parameter can be disconnected. Assume only that condition (9.31) holds, $d_0 \neq 0$, and the boundary problem (8.1) is elliptic with parameter along the rays forming the boundary of the angle (9.30) (then the conditions of ellipticity with parameter hold in some angular neighborhoods of these rays). In this case relation (9.38) holds if $N(\lambda)$ is understood as the distribution function for eigenvalues of A_B outside $\mathcal{L}$.

Theorem 9.3.5. *Let $a_0(x,\xi) > 0$, and let the boundary problem* (8.1) *be elliptic with parameter along the rays $\{\lambda : \arg\lambda = \theta\}$ for $0 < |\theta| < \varepsilon$ with some $\varepsilon > 0$. Then for the distribution function of the moduli of eigenvalues of A_B in arbitrarily narrow angle $\{\lambda : |\arg\lambda| \le \delta\}$ with fixed $\delta \in (0,\varepsilon)$, we have*

$$N(\lambda) = d_0\lambda^{n/m} + o(\lambda^{n/m}) \qquad (\lambda \to +\infty) . \tag{9.39}$$

b. The results presented above in this section are easily generalized to the case of matrix boundary problems elliptic in the usual sense (see Sect. 6.1) with $g = 0$ and $r_j < m$. It is only necessary to make the following replacements in the statements: $H_0(M)$ is replaced by $[H_0(M)]^p$, in (9.32) $a_0^{-n/m}$ is replaced by $\operatorname{tr} a_0^{-n/m}$, and in (9.33) $|a_0|^{-n/m}$ is replaced by $\operatorname{tr}(a_0^* a_0)^{-n/m}$. In addition, in the matrix case the set of directions containing the eigenvalues of the principal symbol $a_0(x,\xi)$ can be disconnected. In this case let us define d_0 by a formula of the form (9.32) with replacement of $a_0^{-n/m}$ by the sum of eigenvalues of the matrix $a_0^{-n/m}$ for all the eigenvalues of a_0 lying in the complement of the angle (9.30). If this quantity d_0 is different from zero and the boundary problem is elliptic with parameter along the rays forming the boundary of this angle, then we again obtain relation (9.38) for the distribution function of the moduli of eigenvalues of A_B outside $\mathcal{L}$. Theorem 9.3.5 extends similarly. Namely, instead of $a_0 > 0$ we can assume that the ray $\mathbb{R}_+$ is a connected component in the set of directions of ellipticity with parameter. In (9.39), d_0 is defined by a formula of the form (9.32) with replacement of $a_0^{-n/m}$ by the sum of the eigenvalues of the matrix $a_0^{-n/m}$ that correspond to positive eigenvalues of a_0.

c. The next step of possible generalizations concerns the matrix boundary problems elliptic in the sense of Agmon, Douglis, and Nirenberg with an angle, or angles, of ellipticity with parameter (see Sect. 6.1). Results similar to (9.36)–(9.39) are formulated as in the case of systems elliptic in the usual sense, and we do not dwell on this here (see (Agranovich 1992)).

The question of completeness is somewhat more complicated. We associate with the boundary problem the operator A_B in $H_{\{s-l_j\}}(M)$ acting according to the formula $A_B u = Au$, with domain $D(A_B) \subset H_{\{s+m_k\}}(M)$ defined by the boundary conditions $Bu = 0$. Here s satisfies inequalities (6.28); we also recall that $m_k + l_k \equiv m$. Generally, some boundary conditions can preserve their sense in $H_{\{s-l_j\}}(M)$. Then the system of root functions cannot be complete in this space, since these functions are subjected to these boundary conditions.

The same situation is possible even in the case of scalar elliptic boundary problems (this remark is due to S.Ya. Yakubov). If the operator A_B is considered as acting in $H_s(M)$ with $s > 0$ instead of $s = 0$ (for example, if $r_j > m$ for some j), then its system of root functions can turn out to be incomplete in $H_s(M)$. Corollary 9.3.2 suggests that one can expect completeness in subspaces of $H_s(M)$ defined by boundary conditions.

The following abstract result permit us to find such subspaces.

Theorem 9.3.6. *Let A be a closed operator in a separable Hilbert space H. Assume that A has a nondense domain and nonempty resolvent set, and let the resolvent $R_A(\lambda) = (A - \lambda I)^{-1}$ belong to the Neumann–Schatten class $\mathfrak{S}_\rho$ for some $\rho > 0$. Assume that along some rays $\Gamma_1, \ldots, \Gamma_r$ issuing from the origin, $R_A(\lambda)$ exists for sufficiently large $|\lambda|$ and satisfies the inequality*

$$\|R_A(\lambda)\| \leq C|\lambda|^\sigma \tag{9.40}$$

with some integer $\sigma \geq -1$. Let the angles between adjacent rays Γ_j be less than π/ρ. Then the system of root vectors of A is complete in the closure $\overline{D(A^{\sigma+2})}$ in H of the domain $D(A^{\sigma+2})$ of the operator $A^{\sigma+2}$.

This theorem follows from Theorem 29 in (Dunford and Schwartz 1963, Chapter XI, Sect. 9).

From Theorem 6.4.1 it is clear that such an estimate holds for the resolvent of the operator A_B in $H_{\{s-l_j\}}(M)$, which corresponds to the matrix boundary problem, along the rays of ellipticity with parameter. The set of directions of these rays is open. The boundary conditions that define the domain of $A_B^{\sigma+2}$ have the form

$$B_j A^k u = 0 \quad \text{on} \quad \Gamma \tag{9.41}$$

with some $k = 0, 1, \ldots$, where B_j are the rows of the matrix B. However here we have to preserve only those of conditions (9.41) that make sense in $H_{\{s-l_j\}}(M)$; exactly for these conditions we have $\min_j(s - r_j - m(k+1)) > 1/2$. This leads to the following result (Agranovich 1990b):

Theorem 9.3.7. *Let the boundary problem* (6.37) *be elliptic with parameter along the rays $\Gamma_1, \ldots, \Gamma_r$, and let the angles between the adjacent rays be not greater than $m\pi/n$. Assume that s satisfies conditions* (6.28). *Then the system of root functions of the operator A_B corresponding to this boundary problem with $g = 0$ is complete in the subspace of $H_{\{s-l_j\}}(M)$ defined by all boundary conditions of the form* (9.41) *with* $\min_j(s - r_j - m(k+1)) > 1/2$.

As a corollary, under the same conditions we obtain the completeness of the root functions in $[H_0(M)]^p$.

9.4. Boundary Problems with Spectral Parameter only in Boundary Conditions.

a. The simplest boundary problem of this sort is well known:

$$\Delta u + a(x)u = 0 \quad \text{in } G\,, \quad \gamma_0 \partial_\nu u = \lambda \gamma_0 u \quad \text{on } \Gamma\,. \tag{9.42}$$

Taking it as a model, we first consider the following boundary problem, for simplicity scalar:

$$Au = 0 \quad \text{on} \quad M\,, \quad B_{j,1}u = \lambda B_{j,0}u \quad (j = 1, \ldots, q) \quad \text{on} \quad \Gamma\,. \tag{9.43}$$

Here A is a (properly) elliptic partial differential operator of order $m = 2q$ on M, and $B_{j,k}$ ($j = 1, \ldots, q$; $k = 0, 1$) are boundary operators of orders $r_{j,k}$, respectively. Assume that the following conditions hold:

1) The boundary problem

$$Au = 0 \quad \text{on} \quad M\,, \quad B_{j,0}u = g_j \quad (j = 1, \ldots, q) \quad \text{on} \quad \Gamma \tag{9.44}$$

is elliptic and, for simplicity, has one and only one solution $u \in H_m(G)$ for any $g_j \in H_{m-r_{j,0}-\frac{1}{2}}(\Gamma)$.

2) The difference $\tau = r_{j,1} - r_{j,0}$ does not depend on j and is positive. (This condition permits us to attribute a weight τ to the parameter λ with respect to the differentiation. We restrict ourselves to this case.)

Let $\mathcal{P} = (\mathcal{P}_1, \ldots, \mathcal{P}_q)$ be the Poisson operator of the boundary problem (9.44) (see Subsect. 5.2c). The substitution

$$u = \mathcal{P}g = \mathcal{P}_1 g_1 + \ldots + \mathcal{P}_q g_q \tag{9.45}$$

reduces the boundary problem (9.43) to the system of pseudodifferential equations on Γ

$$Vg = \lambda g\,, \quad \text{where} \quad V = (V_{j,l})_{j,l=1}^{q}\,, \quad V_{j,l} = B_{j,1}\mathcal{P}_l\,, \tag{9.46}$$

and $\operatorname{ord} V_{j,l} \leq r_{j,1} - r_{l,0}$. This system is equivalent to (9.43). From the considerations in Sect. 5.2 it is clear that ellipticity (in the Douglis–Nirenberg sense) of the operator V is equivalent to the following condition:

3) The boundary problem

$$Au = f \quad \text{on} \quad M\,, \quad B_{j,1}u = g_j \quad (j = 1, \ldots, q) \quad \text{on} \quad \Gamma \tag{9.47}$$

is elliptic.

We assume that this condition is also satisfied.

At least theoretically, the calculation of the local complete symbol of V is available, and this permits us to apply the assertions on spectral properties of pseudodifferential operators from (Agranovich 1990a, § 6) to V. However, for convenience, we first transform V into a matrix pseudodifferential operator with equal orders of elements. Let Λ be an invertible scalar elliptic pseudodifferential operator of order 1 on Γ. We set

$$U = \operatorname{diag}(\Lambda^{r_{1,1}}, \ldots, \Lambda^{r_{q,1}})\,, \quad g = Uh\,.$$

Instead of (9.46) we now have the equation

$$Wh = \lambda h\,, \quad \text{where} \quad W = U^{-1}VU = (W_{j,l})_{j,l=1}^{q}\,, \tag{9.48}$$

and here $\operatorname{ord} W_{j,l} \leq \tau$ for all j and l. We can consider the operator W as acting in $[H_s(\Gamma)]^q$ and having the domain $[H_{s+\tau}(\Gamma)]^q$, for example, with $s = 0$.

Returning to (9.43), we agree to consider the spectrum of W as the spectrum of this boundary problem. If the resolvent set $\rho(W)$ is nonempty, then this spectrum is discrete and the root functions of W belong to $[H_\infty(\Gamma)]^q$, since W is elliptic. If h is an eigenfunction (or a root function) of W corresponding to the eigenvalue λ, then the function $u = \mathcal{P}Uh$ (it clearly belongs to $H_\infty(\Gamma)$) will be called the eigenfunction (accordingly, the root function) of the boundary problem (9.43). To clarify this definition, we note that if, say,

$$Wh = \lambda h \quad (h \neq 0) \quad \text{and} \quad W\tilde{h} = \lambda\tilde{h} + h\,,$$

then for $u = \mathcal{P}Uh$ and $\tilde{u} = \mathcal{P}U\tilde{h}$ we have

$$A\tilde{u} = 0 \quad \text{on} \quad M_+\,, \quad B_{j,1}u = \lambda B_{j,0}u\,, \quad B_{j,1}\tilde{u} = \lambda B_{j,0}\tilde{u} + B_{j,0}u \quad \text{on} \quad \Gamma\,.$$

b. Let $\mathcal{L}$ be a closed angle on the complex plane with vertex at the origin. In Sect. 5.3 we have defined the ellipticity with parameter of the boundary problem (9.43) in $\mathcal{L}$. Under conditions 1) and 2) it is equivalent to ellipticity with parameter of the pseudodifferential operator W in $\mathcal{L}$.

Therefore, for the boundary problem (9.43) we automatically obtain corollaries from theorems on elliptic pseudodifferential operators (see (Agranovich 1990a), especially Sect. 6.4). In particular, if there is an angle $\mathcal{L}$ of ellipticity with parameter, then the spectrum of this boundary problem is discrete and does not contain points in $\mathcal{L}$ with sufficiently large $|\lambda|$. Furthermore, if the boundary problem (9.43) is elliptic with parameter along some rays $\Gamma_1, \ldots, \Gamma_r$ issuing from the origin, and if the largest of the angles between adjacent rays does not exceed $\pi\tau/(n-1)$, then the system of root functions of W is complete in $[H_s(\Gamma)]^q$; moreover, there exists an Abel basis in $[H_s(\Gamma)]^q$ consisting of finite-dimensional subspaces that are invariant with respect to W. Finally, we can obtain a rough asymptotics for the moduli of eigenvalues in some angles between directions of ellipticity with parameter (see exact assumptions in (Agranovich 1990a, Sect. 6.4)). In particular, if there is an isolated ray containing some eigenvalues of the principal symbol of W, then we can obtain the exact asymptotics for the moduli of eigenvalues of W close to this ray.

c. If the principal symbol w_0 of the pseudodifferential operator W is an Hermitian matrix, then we can apply the results presented in (Agranovich 1990a, Sects. 6.1 and 6.2). We can calculate the principal symbol w_0 of W (at least theoretically). We also note that some conditions sufficient for the selfadjointness of W can be obtained from Green's formula; see (Ercolano and Schechter 1965). We restrict ourselves to an elementary example.

Consider the boundary problem (9.42) with a real-valued $a(x)$. Assume that the homogeneous Dirichlet problem for the equation $\Delta u + a(x)u = 0$ does not have nontrivial solutions, and let $\mathcal{P} = \mathcal{P}_D$ be the corresponding Poisson operator. Then the boundary problem(9.42) is equivalent to the equation $Wg = \lambda g$, where $Wg = \gamma_0\partial_\nu\mathcal{P}g$. As we know, this is an elliptic operator of the first order. Let g_1 and g_2 be smooth functions on Γ, and let u_1 and u_2 be two solutions of the equation $\Delta u + a(x)u = 0$ with Dirichlet conditions $\gamma_0 u_j = g_j$ $(j = 1, 2)$. Then from Green's formula (4.9) it follows that

$$(Wg_1, g_2)_\Gamma = (g_1, Wg_2)_\Gamma\,.$$

We see that $W^* = W$. As can be verified, the principal symbol w_0 of W is negative (cf. Subsect. 9.4e below); hence, W is a selfadjoint and semibounded from above operator in $H_0(\Gamma)$ of order 1. A formula for the asymptotics of its eigenvalues holds with Hörmander's estimate of the remainder.

If we replace the boundary condition in (9.42) by $\gamma_0 \partial_\nu u + b(x)\gamma_0 u = \lambda \gamma_0 u$ on Γ with a complex-valued function $b(x)$, then the corresponding operator W is close to a selfadjoint one; more precisely, we have $\operatorname{ord}(W - W^*) = 0$. The reader can deduce corollaries for W from the results presented in (Agranovich 1990a, Sect. 6.2).

If the homogeneous Dirichlet problem for the equation $\Delta u + a(x)u = 0$ has nontrivial solutions, then the boundary problem (9.42) has the "eigenvalue infinity," i.e., if we divide the boundary condition by λ and use the substitution $\lambda = \mu^{-1}$, we obtain the boundary problem with the eigenvalue $\mu = 0$. From this example one can see that the existence of nontrivial solutions to the boundary problem (9.44) is not an insuperable difficulty for the investigation. We do not dwell on this in detail; cf. (Agranovich 1977, §37).

d. Now we consider a scalar boundary problem with boundary conditions depending on the spectral parameter polynomially:

$$Au = 0 \quad \text{in} \quad G\,, \quad \sum_{k=0}^{p} \lambda^{p-k} B_{j,k} u = 0 \quad (j = 1, \ldots, q) \quad \text{on} \quad \Gamma\,. \tag{9.49}$$

Here $B_{j,k}$ are boundary differential operators of orders $r_{j,0} + k\tau$, $\tau \in \mathbb{N}$.

Again we assume that the boundary problem (9.44) is elliptic and uniquely solvable, and let $\mathcal{P}$ be its Poisson operator. The substitution (9.45) transforms the boundary problem (9.49) into the system

$$\sum_{k=0}^{p} \lambda^{p-k} V_k g = 0 \tag{9.50}$$

on Γ, which is equivalent to (9.49). Here V_k are matrix pseudodifferential operators, $V_0 = I$,

$$V_k = (V_{k,j,l})_{j,l=1}^{q}\,, \quad V_{k,j,l} = B_{j,k}\mathcal{P}_l\,, \quad \text{and} \quad \operatorname{ord} V_{j,k,l} = r_{j,0} - r_{l,0} + k\tau\,. \tag{9.51}$$

This system admits the equalization of the orders of matrix elements: using the transformation $g = Uh$, $U = \operatorname{diag}(\Lambda^{r_{1,0}}, \ldots, \Lambda^{r_{q,0}})$, we obtain the system

$$W(\lambda)h = 0\,, \quad \text{where} \quad W(\lambda) = \sum_{k=0}^{p} \lambda^{p-k} W_k\,, \tag{9.52}$$

$W_0 = I$, $W_k = (W_{k,j,l})_{j,l=1}^{q}$, and $\operatorname{ord} W_{k,j,l} = k\tau$ for all j and l. The principal symbols of W_k can be calculated in terms of the corresponding Lopatinskij matrices; in principle, it is possible to calculate the local complete symbols.

Now we define the spectrum of the boundary problem (9.49) as the spectrum of the pencil $W(\lambda)$. (See definitions e.g. in (Agranovich 1990a, Subsect. 6.4b).) We also define the eigenfunctions and the root functions of this boundary problem by the formula $u = \mathcal{P}Uh$, where h stands for eigenfunctions and root functions of the pencil $W(\lambda)$ that correspond to the eigenvalue λ.

The ellipticity with parameter of the boundary problem (9.49) in an angle $\mathcal{L}$ is equivalent to the ellipticity with parameter of the pencil $W(\lambda)$ in $\mathcal{L}$. Therefore, ellipticity with parameter in $\mathcal{L}$ of the boundary problem (9.49) implies the discreteness of its spectrum and the absence of eigenvalues in $\mathcal{L}$ sufficiently far from the origin. If there are rays $\Gamma_1, \dots, \Gamma_r$ of ellipticity with parameter and the angles between the adjacent rays do not exceed $\pi\tau/(n-1)$, then the system of root functions of the pencil $W(\lambda)$ is p-fold complete in the space

$$[H_{s+(p-1)\tau})\Gamma)]^q \times \dots \times [H_s(\Gamma)]^q$$

for any s (cf. (Agranovich 1990a, Subsect. 6.4b)). Moreover, using the standard procedure of linearization in λ (see (Agranovich 1990a, Sect. 4.3)), we can find a rough asymptotics for the moduli of eigenvalues lying in an angle bounded by two directions of ellipticity with parameter, and if there is an isolated direction of ellipticity with parameter, than we can obtain the exact asymptotics of eigenvalues close to it.

e. Now we mention some other spectral problems similar to those discussed above. First, what we have discussed can be extended to matrix boundary problems.

Second, we can consider the case in which only some of the boundary conditions contain λ. In particular, the boundary problem can be of the form

$$\begin{aligned} &Au = 0 \quad \text{on} \quad M\ ; \quad B_j u = 0 \quad (j = 1, \dots, s) \quad \text{and} \\ &B_{j,1}u = \lambda B_{j,0}u \quad (j = s+1, \dots, q) \quad \text{on} \quad \Gamma \end{aligned} \tag{9.53}$$

($1 \le s < q$); such a boundary problem can be investigated by means of tools similar to those used above. We must omit the details.

Furthermore, there are transmission problems (see Sect. 7.3) with spectral parameter entering in the corresponding boundary conditions, and also exterior boundary problems (see Sect. 7.4) with spectral parameter in boundary conditions on the compact boundary of an unbounded domain.

Here we briefly discuss some spectral problems for the Helmholtz equation. They were formulated by the physicists Katsenelenbaum, Sivov, and Voĭtovich: see their book (Voĭtovich et al. 1977) with the Supplement (Agranovich 1977) and (Golubeva 1976). For definiteness, we assume that $n = 3$. We use the notation and formulas of Subsect. 5.4a.

Boundary Problem 9.4.1. Find solutions of the Helmholtz equation (5.51) in $G_+ \cup G_-$ with radiation condition at infinity and the conditions

$$u^+ = u^- \quad \text{and} \quad \lambda[\partial_\nu u^+ - \partial_\nu u^-] - u = 0 \quad \text{on} \quad \Gamma \tag{9.54}$$

where $u = u^{\pm}$ on Γ. Setting

$$\varphi = \partial_\nu u^+ - \partial_\nu u^- \tag{9.55}$$

and using relations (5.55), we obtain the equation

$$T_1\varphi = \lambda\varphi . \tag{9.56}$$

Conversely, if φ is a (smooth) solution of this equation for some λ and we set

$$u(x) = \int_\Gamma \Phi(s-y)\varphi(y)\, dS_y , \tag{9.57}$$

then conditions (9.54) hold and the function $u(x)$ is the solution to the boundary problem 9.4.1.

Recall that T_1 is a polyhomogeneous elliptic pseudodifferential operator of order -1 with negative principal symbol, and that this operator is infinitely close to a selfadjoint operator $\operatorname{Re} T_1$ in the following sense: the order of the operator $\operatorname{Im} T_1$ is $-\infty$ (see (Agranovich 1990a, Example 2.2.3), where this operator is denoted by A). From the results presented in (Agranovich 1990a, Sect. 6.2) it follows that for the eigenvalues of T_1 there is an asymptotic formula with Hörmander's estimate of the remainder, and that very strong assertions on basic properties of the system of root functions of this operator are true. By means of Green's formula it can be verified that

$$\operatorname{Im}(T_1\varphi,\varphi)_\Gamma > 0 \quad \text{if} \quad T_1\varphi \neq 0 ;$$

in particular, T_1 is a dissipative operator. Note that $\operatorname{Ker} T_1$ consists of the values $\partial_\nu u^+$ for nontrivial solutions to the Dirichlet boundary problem

$$\Delta u + k^2 u = 0 \quad \text{in} \quad G_+ , \quad u^+ = 0 \quad \text{on} \quad \Gamma \tag{9.58}$$

if they exist for the given k.

Boundary Problem 9.4.2. Find solutions of the Helmholtz equation (5.51) in G_+ with the condition

$$u^+ = \lambda\partial_\nu u^+ \quad \text{on} \quad \Gamma . \tag{9.59}$$

Boundary Problem 9.4.3. Find solutions of equation (5.51) in G_- with radiation condition at infinity and the condition

$$-u^- = \lambda\partial_\nu u^- \quad \text{on} \quad \Gamma . \tag{9.60}$$

Boundary Problem 9.4.4. Find solutions of equation (5.51) outside Γ with radiation condition at infinity and the conditions

$$\partial_\nu u^+ = \partial_\nu u^- \quad \text{and} \quad \frac{1}{2}[u^+ - u^-] = \lambda\partial_\nu u \quad \text{on} \quad \Gamma , \tag{9.61}$$

where $\partial_\nu u = \partial_\nu u^{\pm}$ on Γ.

We accept two assumptions to simplify the reduction of these boundary problems to equations on Γ.

1°. The interior homogeneous Dirichlet problem (9.58) has no nontrivial solutions for the given k.

2°. The interior homogeneous Neumann problem

$$\Delta u + k^2 u = 0 \quad \text{on} \quad G_+ \,, \quad \partial_\nu u^+ = 0 \quad \text{on} \quad \Gamma \tag{9.62}$$

has no nontrivial solutions for the given k.

Each of these assumptions excludes from consideration a sequence of values of k tending to infinity. Namely, these values are square roots of eigenvalues of the operator $-\Delta$ under the corresponding boundary condition.

A not complicated analysis (see e.g. (Colton and Kress 1983, Chapter 3)) shows that conditions 1° and 2° are equivalent to the invertibility of the operators $I - 2T_2$ and $I + 2T_2$, respectively. Setting, under these conditions,

$$T_- = 2(I - 2T_2)^{-1}T_1 \quad \text{and} \quad T_+ = 2(I + 2T_2)^{-1}T_1 \,, \tag{9.63}$$

we obtain

$$u^+ = T_+\partial_\nu u^+ \,, \quad \text{and} \quad u^- = -T_-\partial_\nu u^- \tag{9.64}$$

(cf. (5.55)); these relations explain the roles of the operators T_- and T_+ in the boundary problems for the Helmholtz equation. Setting, now,

$$\varphi = \partial_\nu u^+ \,, \quad \varphi = \partial_\nu u^- \,, \quad \text{and} \quad \varphi = \partial_\nu u \tag{9.65}$$

in the cases of boundary problems 9.4.2, 9.4.3, and 9.4.4, respectively, and

$$T = \frac{1}{2}(T_- + T_+) \,, \tag{9.66}$$

we easily reduce these boundary problems to the equations

$$T_+\varphi = \lambda\varphi \,, \quad T_-\varphi = \lambda\varphi \,, \quad \text{and} \quad T\varphi = \lambda\varphi \,. \tag{9.67}$$

As we noted in Subsect. 5.4a, T_2 is a polyhomogeneous pseudodifferential operator of order not greater than -1. Therefore, T_+, T_-, and T are polyhomogeneous elliptic pseudodifferential operators with principal symbol equal to the principal symbol of T_1 multiplied by 2. The first operator, T_+, is self-adjoint (which can be easily verified by means of Green's formula), while the second and the third possess the symmetry property, $T_-^* = \overline{T}_-$, $T^* = \overline{T}$, are dissipative, and are infinitely close to the pseudodifferential operators $\operatorname{Re} T_-$ and $\operatorname{Re} T$, respectively, i.e. differ from them by terms of order $-\infty$. As in the case of T_1, it is possible to obtain the asymptotics of eigenvalues, and for the systems of root functions of T_- and T very strong assertions on basic properties are true.

Similar assertions hold for two-dimensional analogs of the boundary problems 9.4.1–9.4.4. In this case it is possible to find some first terms of the complete asymptotic expansion for the eigenvalue λ_n in powers of $1/n$. See Example 2.4.8 in (Agranovich 1990a) and also formulas (6.98)–(6.100) there.

Further details and some generalizations can be found in (Agranovich 1977).

We also mention that it is possible to consider some spectral problems for the stationary Maxwell system

$$\operatorname{rot} H - ikE = 0\,, \quad \operatorname{rot} E + ikH = 0$$

with spectral parameter in the boundary conditions or transmission conditions, and the radiation conditions at infinity. The eigenvalues λ_j of these boundary problems accumulate at zero and ∞, and the expressions $\lambda_j - \lambda_j^{-1}$ are eigenvalues of some elliptic pseudodifferential operators on Γ close to self-adjoint ones. See (Agranovich and Golubeva 1976) and (Agranovich 1977, §40).

In the recent paper (Kozhevnikov and Yakubov 1995) spectral boundary problems of the form (9.53) are considered without the assumption that the spectral parameter λ has a definite weight with respect to differentiation. The results relate to completeness and Abel summability. The investigation is based on the results for systems of pseudodifferential operators on a closed manifold; these results were obtained by the first author and are presented in the Supplement to this paper. Here for a system of pseudodifferential operators with "elliptic principal minors" and parameter that has no definite weight a notion of ellipticity with parameter is introduced, an it is shown that this property keeps under the procedure of "almost block diagonalization" of the system. (Cf. (Kozhevnikov 1973)).

9.5. Boundary Problems with Spectral Parameter in Equation and Boundary Conditions. Consider the scalar boundary problem

$$A(\lambda)u = 0 \quad \text{on} \quad M_+\,, \quad B_j(\lambda)u = 0 \quad \text{on} \quad \Gamma \quad (j = 1, \ldots, q)\,, \tag{9.68}$$

where $A(\lambda)$ and $B_j(\lambda)$ have the same form as in (3.2):

$$A(\lambda) = \sum_{0 \le l\tau \le m} \lambda^l A_{m-l\tau} \quad \text{and} \quad B_j(\lambda) = \sum_{0 \le l\tau \le r_j} \lambda^l B_{j,r_j-l\tau}\,. \tag{9.69}$$

Here A_s is a differential operator on M of order s and $B_{j,s}$ are boundary operators of orders s. Additionally, we assume that $r_j < \tau p = 2q = m$. The substitution

$$U = (u, \lambda u, \ldots, \lambda^{p-1}u)' \tag{9.70}$$

leads to the matrix boundary problem

$$(A - \lambda I)U = 0 \quad \text{on} \quad M\,, \quad B_j U = 0 \quad \text{on} \quad \Gamma \quad (j = 1, \ldots, q)\,, \tag{9.71}$$

which has the same structure as in (6.37), with $l_j = -\tau(p - j - 1)$ and $m_k = \tau(p - k)$; hence, $l_k + m_k \equiv \tau$, and the spectral parameter can be considered as having weight τ. The results discussed in Subsect. 9.3c can be applied to the boundary problem (9.71). First, we can apply Theorem 9.3.7 on the completeness of the root functions, and we obtain p-fold completeness for the boundary problem (9.68). Namely, let the boundary problem (9.68) be elliptic with parameter along the rays $\Gamma_1, \ldots, \Gamma_r$ with the angles between the adjacent rays not greater than $\pi\tau/n$. Then, for the root functions of (9.68), p-fold completeness holds in a subspace of

$$H_{s+\tau(p-1)}(M) \times \ldots \times H_s(M) \tag{9.72}$$

($s \geq 0$) defined by certain boundary conditions. In terms of the boundary problem (9.71), these boundary conditions have the form $B_j A^k U = 0$ on Γ, where k are nonnegative integers for which $B_j A^k U|_\Gamma$ makes sense in (9.72). As a corollary, p-fold completeness holds in $[H_0(M)]^p$.

Second, for the distribution function $N(\lambda)$ of the moduli of eigenvalues of the boundary problem (9.68) in the angle $\mathcal{L}$ between two rays of ellipticity with parameter, we obtain the relation $N(\lambda) \asymp \lambda^{n/\tau}$ if the corresponding number d_0 (see Sect. 9.3) is different from zero. If, in addition, $\mathcal{L}$ contains only one direction without ellipticity with parameter, so that we can take the sides of $\mathcal{L}$ arbitrarily close to this direction, then $N(\lambda) \sim d_0 \lambda^{n/\tau}$.

These results can be generalized to matrix boundary problems of the form (9.68) that are elliptic in the sense of Agmon, Douglis, and Nirenberg.

9.6. Spectral Properties of General Realizations. Here we present some information about spectral properties of the realizations defined in Sect. 7.7. As in Sect. 7.7, we assume that the principal symbol of $A(x, D)$ is positive.

a. From Assertion 6 in Sect. 7.7 with $p = \infty$ we see that the realization $\mathbb{A}_C$ has a discrete spectrum if and only if the operator C is compact. We assume that this condition holds and that $C = C^*$, so that $\mathbb{A}_C$ is a selfadjoint operator: see Assertion 3 in Sect. 7.7. As was noted in (Birman and Solomyak 1977b), the distribution functions $N_+(\lambda)$ and $N_-(\lambda)$ for positive and negative eigenvalues of $\mathbb{A}_C$ can generally have arbitrary power asymptotics

$$N_+(\lambda) \sim C_+ \lambda^{\alpha_+} \quad \text{and} \quad N_-(\lambda) \sim C_- \lambda^{\alpha_-} \qquad (\lambda \to \infty) \tag{9.73}$$

with $C_\pm > 0$ and $\alpha_\pm \geq n/m$; a non-power growth of these functions is also possible.

b. We set

$$n(\lambda) = \operatorname{card}\{k : |\lambda_k(C)| \geq \lambda^{-1}\}\,. \tag{9.74}$$

Theorem 9.6.1. (Mikhailets 1982, 1990.) *The behavior of $N(\lambda)$ is connected with the behavior of $n(\lambda)$ as follows:*

$$n(\lambda) = O(\lambda^\delta) \iff N(\lambda) = \begin{cases} O(\lambda^\delta) & \text{if } \delta \geq \frac{n}{m}, \\ d_0\lambda^{n/m} + O(\lambda^\delta) & \text{if } \delta \in (\frac{n-1}{m}, \frac{n}{m}), \end{cases}$$

$$n(\lambda) = o(\lambda^\delta) \iff N(\lambda) = \begin{cases} o(\lambda^\delta) & \text{if } \delta > \frac{n}{m}, \\ d_0\lambda^{n/m} + o(\lambda^{n/m}) & \text{if } \delta = \frac{n}{m}, \\ d_0\lambda^{n/m} + o(\lambda^\delta) & \text{if } \delta \in (\frac{n-1}{m}, \frac{n}{m}). \end{cases} \tag{9.75}$$

Here d_0 is the quantity (9.5).

If, in addition, $\mathbb{A}_C$ is not bounded from below, then

$$N_-(\lambda) = \begin{cases} O(\lambda^\delta) & \text{if } n(\lambda) = O(\lambda^\delta),\ \delta \geq 0, \\ o(\lambda^\delta) & \text{if } n(\lambda) = o(\lambda^\delta),\ \delta > 0, \end{cases} \tag{9.76}$$

and $N_+(\lambda)$ satisfies relations similar to (9.75) for $N(\lambda)$.

c. Denote by $\langle A, H_0, H_s\rangle$ the class of all selfadjoint realizations $\mathbb{A}_C$ with $D(\mathbb{A}_C) \subset H_s(G)$, $s \in [0, m]$. This class expands when s decreases. Realizations from a given class have similar spectral properties. If $\mathbb{A}_C \in \langle A, H_0, H_s\rangle$, then $N_\pm(\lambda) = O(\lambda^{n/s})$. These relations can be sharpened. In Subsect. 9.1a we mentioned results of Agmon and Kannai, Agmon, and Brüning. These results relate to the case $s = m = 2q$ if $m > n$, while if $m < n$, then it is assumed that $(\mathbb{A}_C)^k \in \langle A^k, H_0, H_{mk}\rangle$ with some k such that $mk > n$. Now we present a theorem from (Mikhailets 1982, 1989).

Theorem 9.6.2. *Let $\mathbb{A}_C \in \langle A, H_0, H_s\rangle$. Then*

$$N_+(\lambda) = \begin{cases} O(\lambda^{\frac{n-1}{s}}) & \text{if } s \in (0, s_0], \\ d_0\lambda^{\frac{n}{m}} + O(\lambda^{\frac{n-1}{s}}) & \text{if } s \in (s_0, m], \end{cases} \tag{9.77}$$

where $s_0 = m(n-1)/n$, while

$$N_-(\lambda) = O(\lambda^{\frac{n-1}{s}})\,. \tag{9.78}$$

In general, in (9.77) it is impossible to reduce the degree under the O-sign for any $s \in (0, m]$. The same is true for (9.78) if $s = m$. In the latter case in (9.77) it is impossible to replace O by o. This follows from Theorem 9.1.2 for differential boundary problems and a result of Métivier (Métivier 1983, Theorem 6).

For positive selfadjoint realizations the condition $D(\mathbb{A}_C) \subset H_s(G)$ can be replaced by $D((\mathbb{A}_C)^{1/2}) \subset H_{s/2}(G)$. Formulas (9.77) remain true in this case (see Mikhailets 1982) and (Boĭmatov and Kostyuchenko 1988).

The results presented here show that to a large extent the distribution of eigenvalues of a selfadjoint elliptic operator is defined by the smoothness of functions from the domain of the operator and are not affected by the concrete form of (homogeneous) boundary conditions.

These results can be generalized to the case in which the operator $A(x, D)$ is not formally selfadjoint, to the case of matrix elliptic operators, and, finally, to the case of spaces L_p instead of L_2.

9.7. Boundary Problems Elliptic in a Subspace.

As usual, here we have in mind variational spectral problems in which the functions from the domain of the given quadratic form are subjected to some additional differential equations, the "constraints." It is convenient to use the term *variational triple* (H, V, a). Here H is the basic Hilbert space, V is a dense linear manifold in H, and $a = a[u]$ is a real-valued quadratic form that is semibounded from below, with domain V. The form $a[u]$ defines the sesquilinear form

$$a[u,v] = \{a[u+v] - a[u-v] + ia[u+iv] - ia[u-iv]\}/4 \,.$$

In its turn, $a[u,v]$ defines an operator in H (cf. Sect. 7.2); this operator is selfadjoint and semibounded from below. Its spectrum is called the *spectrum of the variational triple* (H, V, a). In the same sense, one can speak about the eigenvalue distribution function of the variational triple, etc.[17]

Below for simplicity we consider boundary problems in a bounded domain $G \subset \mathbb{R}^n$, and we assume that the boundary $\Gamma = \partial G$ is C^∞ and that all the coefficients in equations and quadratic forms belong to $C^\infty(\overline{G})$.

a. We consider two typical examples of variational problems on subspaces.

Example 9.7.1. (The Stokes boundary problem.) Let

$$\begin{aligned} H &= \{u = (u_1, \ldots, u_n)' \in [H_0(G)]^n : \operatorname{div} u = 0\} \,, \\ V &= \{u \in (\overset{\circ}{H}_1(G))^n : \operatorname{div} u = 0\} \,, \\ a[u] &= \sum_{k=1}^{n} \int_G |\nabla u_k|^2 \, dx \,. \end{aligned}$$

The spectrum of this variational triple corresponds to the Stokes problem

$$-\Delta u + \operatorname{grad} p = \lambda u \,, \quad \operatorname{div} u = 0 \quad \text{in} \quad G \,, \quad u = 0 \quad \text{on} \quad \Gamma \tag{9.79}$$

(cf. Subsect. 6.2c). The function $p \in H_1(G)$ that is absent in the original setting of the problem appears in (9.79) in view of the Euler–Lagrange equation. Indeed, the difference $-\Delta u - \lambda u$ must be orthogonal in $[H_0(G)]^n$ to the linear subset $\{u : \operatorname{div} u = 0, u|_\Gamma = 0\}$. (If we remove the constraint $\operatorname{div} u = 0$ from the original setting, then the problem splits into n scalar spectral problems for the operator $-\Delta_D$.)

Example 9.7.2. Let

$$\begin{aligned} H &= \left\{u \in H_0(G) : \Delta u = 0, \int_G u \, dx = 0\right\} \,, \\ V &= \left\{u \in H_1(G) : \Delta u = 0, \int_G u \, dx = 0\right\} \,, \end{aligned}$$

[17] For the problems of Sect. 7.1 the variational triple can be defined if the form $a[u,v]$ is Hermitian: it is $(H_0(G), \mathcal{H}, a[u,u])$.

$$a[u] = \int_G |\nabla u|^2 \, dx \, .$$

The spectrum of this variational triple consists of the numbers λ_k^{-1}, where $\{\lambda_k\}$ is the spectrum of the operator $\Delta_D^{-1} - \Delta_N^{-1}$, the difference of inverse Dirichlet and Neumann Laplacians (Birman 1956). (If we remove the constraint $\Delta u = 0$, then we get the operator $-\Delta_N^{-1}$.)

There is an essential difference between Examples 9.7.1 and 9.7.2. In the first of them the number 1 of the constraints is less than the "vectorial dimension" n of the boundary problem, while in the second example the corresponding numbers are equal. Such problems are called problems with an *incomplete system of constraints* or a *complete system of constraints*, respectively. Below we will see that this difference is reflected in formulas for spectral asymptotics.

b. The main result for boundary problems of the first type was obtained in (Métivier 1978). Let $p > r \geq 1$, $q \geq 1$, and $0 \leq l_j \leq q$, $j = 1, \dots, r$. Introduce the spaces

$$W(G) = [H_q(G)]^p \quad \text{and} \quad W_0(G) = [\overset{\circ}{H}_q(G)]^p \, .$$

Let C_j be differential operators in $\overline{G}$ of orders l_j, $j = 1, \dots, r$, that act on vector-valued functions $u = (u_1, \dots, u_p)' \in W(G)$:

$$C_j u = C_j(x, D)u = \sum_{k=1}^{p} \sum_{|\alpha| \leq l_j} c_{j,k}^{\alpha}(x) D^\alpha u_k \, . \tag{9.80}$$

Denote by c_0 the matrix principal symbol of the operator $C = (C_1, \dots, C_r)'$,

$$c_0(x, \xi) = \left\{ \sum_{|\alpha| = l_j} c_{jk}^{\alpha}(x) \xi^\alpha \right\} \qquad (1 \leq j \leq r, \ 1 \leq k \leq p) \, ,$$

and also the corresponding linear mapping from $\mathbb{C}^p$ to $\mathbb{C}^r$. Assume that on $W(G)$ a quadratic form

$$a[u] = \sum_{1 \leq j,k \leq p} \sum_{|\alpha|, |\beta| \leq q} \int_G a_{j,k}^{\alpha,\beta}(x) D^\alpha u_j \overline{D^\beta u_k} \, dx \tag{9.81}$$

is given, with $a_{j,k}^{\alpha,\beta} = \overline{a_{k,j}^{\beta,\alpha}}$. Under this condition the form (9.81) is real-valued, and the principal symbol

$$a_0(x, \xi) = \left\{ \sum_{|\alpha|, |\beta| = q} a_{j,k}^{\alpha,\beta}(x) \xi^{\alpha+\beta} \right\} \qquad (1 \leq j, k \leq p) \tag{9.82}$$

is an Hermitian matrix for all $x \in \overline{G}$, $\xi \in \mathbb{R}^n$.

Let $\mathcal{H} \supset W_0(G)$ be a (closed) linear subspace in $W(G)$; its choice corresponds to the assignment of the "stable" boundary conditions on Γ (cf. Sect. 7.1). Consider the triple

$$(H, V, a): \quad V = \mathcal{H} \cap \operatorname{Ker} C\,, \quad a[u] \text{ is the form (9.81)}\,; \quad H \text{ is the closure of } V \text{ in } [H_0(G)]^p\,. \tag{9.83}$$

Introduce the family of finite-dimensional subspaces $V(x,\xi) = \operatorname{Ker} c_0(x,\xi)$, where $(x,\xi) \in \overline{G} \times \dot{\mathbb{R}}_n$, $\dot{\mathbb{R}}_n = \mathbb{R}_n \setminus \{0\}$. On $V(x,\xi)$ consider the quadratic form

$$a(x,\xi)[h] = a_0(x,\xi)h \cdot \overline{h}\,, \qquad h \in V(x,\xi)\,, \tag{9.84}$$

where $g \cdot \overline{h}$ is the standard scalar product in $\mathbb{C}^p$.

Denote by $n(\lambda, x, \xi)$ the eigenvalue distribution function for the operator in $V(x,\xi)$ generated by the form (9.84), and set $\dot{\mathbb{C}}^n = \mathbb{C}^n \setminus \{0\}$.

Theorem 9.7.3. (Métivier 1978.) *Assume that the following conditions hold:*

$$\operatorname{rank} c_0(x,\xi) = r \quad \text{if } (x,\xi) \in \overline{G} \times \dot{\mathbb{R}}_n\,, \tag{9.85}$$

$$\operatorname{rank} c_0(x,\xi) = r \quad \text{if } (x,\xi) \in \Gamma \times \dot{\mathbb{C}}_n\,, \tag{9.86}$$

$$a(x,\xi)[h] > 0 \quad \text{if } (x,\xi) \in \overline{G} \times \dot{\mathbb{R}}_n,\ h \in V(x,\xi) \setminus \{0\}\,. \tag{9.87}$$

Then (9.83) *is a variational triple, and for the corresponding eigenvalue distribution function the following formula holds:*

$$N(\lambda) = \frac{\lambda^{n/m}}{(2\pi)^n} \int_{G \times \mathbb{R}_n} n(1, x, \xi)\, dx\, d\xi + o(\lambda^{n/m})\,, \qquad \lambda \to \infty\,. \tag{9.88}$$

In Example 9.7.1, $p = n$, $r = 1$, $q = 1$, $l = 1$, $\mathcal{H} = W_0(G) = (\overset{\circ}{H}_1(G))^n$, and

$$a_0(x,\xi)h \cdot \overline{h} = |\xi|^2 |h|^2\,, \qquad c_0(\xi) = (\xi_1, \dots, \xi_n)\,.$$

Obviously, the conditions of Theorem 9.7.3 are satisfied. According to formula (9.88),

$$N(\lambda) \sim (n-1)(2\pi)^{-n} V_n \operatorname{vol} G \cdot \lambda^{n/2}\,,$$

where V_n is the volume of the unit ball in $\mathbb{R}_n$.

We make three remarks concerning Theorem 9.7.3.

1. Condition (9.86) can be replaced by the assumption of closedness of the linear manifold $C(x,D)\mathcal{H}$ in $\prod_{j=1}^{r} H_{q-l_j}(G)$. Condition (9.86) ensures the closedness of it for any $\mathcal{H}$ such that $W_0(G) \subset \mathcal{H} \subset W(G)$.

2. The differential operator $A(x,D)$ formally corresponding to the form (9.81) need not be elliptic. Condition (9.87) means ellipticity of the symbol (9.82) only on the family of the subspaces $V(x,\xi)$. It is this circumstance that is reflected in the term "ellipticity in a subspace."

3. Assume, however, that the operator $A(x,D)$ is elliptic and semibounded from below on the whole of $\mathcal{H}$. Then the principal term of the asymptotics

of its spectrum can be calculated by the general formula (see Sect. 9.1a). We see that the inclusion of a differential constraint $Cu = 0$ leads only to the alteration of the coefficient in that principal term. The order $\lambda^{n/m}$ remains the same.

c. The setting of the problem and Theorem 9.7.3 can evidently be carried over to the case of operators on a smooth compact manifold M with boundary and given positive smooth density. In statements, $\overline{G} \times \dot{\mathbb{R}}_n$ should be replaced by $T^*M\setminus 0$, and in condition (9.86) the restriction of the bundle $T^*M\setminus 0$ to the boundary $\Gamma = \partial M$ appears. The problem remains meaningful for manifolds without boundary; in this case condition (9.86) is omitted.

Now we briefly discuss a useful variant of Theorem 9.7.3 for the case $\partial M = \varnothing$. Let $\mathcal{H} \subset [H_q(M)]^p$ be a (closed) subspace with the following property: in $[H_q(M)]^p$ there exists a pseudodifferential projector P (i.e. $P^2 = P$) such that the range $R(P)$ coincides with $\mathcal{H}$. It is obvious that the principal symbol $p_0(x,\xi)$ of P is a projector in $\mathbb{C}^p$. Theorem 9.7.3 can be carried over to this case without essential changes, and here the subspaces $R(p_0(x,\xi))$ play the role of $V(x,\xi)$.

Note that each subspace $\operatorname{Ker} C(x,D)$, where $C(x,D)$ is a differential or pseudodifferential operator satisfying a condition of the form (9.85), admits a pseudodifferential projector. The converse is not true.

The results mentioned here were obtained in (Birman and Solomyak 1982a,b), where extensions to the case of operators acting in sections of bundles were also presented.

d. For boundary problems with a complete system of constraints the statement of general results is rather complicated. We restrict ourselves to a particular but typical case.

Assume that the quadratic form

$$a[u] = \sum_{|\alpha|=|\beta|\leq q} \int_G a_{\alpha\beta}(x) D^\alpha u \, \overline{D^\beta u} \, dx \, , \qquad a_{\alpha\beta} = \overline{a_{\beta\alpha}} \, ,$$

acting on scalar functions $u \in H_q(G)$ is given. This is a form of the type (9.81) with $p = 1$. In addition, assume that an elliptic differential operator $C = C(x,D)$ of order $2l$ is given in $\overline{G}$. Set

$$H_s[G;C] = \{u \in H_s(G) : Cu = 0\}, \qquad s \geq 0$$

(if $s < 2l$, then the equation $Cu = 0$ is satisfied in the sense of distributions). Clearly $H_s[G;C]$ is a (closed) subspace of $H_s(G)$.

We discuss the conditions for

$$(H_0[G;C], H_q[G;C], a) \tag{9.89}$$

to be a variational triple, and we present the formula for the principal term of its spectral asymptotics.

Let $\nu(x)$ be the unit vector of the inner normal to Γ at the point $x \in \Gamma$. The set

$$\{\eta \in \mathbb{R}_n : \nu(x) \cdot \eta = 0\}$$

is naturally identified with $T_x^*\Gamma$. Each vector $\xi \in \mathbb{R}_n$ has a unique representation of the form

$$\xi = \eta + \tau\nu(x) , \qquad \tau \in \mathbb{R}, \quad \eta \in T_x^*\Gamma .$$

With each point $(x, \eta) \in T_x^*\Gamma\backslash 0$ we associate the ordinary differential equation

$$c_0(x, \eta + \nu(x)D_t)v(t) = 0 , \tag{9.90}$$

where c_0 is the principal symbol of $C(x, D)$. Let $\mathcal{F}(x,\eta)$ be the space of solutions to the equation (9.90) with $|v(t)| \to 0$ as $t \to +\infty$. Assume that C is properly elliptic, i.e. $\dim \mathcal{F}(x, \eta) = l$. We consider $\mathcal{F}(x, \eta)$ as an l-dimensional subspace in $H_0(\mathbb{R}_+)$. On $\mathcal{F}(x, \eta)$ we consider the quadratic form

$$a(x,\eta)[v] = \sum_{|\alpha|=|\beta|=q} a_{\alpha\beta}(x) \int_{\mathbb{R}_+} (\eta + \nu(x)D_t)^\alpha v(t)\overline{(\eta + \nu(x)D_t)^\beta v(t)}\, dt . \tag{9.91}$$

Obviously the form (9.91) is real-valued. Let $n(\lambda, x, \eta)$ be the eigenvalue distribution function for the selfadjoint operator defined by this form in the space $\mathcal{F}(x, \eta)$.

Theorem 9.7.4. (Birman and Solomyak 1979b, 1982c.) 1) *Assume that for any* $(x, \eta) \in T^*\Gamma \setminus 0$ *the form* (9.91) *is positive. Then* (9.89) *is a variational triple with a discrete spectrum. For its eigenvalue distribution function* $N(\lambda)$ *the following relation holds*:

$$N(\lambda) = \lambda^{\frac{n-1}{m}}(2\pi)^{1-n} \int_{T^*\Gamma} n(1; x, \eta)\, dx\, d\eta + o(\lambda^{\frac{n-1}{m}}) , \qquad \lambda \to \infty . \tag{9.92}$$

2) *Conversely, if* (9.89) *is a variational triple, then the form* (9.91) *is positive for any* $(x, \eta) \in T^*\Gamma \setminus 0$.

For Example 9.7.2 formula (9.92) gives

$$N(\lambda) \sim (2\pi)^{1-n} V_{n-1} \operatorname{vol}(\Gamma)(\lambda/2)^{\frac{n-1}{2}}$$

(the additional condition $\int_G u\, dx = 0$ in Example 9.7.2 does not affect the spectral asymptotics).

We see that in comparison with problems without differential constraints or with an incomplete system of constraints, here an essential distinction arises: the order of $N(\lambda)$ changes, and the asymptotic coefficient is expressed in the form of an integral along $T^*\Gamma$.

The basic scheme of the proof of Theorem 9.7.4 is rather clear. The solutions of the elliptic equation $Cu = 0$ are expressed in terms of the boundary values $\gamma_0 u, \ldots, \gamma_{l-1}u$ by means of the Poisson operators (see Sect. 5.2). Substituting these expressions in $\int_G |u|^2\,dx$ and $a[u]$ and omitting the lower-order terms, we get an eigenvalue problem for a variational triple in the space of l-component vector-valued functions on Γ. The operator corresponding to this triple is pseudodifferential, and under the conditions of the theorem it is elliptic. Using the well-known asymptotic formula for such operators (see e.g. (Agranovich 1990, Sect. 6.1)), we obtain (9.92).

The class of problems that admit investigation with the same tools can be considerably extended. In particular, integrals along the boundary can be included in the form a, and the functions from $H_q[G;C]$ can be subjected to some boundary conditions (the number of them must be less than l). In contrast to problems with incomplete systems of constraints, all this does affect the principal term of the asymptotics of $N(\lambda)$. The results are generalized to the case of operators on vector-valued functions and on sections of vector bundles. In these more general situations the results briefly discussed in Subsect. 9.7c are used; see (Birman and Solomyak 1982c).

9.8. Boundary Problems in Nonsmooth Domains.

a. At first we consider operators $A = A_D, A_N$, and A_V that correspond to variational problems in a bounded domain $G \subset \mathbb{R}^n$ (see Sect. 7.2). In addition, we first assume that these operators are selfadjoint.

In (Birman and Solomyak 1972, 1973) formula (9.4) for $N_A(\lambda)$ is obtained in the cases 1) of $A = A_D$ and any bounded domain and 2) of $A = A_N$ and a Lipschitz domain by means of the variational method. The same method is used in (Métivier 1977) to obtain the formula

$$N_A(\lambda) = d_0\lambda^{n/m} + O(\lambda^{(n-\frac{1}{2})/m}) \tag{9.93}$$

for $A = A_D$, A_N, and A_V in a Lipschitz domain in the general case and any $O(\lambda^{(n-1)/m}\log\lambda)$ in the case of constant coefficients $a_{\alpha,\beta}$ in the principal symbol.[18]

There are some other papers in which the estimate of the remainder is strengthened under additional assumptions; see, in particular, the paper (Boĭmatov and Kostyuchenko 1988) about the Dirichlet problem. In (Vassiliev 1986b) the boundary is assumed to be piecewise-smooth (i.e. G is a "curvilinear polyhedron"), and formula (9.6) is established. See also references in Sect. 9.6.

b. The operator A defined by the form $a[u,v]$ is selfadjoint if and only if this form is Hermitian. We now do not assume this and introduce the Hermitian forms

$$b[u,v] = \frac{1}{2}\{a[u,v] + \overline{a[v,u]}\} \quad \text{and} \quad c[u,v] = \frac{1}{2i}\{a[u,v] - \overline{a[v,u]}\} \tag{9.94}$$

[18] Birman and Solomyak consider more general problems, roughly speaking, for equations of the form $Au = \lambda A_1 u$, where $\operatorname{ord} A_1 < \operatorname{ord} A$, and nonsmooth coefficients. Métivier also considers nonsmooth coefficients. In addition, along with Lipschitz domains, they consider somewhat more general domains.

on $\mathcal{H}$ called the *real part* and the *imaginary part* of $a[u,v]$. The form $b[u,v]$ is strongly elliptic simultaneously with $a[u,v]$. Let B be the operator defined by the form $b[u,v]$. If the boundary Γ is sufficiently smooth, then $D(A) = D(B)$, and $A = B + iC$, where C is the operator corresponding by the form $c[u,v]$:

$$(Cu, v) = c[u,v] \quad \text{for} \quad u \in D(A),\ v \in \mathcal{H}\,.$$

In this case we can consider the operator A as a weak perturbation of the selfadjoint operator B, assuming e.g. that the operator CB^{-q} is bounded for some $q < 1$ (see (Agranovich 1990a, Sect. 6.2) and Sect. 9.2 in the present paper). However, in general $D(A) \neq D(B)$ if Γ is nonsmooth, and then we cannot use abstract theorems on spectral properties of weak perturbations of selfadjoint operators.

Instead, we can compare the forms $b[u,v]$ and $c[u,v]$. In (Markus and Matsaev 1981), in particular, the following abstract condition is proposed:

$$|c[u,v]| \leq \gamma_1 b[u,v]^q \|u\|_H^{2-2q} \qquad (u \in \mathcal{H})\,, \tag{9.95}$$

where $\gamma_1 > 0$, $0 \leq q < 1$, and $b[u,v]$ is assumed to be positive:

$$\varepsilon \|u\|_{\mathcal{H}}^2 \leq b[u,u] = \operatorname{Re} a[u,u] \qquad (u \in \mathcal{H}) \tag{9.96}$$

for some $\varepsilon > 0$. Under these assumptions, Markus and Matsaev prove that the eigenvalues of A lie in a domain of the form

$$\{\lambda :\ \operatorname{Re}\lambda > 0,\ |\operatorname{Im}\lambda| \leq \gamma_2 (\operatorname{Re}\lambda)^q\}$$

with some positive γ_2, and the asymptotics of the distribution function $N_A(\lambda)$ of the real parts of the eigenvalues of A coincides with that of $N_B(\lambda)$. We do not present the exact formulation of this abstract theorem but indicate a corollary for our differential forms: if the maximal order s of members in $c[u,v]$ is less than m, and

$$N_B(\lambda) = d_0 \lambda^{n/m} + o(\lambda^{n/m}) \quad \text{or} \quad N_B(\lambda) = d_0 \lambda^{n/m} + O(\lambda^{(n-\theta)/m})$$

with some $\theta \in (0,1]$, then the same is true for $N_A(\lambda)$.

In (Agranovich 1994b) an abstract theorem on basic properties of root functions is proved. Let conditions (9.96) and (9.95) be satisfied and

$$\lambda_j(B) = c_0 j^p + o(j^p)\,. \tag{9.97}$$

Then the root functions of A form a complete set in the corresponding Hilbert space H. Moreover, there exists a system of finite-dimensional subspaces in H that are invariant with respect to A and form a Bari basis, Riesz basis, or Abel basis if $p(1-q) > 1$, $p(1-q) = 1$, or $p(1-q) < 1$, respectively. In the latter case, the order of the corresponding Abel's method is $p^{-1} - (1-q) + \varepsilon$, where ε is an arbitrary sufficiently small positive number. If, in (9.97), we have an estimate of the remainder of the form $O(j^{p-r})$ with $r > 0$, then, additionally, we can control the positions of brackets in the Fourier series with respect to the root functions of A (as in (Agranovich 1990a, Sect. 6.2)).

To apply this result to our differential forms and $H = H_0(G)$, we can take the maximal order s of members in $c[u, v]$ into account. The number $p(1-q)$ is equal to $(m-s)/n$.

c. Now we consider the case in which condition (9.95) is not satisfied, so that the form $a[u, v]$ is "far" from the Hermitian form $b[u, v]$. Consider first the operator A_D in an arbitrary bounded domain G. Assume that condition (7.18) for the principal symbol of the form $a[u, v]$ is satisfied. Then the following assertion is true: if $\theta < \pi/2$ and $\theta < \pi m/n$, then the set of root functions of A is complete in $H_0(G)$ and there is an Abel's basis with brackets in $H_0(G)$ consisting of the root functions, of order $n/m + \varepsilon$, where ε is positive and sufficiently small. (In the matrix case, a condition of the form (7.18) is imposed on the eigenvalues of the principal symbol $a_0(x, \xi)$.) The proof uses estimate (7.19) of the resolvent $R_A(\lambda)$ and the estimate $|s_j(R_A)(\lambda)| \leq Cj^{-m/n}$ for its s-numbers; the latter follows from the asymptotics of $N_B(\lambda)$. Similar results hold for A_N and A_V in a Lipschitz domain, but in this case inequality (7.18) is replaced by a similar inequality for the generalized symbol $a(x, \zeta)$. See e.g. (Agranovich 1994a).

These formulations are similar to those in (Agranovich 1990a, Sect. 6.3) but contain an additional assumption $\theta < \pi/2$ (for strong ellipticity or coerciveness). Some steps toward the liberation from this condition are made in (Boĭmatov 1993, 1995), where elliptic equations with possible degeneracy are considered; see also references in these papers.

d. The results outlined above in this section relate to boundary problems with spectral parameter in elliptic equations in G. It remains to discuss some questions related to boundary problems with spectral parameter in boundary conditions.

In (Agranovich and Amosov 1996) integral operators of potential type

$$T\varphi(x) = \int_\Gamma \alpha(x)\, K(x, y)\, \beta(x)\varphi(y)\, dS_y \qquad (x \in \Gamma) \tag{9.98}$$

are considered. Here Γ is a Lipschitz $(n-1)$-dimensional compact closed surface in $\mathbb{R}^n$, and $\alpha(x)$, $\beta(x)$ are measurable functions on Γ with $|\alpha(x)| \leq 1$ and $|\beta(x)| \leq 1$. The function $K(x, y)$ can be the restriction to $\Gamma \times \Gamma$ of the kernel of a polyhomogeneous pseudodifferential operator of order $-m-1$ with $m > 0$ in $\mathbb{R}^n$ or in a neighborhood $O(\Gamma)$ of Γ: see the description of such kernels in (Seeley 1965) or (Agranovich 1990a, Sect. 1.6).[19] In the case of a smooth Γ, A is a pseudodifferential operator of negative order $-m$ on S.

[19] Actually, the function $K(x, y)$ can be more general, in particular, it can have a finite smoothness.

In particular, $K(x,y)$ can be a fundamental solution of an elliptic equation of order m in $\mathbb{R}^n$, and then (9.98) is an operator of the single layer potential type. However, in general no ellipticity is assumed.

For s-numbers of T, the following estimate is obtained:

$$s_j(T) \leq Cd(\alpha,\beta)j^{-m/(n-1)} \,. \tag{9.99}$$

(Here the degree of j is the same as in the case of a smooth Γ.) The quantity $d(\alpha,\beta)$ is expressed in terms of $(n-1)$-dimensional Lebesgue measures $m(\operatorname{supp}\alpha)$ and $m(\operatorname{supp}\beta)$ on Γ and is small if at least one of these measures is small.

Earlier, estimates of s-numbers for very general integral operators in domains of $\mathbb{R}^n$ were obtained by Birman and Solomyak (see their survey (Birman and Solomyak 1977a)) and Kostometov (see (Kostometov 1974)). They proved and used the existence of appropriate finite-dimensional approximations of the given operator that have piecewise-polynomial kernels. A smoothness of the given kernel is necessary in this approach for large m.

Clearly the kernel $K(x,y)$ loses its smoothness when we restrict it to $\Gamma\times\Gamma$. To avoid this difficulty, Agranovich and Amosov construct approximations $K_N(x,y)$ of $K(x,y)$ in $O(\Gamma)\times O(\Gamma)$ and then restrict both kernels, K and K_N, to $\Gamma\times\Gamma$. In addition, the piecewise-polynomial approximations $K_N(x,y)$ of $K(x,y)$ are constructed explicitly and very simply, by means of Taylor expansions for $K(x,y)$ as a function of x or y.

Let us call the Lipschitz surface Γ *almost smooth* if it is C^∞ outside a closed "singular subset" $\Gamma_1\subset\Gamma$ with zero surface Lebesgue measure $m(\Gamma_1)$. Assume that the operator T is selfadjoint: $K(x,y)=\overline{K(y,x)}$ and $\alpha(x)=\beta(x)$ on Γ. In this case, Agranovich and Amosov obtain an asymptotic formula for the eigenvalues $\lambda_j(T)$ that generalizes the corresponding result for pseudodifferential operators of negative order in a domain of Euclidean space obtained in (Birman and Solomyak 1977c, 1979a). In the proof, this result is combined with (9.99). A generalization to the matrix case (with eigenvalues of both signs) is also obtained.

f. Possible applications of these results to boundary equations and systems of second order with the spectral parameter in boundary conditions on nonsmooth boundaries strongly depend on the progress in the corresponding potential theory for these equations and systems (see Subsect. 7.2c). Recently, Agranovich considered the boundary problems 9.4.1–9.4.4 for the Helmholtz equation in the case of a Lipschitz surface Γ. He obtained the completeness of the root functions in $L^2(\Gamma)$, the Abel summability of Fourier series with respect to these functions (for boundary problems 9.4.1, 9.4.3, and 9.1.4), and, in the case of an almost smooth Lipschitz surface Γ, the asymptotics of the eigenvalues.

Remarks and Bibliographical Notes

§ 1 and § 2: The conditions of ellipticity of a boundary problem were first formulated in (Shapiro 1953) and (Lopatinskij 1953), in the latter paper in full generality as conditions of reducibility of the boundary problem to Fredholm equations on the boundary. The theory of Sobolev spaces of non-integer positive order, with applications to elliptic boundary problems, was developed in papers of Aronszajn and Slobodetskij ; see, in particular, (Aronszajn 1955) and (Slobodetskij 1958, 1960). More complete references can be found e.g. in (Triebel 1978). A priori estimates for elliptic boundary problems of general form in L_p-Sobolev spaces and Hölder spaces $C^{(\alpha)}$ were obtained in (Agmon et al. 1959, 1964). More old results are described in these papers. To finish the proof of the general theorem "ellipticity implies Fredholm property," it was sufficient to construct a parametrix; this was done in (Browder 1959) and (Dynin 1961), in the last paper for boundary problems with pseudodifferential boundary conditions. Earlier, this theorem was proved in (Schechter 1959a,b), where, instead of the parametrix, the adjoint boundary problem was used, and for general elliptic boundary problems in plane domains in (Vol'pert 1961) (see also references in this paper) by means of the reduction of the problem to equations on the boundary; Vol'pert also obtained a formula for the index. In (Volevich 1965) and (Solonnikov 1964, 1966, 1967) the proof of the Fredholm property was finished for boundary problems elliptic in the Agmon–Douglis–Nirenberg sense in L^p spaces and (in papers of Solonnikov) $C^{(\alpha)}$ spaces.

Subsect. 2.1a: Earlier, an extension operator for a bounded interval of s-axis was used. It was constructed by Hesteness (1941) for C^m-functions and was adapted by Babich (1953) for H_m-functions with natural m and by Slobodetskij for H_s-functions with any $s > 0$. See (Slobodetskij 1958).

§ 3: The conditions of ellipticity with parameter appeared in (Agmon 1961a, 1962), see also (Agmon 1965), for boundary problems of the form $(A - \lambda I)u = f$ in G, $B_j u = g_j$ on Γ, with order of the boundary conditions less than the order of A. The a priori estimate was proved using elliptic equations without parameter but with an additional variable t in a cylindrical domain. In the proof of solvability sketched by Agmon, he used the adjoint boundary problem. This approach and the results were generalized to boundary problems polynomial in λ in (Agmon and Nirenberg 1963). In (Agranovich and Vishik 1963, 1964) the boundary problems polynomial in λ were considered without a restriction on the orders of boundary operators. The direct approach, without introduction of an additional variable, was used. In particular, the right inverse operator was constructed similarly to the parametrix for elliptic boundary problems without a parameter. In §3 we follow these two papers. As to the terminology, for boundary problems linear with respect to parameter Agmon used the term "rays of minimal growth of the resolvent." For the condition of ellipticity with parameter, Seeley used the term "Agmon's condition" (see (Seeley 1969)). As we noted in Sect. 7.8, now a general calculus of boundary problems elliptic with parameter exists (Grubb 1986). Grubb used the term "parameter ellipticity."

§ 4: The construction and investigation of the adjoint boundary problem is not simple if we do not assume that the boundary conditions are normal and are of orders $r_j < m$. See (Dikanskij 1973). Special questions arise for boundary problems elliptic in Agmon, Douglis and Nirenberg sense; see (Grubb 1977a) and (Roĭtberg 1970, 1975, 1991). Apparently, the framework of the calculus of pseudodifferential problems is most convenient for consideration of adjoint problems (see Rempel and Schulze 1982) and the survey of Brenner and Shargorodsky in this volume.

Sect. 6.6: See also the paper (Brenner and Shubin 1991) about elliptic complexes on a manifold with boundary.

Subsect. 7.1b: Grubb first founds conditions of "weak semiboundedness", under which the form $(A_B u, v)$ can be transformed into a form $a_B(u, v)$ that in general contains boundary terms but is bounded on the subspace of $H_{\{m_t\}}(G)$ defined by the stable bounded conditions. The conditions for coerciveness of a_B on this subspace coincide with those for coerciveness of $(A_B u, v)$ on $D(A_B)$.

Sect. 7.9: Elliptic problems in $C^{(\alpha)}$ were considered e.g. in (Agmon et al. 1959, 1964), (Solonnikov 1964, 1966), and (Rempel and Schulze 1982).

§ 8: The results discussed here stem from the classical paper (Minakshisundaram and Pleijel 1949) on the Beltrami-Laplace equation. Investigations of determinants of elliptic operators (see e.g. (Friedlander 1989) and (Burgelea et al. 1992)) are close to this topic. The zeta-function and the exponential function for an elliptic operator with homogeneous boundary conditions on submanifolds of various dimensions were considered in (Shatalov 1977).

See also (Grubb 1986).

Subsects. 9.1a and b: See also the papers (Melrose 1980, 1983) about two–term asymptotics and the treatment of one-term and two-term asymptotics in (Hörmander 1985b).

Sects. 9.2 and 9.3: See also the paper (Lidskij 1962) about basic properties of the root functions of the Dirichlet problem in a plane domain for a second order elliptic equation with real leading coefficients. A general theorem on completeness for boundary problems elliptic with parameter was proved in (Agmon 1962). See also (Agmon and Nirenberg 1963). As we mentioned in (Agranovich 1990a), the notion of Abel summability of Fourier series with respect to root functions was introduced by Lidskij. More complete references to the literature on abstract nonselfadjoint operators can be found in (Agranovich 1990a, Sect. 6). See also (Geymonat and Grisvard 1985, 1991). Theorem 9.3.5, for scalar boundary problems, is contained in (Mizohata 1965).

Sect. 9.4: See also (Fedotov 1976).

Sect. 9.5: The question on multiple completeness in subspaces for scalar boundary problems polynomial in parameter was considered in (Yakubov 1986, 1989, 1994) and (Shkred 1989) and, earlier, for ordinary differential equations in (Shkalikov 1983). The asymptotics of eigenvalues of such boundary problems close to an isolated ray was obtained in (Boĭmatov and

Kostyuchenko 1991), see also the references in this paper. These authors mainly used the investigation of the resolvent of such boundary problems. Instead, in (Agranovich 1990, 1992) a linearization in λ was used, which leads to a boundary problem elliptic in the sense of Agmon, Douglis, and Nirenberg. This approach naturally yields the most general results.

Equations and boundary problems with parameter that has no definite weight were also considered in (Kotko and Kreĭn 1976) and (Boĭmatov 1992).

Subsects. 9.8a,b: See also (Maruo and Tanabe 1971) and (Maruo 1972). In these papers the resolvent method was used in the case $2q > n$ to obtain asymptotic formulas with some remainder estimates for eigenvalue distribution functions for operators A_D, A_N, and A_V with nonsmooth coefficients in Lipschitz domains by the resolvent method on the base of Agmon's results for smooth domains. The references to papers of Métivier published in 1973 and 1974 see in (Métivier 1977).

See also investigations of selfadjoint spectral problems in domains with singularities and spectral problems with degeneracies in (Ivrii 1990–1992).

Unfortunately, we could not dwell on numerous results for $-\Delta_D$ in domains G with fractal boundaries. In this case the spectral asymptotics can have the form $N(\lambda) = d_0\lambda^{n/2} + O(\lambda^{d/2})$, where d, $n-1 < d < n$, is the "Minkowski dimension" of the boundary. In some cases the second term $d_{-1}\lambda^{d/2}$ can be singled out from the remainder. See e.g. (Lapidus 1991) and (Levitin and Vassiliev 1996), where definitions and further references can be found.

In conclusion, we mention the paper (Shubin 1992) on spectral properties of elliptic operators on noncompact manifolds.

Last remark. In this paper we wrote "boundary problems" instead of "boundary value problems" following Hörmander.

References[20]

Adams, R.A. (1975): Sobolev spaces. Academic Press, New York San Francisco London. Zbl. 314.46030

Agmon, S. (1958): The coerciveness problem for integro-differential forms. J. Analyse Math. *6*, 183–223. Zbl. 119, 323

Agmon, S. (1961a): The angular distribution of eigenvalues of non-self-adjoint elliptic boundary value problems of higher order. Proc. Int. Conf. on Partial Differ. Equations and Continuum Mech., Univ. of Wisconsin Press, 9–18. Zbl. 111, 95

Agmon, S. (1961b): Remarks on self-adjoint and semi-bounded elliptic boundary value problems. Proc. Int. Symp. on Linear Spaces, Jerusalem, Israel Acad. Sci. and Humanities, 1–13. Zbl. 123, 73

[20]For the convenience of the reader, references to reviews in Zentralblatt für Mathematik (Zbl), compiled by means of the MATH databaze, have, as far as possible, been included in this bibliography.

Agmon, S. (1962): On the eigenfunctions and on the eigenvalues of general elliptic boundary value problems. Commun. Pure Appl. Math. *15*, 119–147. Zbl. 109, 237

Agmon, S. (1965a): Lectures on Elliptic Boundary Value Problems. Van Nostrand Math. Studies, Van Nostrand, Princeton etc. Zbl. 142, 374

Agmon, S. (1965b): On kernels, eigenvalues, and eigenfunctions of operators related to elliptic problems. Commun. Pure Appl. Math. *18*, 627–663. Zbl. 151, 202

Agmon, S. (1968): Asymptotic formulas with remainder estimates for eigenvalues of elliptic operators. Arch. Ration. Mech. Anal. *28*, No. 3, 165–183. Zbl. 159, 159

Agmon, S., Douglis, A., Nirenberg, L. (1959, 1964): Estimates near the boundary for solutions of elliptic partial differential equations satisfying general boundary conditions. I. Commun. Pure Appl. Math. *12*, 623–727. Zbl. 93, 104. II. Commun. Pure Appl. Math. *17*, 35–92. Zbl. 123, 287

Agmon, S., Kannai, Y. (1967): On the asymptotic behavior of spectral functions and resolvent kernels of elliptic operators. Isr. J. Math. *5*, 1–30. Zbl. 148, 130

Agmon, S., Nirenberg, L. (1963): Properties of solutions of ordinary differential equations in Banach spaces. Commun. Pure Appl. Math. *16*, No. 2, 121–239. Zbl. 117, 110

Agranovich, M.S. (1962): On the problem of index of elliptic operators. Dokl. Akad. Nauk SSSR *142*, No. 5, 983–995. English transl.: Sov. Math., Dokl. *3*, 194–197 (1962). Zbl. 122, 103

Agranovich, M.S. (1964): General boundary value problems for integro–differential elliptic systems. Dokl. Akad. Nauk SSSR *155*, No. 3, 495–498. English transl.: Sov. Math., Dokl. *5*, 419–423 (1964). Zbl. 132, 355

Agranovich, M.S. (1965): Elliptic singular integro-differential operators. Usp. Mat. Nauk *20*, No. 5, 3–120. English transl.: Russ. Math. Surv. *20*, No. 5, 1–121 (1965). Zbl. 149, 361

Agranovich, M.S. (1977): Spectral properties of diffraction problems. Supplement to (Vojtovich et al. 1977), 289–416 (Russian). Zbl. 384.35002

Agranovich, M.S. (1990a): Elliptic operators on closed manifolds. Itogi Nauki Tekh., Ser. Sovrem. Probl. Mat., Fundam. Napravleniya *63*, VINITI, Moscow, 5–129. English transl.: Encycl. Math. Sci. *63*, Springer, Berlin Heidelberg New York, 1–130, 1994. Zbl. 802.58050

Agranovich, M.S. (1990b): Non-self-adjoint problems with a parameter that are elliptic in the sense of Agmon–Douglis–Nirenberg. Funkts. Anal. Prilozh. *24*, No. 1, 59–61. English transl.: Funct. Anal. Appl. *24*, No. 1, 50–53 (1990). Zbl. 718.35032

Agranovich, M.S. (1992): On moduli of eigenvalues for non-self-adjoint Agmon–Douglis–Nirenberg elliptic boundary problems with a parameter. Funkts. Anal. Prilozh. *26*, No. 2, 51–55. English transl.: Funct. Anal. Appl. *26*, No. 2, 116–119 (1992). Zbl. 795.58047

Agranovich, M.S. (1994a): Nonselfadjoint elliptic operators on nonsmooth domains. Russ. J. Math. Phys. *2*, No. 2, 139–148.

Agranovich, M.S. (1994b): On series with respect to root vectors of operators associated with forms having symmetric principal part. Funkts. Anal. Prilozh. *28*, No. 3, 1–21. English transl.: Funct. Anal. Appl. *28*, No. 3, 151–167 (1994). Zbl. 819.47025

Agranovich, M.S., Amosov, B.A. (1996): Estimates of s–numbers and spectral asymptotics for integral operators of potential type on nonsmooth surfaces. Funkts. Anal. Prilozh. *30*, No. 2, 1–18. English transl.: Funct. Anal. Appl. *30*, No. 2 (1996).

Agranovich, M.S., Dynin, A.S. (1962): General boundary problems for elliptic systems in multidimensional domains. Dokl. Akad. Nauk SSSR *146*, No. 3, 511–514. English transl.: Sov. Math., Dokl. *3*, 1323–1327 (1962). Zbl. 132, 354

Agranovich, M.S., Golubeva, Z.N. (1976): Some problems for Maxwell's system with a spectral parameter in the boundary condition. Dokl. Akad. Nauk SSSR *231*, No. 4, 777–780. English transl.: Sov. Math., Dokl. *17*, No. 6, 1614–1619 (1977). Zbl. 353.35074

Agranovich, M.S., Markus, A.S. (1989): On spectral properties of elliptic pseudo–differential operators far from self-adjoint ones. Z. Anal. Anwend. *8*, No. 3, 237–260. Zbl. 696.35123

Agranovich, M.S., Vishik, M.I. (1963): Elliptic boundary value problems with a parameter. Dokl. Akad. Nauk SSSR *149*, No. 2, 223–226. English transl.: Sov. Math., Dokl. 325–329 (1963). Zbl. 145, 147

Agranovich, M.S., Vishik, M.I. (1964): Elliptic problems with parameter and parabolic problems of general form. Usp. Mat. Nauk *19*, No. 3, 53–161. English transl.: Russ. Math. Surv. *19*, No. 3, 53–157 (1964). Zbl. 137, 296

Aronszajn, N. (1955): On coercive integro–differential quadratic forms. Tech. Report No. 14, Univ. of Kansas, 94–106. Zbl. 67, 327

Aronszajn, N. (1957): A unique continuation theorem for solutions of elliptic partial differential equations or inequalities of second order. J. Math. Pures Appl. *36*, 235–249. Zbl. 84, 304

Aronszajn, N., Milgram, A.N. (1953): Differential operators on Riemannian manifolds. Rend. Circ. Mat. Palermo *2*, No. 3, 266–325. Zbl. 53, 65

Atiyah, M.F., Bott, R. (1964): The index problem for manifolds with boundary. Proc. Symp. Differ. Analysis, Bombay Coll., Oxford Univ. Press, London, 175–186. Zbl. 163, 346

Bagirov, L.A., Kondrat'ev, V.A. (1975): On elliptic equations in $\mathbb{R}^n$. Differ. Uravn. *11*, No. 3, 498–504. Zbl. 324.35032. English transl.: Differ. Equations *11*, 375–379 (1976).

Berezanskij, Yu.M. (1965): Expansions in Eigenfunctions of Selfadjoint Operators. Naukova Dumka, Kiev. Zbl.142, 372. English transl.: Transl. Math. Monographs *17*, Am. Math. Soc., Providence, 1968.

Berezanskij, Yu.M., Kreĭn, S.G., Roĭtberg, Ya.A. (1963): A theorem on homeomorphisms and local increase in smoothness up to the boundary of solutions to elliptic equations. Dokl. Akad. Nauk SSSR *148*, 745–748. English transl.: Sov. Math., Dokl. *4*, 152–155 (1963). Zbl. 156, 344

Berezanskij, Yu.M., Roĭtberg, Ya.A. (1967): A theorem on homeomorphisms and Green's function for general elliptic boundary problems. Ukr. Mat. Zh. *19*, No. 5, 3–32. Zbl. 163, 348. English transl.: Ukr. Math. J. *19*, 509–530 (1967).

Birman, M.S. (1956): To the theory of selfadjoint extensions of positive definite operators. Mat. Sb. *38*, No. 4, 431–450 (Russian). Zbl. 70, 121

Birman, M.S., Solomyak, M.Z. (1972, 1973): Spectral asymptotics of nonsmooth elliptic operators. I, II. Tr. Mosk. Matem. O-va *27*, 3–52 and *28*, 3–34. Zbl. 251.35075 and 281.35079. English transl.: Trans. Moscow Math. Soc. *27*, 1–52 (1975) and *28*, 1–32 (1975).

Birman, M.S., Solomyak, M.Z. (1977a): Estimates of singular numbers of integral operators. Usp. Mat. Nauk *32*, No. 1 (193), 17–84. Zbl. 344.47021. English transl.: Russ. Math. Surv. *32*, 15–89 (1977).

Birman, M.S., Solomyak, M.Z. (1977b): Asymptotics of the spectrum of differential equations. Itogi Nauki Tekh., Ser. Mat. Anal., VINITI, Moscow *14*, 5–58. Zbl. 417.35061. English transl.: J. Sov. Math. *12*, 247–283 (1979).

Birman, M.S., Solomyak, M.Z. (1977c, 1979a): Asymptotics of the spectrum of pseudodifferential operators with anisotropic–homogeneous symbols. I, II. Vestn. Leningr. Univ., Mat. Mekh. Astron. No. 13, 1–21 (1977) and No. 13, 5–10 (1979). Zbl. 377.47033 and 439.47040 (1979). English transl.: Vestn. Leningr. Univ., Math. *10*, 237–247 (1982) and *12*, 155–161 (1980).

Birman, M.S., Solomyak, M.Z. (1979b): Asymptotics of the spectrum of variational problems on solutions of elliptic equations. Sib. Mat. Zh. *20*, No. 1, 3–22, Zbl. 414.35059. English transl.: Sib. Math. J. *20*, 1–15 (1979).

Birman, M.S., Solomyak, M.Z. (1982a): On subspaces that admit a pseudodifferential projector. Vestn. Leningr. Univ., Mat. Mech. Astron. No.1, 18–25. Zbl. 484.35085. English transl.: Vestn. Leningr. Univ., Math. *15*, 17–27 (1983).

Birman, M.S., Solomyak, M.Z. (1982b): Asymptotics of the spectrum of pseudodifferential variational problems with constraints. Probl. Mat. Fiz. *10*, 20–36. English transl.: Sel. Math. Sov. *5*, 245–256 (1986). Zbl. 602.35086

Birman, M.S., Solomyak, M.Z. (1982c): Asymptotics of the spectrum of variational problems on solutions of elliptic systems. Zap. Nauchn. Semin. Leningr. Otd. Mat. Inst. Steklova *115*, 23–39. Zbl. 501.35065. English transl.: J. Sov. Math. *28*, 633–644 (1985).

Bitsadze, A.V. (1966): Boundary Value Problems for Elliptic Equations of Second Order. Nauka, Moscow. Zbl. 148, 89. English transl.: North–Holland, Amsterdam (1968).

Bitsadze, A.V., Samarskij, A.A. (1969): On some simple generalizations of linear elliptic boundary value problems. Dokl. Akad. Nauk SSSR *185*, No. 4, 739–740. English transl.: Sov. Math., Dokl. *10*, 427–431 (1969). Zbl. 187, 355

Boĭmatov, K.Kh. (1992): Spectral asymptotics for linear and nonlinear pencils of pseudodifferential operators that are elliptic in the Douglis–Nirenberg sense. Dokl. Akad. Nauk SSSR *322*, No. 3, 441–445. English transl.: Sov. Math., Dokl. *45*, No. 1, 99–104 (1992). Zbl. 798.58072

Boĭmatov, K.Kh. (1993): Generalized Dirichlet problem associated with a noncoercive bilinear form. Dokl. Ross. Akad. Nauk *330*, No. 3, 285–290. English transl.: Ross. Acad. Sci., Dokl., Math. *47*, No. 3. 455–463 (1993). Zbl. 827.35028

Boĭmatov, K.Kh. (1995): Some spectral properties of matrix differential operators far from selfadjoint. Funkts. Anal. Prilozh. *29*, No. 3, 55–58. English transl.: Funct. Anal. Appl. *29*, No. 3 (1995).

Boĭmatov, K.Kh., Kostyuchenko, A.G. (1988): Distribution of eigenvalues of an elliptic operator in a bounded domain. Tr. Semin. Petrovskogo *13*, 3–18. Zbl. 668.35074

Boĭmatov, K.Kh., Kostyuchenko, A.G. (1991): Spectral asymptotics of polynomial pencils of differential operators in a bounded domain. Funkts. Anal. Prilozh. *25*, No. 1, 7–20. English transl.: Funct. Anal. Appl. *25*, No. 1, 5–16 (1991). Zbl.737.47011

Boutet de Monvel, L. (1966): Comportment d'un opérateur pseudo–différentiel sur une variété à bord. J. Anal. Math. *17*, 241–304. Zbl. 161, 79

Boutet de Monvel, L. (1971): Boundary problems for pseudo–differential operators. Acta Math. *126*, No. 1–2, 11–51. Zbl. 206, 394

Boyarskij, B.V. (1960): On the Dirichlet problem for a system of equations of elliptic type in the space. Bull. Acad. Polon. Sci., Ser. Sci. Math. Astron. Phys. *8*, No. 1, 19–23. Zbl. 106, 76

Brenner, A.V., Shargorodsky, E.M. (1996): Boundary value problems for elliptic pseudodifferential operators. Itogi Nauki Tekh., Ser. Sovr. Probl. Mat., Fundam. Napravleniya *79*, VINITI, Moscow, in preparation. English transl.: in this volume.

Brenner, A.V., Shubin, M.A. (1991): The Atiyah–Bott–Lefschetz formula for elliptic complexes on a manifold with boundary. Itogi Nauki Tekh., Ser. Sovrem. Probl. Mat., Novejshie Dostizh. *38*, 119–183. Zbl. 738.58043. English transl.: J. Sov. Math. *64*, 1069–1111 (1993).

Browder, F.E. (1959): Estimates and existence theorems for elliptic boundary value problems. Proc. Natl. Acad. Sci. USA *45*, 365–372. Zbl. 93, 294

Browder, F.E. (1961): A continuity property for adjoints of closed operators in Banach spaces and its applications to elliptic boundary value problems. Duke Math. J. *28*, No. 2, 157–182. Zbl. 102, 315

Brüning, J. (1974): Zur Abschätzung der Spektralfunktion elliptischer Operatoren. Math. Z. *137*, No. 1, 75–85. Zbl. 275.47030

Burgelea, D., Friedlander, L., Kapeller, T. (1992): Mayer–Vietoris type formula for determinants of elliptic differential operators. J. Funct. Anal. *107*, 34–65. Zbl. 759.58043

Calderón, A.P. (1958): Uniqueness in the Cauchy problem for partial differential equations. Am. J. Math. *80*, 16–36. Zbl. 80, 303

Calderón, A.P. (1963): Boundary value problems for elliptic equations. Outlines of the Joint Soviet–American Symposium on Partial Differential Equations, Novosibirsk, 303–304.

Chazarain, J., Piriou, A. (1981): Introduction à la théorie des équations aux dérivées partielles linéaires. Paris, Gauthier–Villars. Zbl. 446.35001

Clark, C. (1967): The asymptotic distribution of eigenvalues and eigenfunctions for elliptic boundary value problems. SIAM Review *9*, No. 4, 627–646. Zbl. 159, 197

Colton, D., Kress, R. (1983): Integral Equation Methods in Scattering Theory. Wiley, Interscience Publishers, New York etc. Zbl. 522.35001

Costabel, M. (1988): Boundary integral operators on Lipschitz domains: elementary results. SIAM J. Math. Anal. *19*, No. 3, 613–626. Zbl. 644.35037

Costabel, M., Wendland, W.L. (1986): Strong ellipticity of boundary integral operators. J. Reine Angew. Math. *372*, 34–63. Zbl. 628.35027

Dahlberg, E.D., Kenig, C.E., Verchota, G.C. (1988): Boundary value problems for the systems of elastostatics in Lipschitz domains. Duke Math. J. *57*, No. 3, 795–818. Zbl. 699.35073

Dieudonné, J. (1978): Eléments d'analyse. VIII. Paris, Gauthier–Villars. Zbl. 393. 35001

Dikanskij, A.S. (1973): Conjugate problems to elliptic differential and pseudodifferential boundary problems in a bounded domain. Mat. Sb. *91*, 62–77. Zbl. 278.35080. English transl.: Math. USSR, Sb. *20*, 67–83 (1974).

Dudnikov, P.I., Samborskij, S.N. (1991): Linear overdetermined systems of partial differential equations. Initial and initial–boundary value problems. Itogi Nauki Tekh., Ser. Sovrem. Probl. Mat., Fundam. Napravleniya *65*, VINITI, Moscow, 5–93. Zbl. 807.35100. English transl.: Encycl. Math. Sci. *65*, Springer, Berlin Heidelberg New York, 1–86, 1996.

Duistermaat, J.J. (1981): On operators of trace class in $L^2(X,\mu)$. Proc. Indian Acad. Sci., Math. Sci. *90*, No. 1, 29–32 Zbl. 482.47012

Dunford, N., Schwartz, J.T. (1963): Linear Operators. II. Interscience Publishers, New York. Zbl. 128, 348

Dynin, A.S. (1961): Multidimensional elliptic boundary problems with one unknown function. Dokl. Akad. Nauk SSSR *141*, No. 2, 285–287. English transl.: Sov. Math., Dokl. *2*, 1431–1433 (1961). Zbl. 125, 348

Egorov, Yu.V. (1984): Linear Differential Equations of Principal Type. Nauka, Moscow. English transl.: Plenum Press, New York London 1986. Zbl. 574.35001

Egorov, Yu.V., Shubin, M.A. (1988a): Linear partial differential equations. Foundations of the classical theory. Itogi Nauki Tekh., Ser. Sovrem. Probl. Mat., Fundam. Napravleniya *30*, VINITI, Moscow. Zbl. 657.35002. English transl.: Encycl. Math. Sci. *30*, Springer, Berlin Heidelberg New York, 1992.

Egorov, Yu.V., Shubin, M.A. (1988b): Linear partial differential equations. Elements of the modern theory. Itogi Nauki Tekh., Ser. Sovrem. Probl. Mat., Fundam. Napravleniya *31*, VINITI, Moscow, 5–125. Zbl. 686.35005. English transl.: Encycl. Math. Sci. *31*, Springer, Berlin Heidelberg New York, 1–120, 1994.

Ercolano, J., Schechter, M. (1965): Spectral theory for operators generated by elliptic boundary problems with eigenvalue parameter in boundary conditions. I, II. Commun. Pure Appl. Math. *18*, 83–105 and 397–414. Zbl. 138, 362

Eskin, G.I. (1973): Boundary Value Problems for Elliptic Pseudodifferential Equations. Nauka, Moscow. Zbl. 262.35001. English transl.: Transl. Math. Monogr.*52*, Am. Math. Soc., Providence, 1981.
Fabes, E.B., Kenig, C.E., Verchota, G.C. (1988): The Dirichlet problem for the Stokes system on Lipschitz domains. Duke Math. J. *57*, 769–793. Zbl. 685.35085
Fedosov, B.V. (1991): Index theorems. Itogi Nauki Tekh., Ser. Sovrem. Probl. Mat., Fundam. Napravleniya *65*, VINITI, Moscow, 165–268. English transl.: Encycl. Math. Sci. *65*, Springer, Berlin Heidelberg New York, 155–252, 1996.
Fedotov, A.G. (1977): On elliptic boundary value problems with a spectral parameter entering polynomially in boundary conditions. Usp. Mat. Nauk *32*, No. 4, 269–270. Zbl. 354.35081
Figueiredo, D.E. de (1963): The coerciveness problem for forms over vector valued functions. Commun. Pure Appl. Math. **16**, 63–94. Zbl. 136, 95
Friedlander, L. (1989): The asymptotics of the determinant function for a class of operators. Proc. Am. Math. Soc. *107*, No. 1, 169–178. Zbl. 694.47036
Gårding, L. (1953): Dirichlet's problem for linear elliptic partial differential equations. Math. Scand. *1*, 55–72. Zbl. 53, 391
Gao, W. (1991): Layer potentials and boundary value problems for elliptic systems in Lipschitz domains. J. Funct. Anal. *95*, 377–399. Zbl. 733.35034
Gel'fand, I.M. (1960): On elliptic equations. Usp. Mat. Nauk *15*, No. 3, 121–132. English transl.: Russ. Math. Surv. *15*, No. 3, 113–123 (1960). Zbl. 95, 78
Geymonat, G., Grisvard, P. (1967): Alcuni risultati di teoria spettrale per i problemi ai limiti lineari ellittici. Rend. Sem. Mat. Univ. Padova *38*, 127–173. Zbl. 182, 185
Geymonat, G., Grisvard, P. (1985): Eigenfunction expansions for non-self-adjoint operators and separation of variables. Lecture Notes in Math. *1121*, 123–136. Zbl. 589.47021
Geymonat, G., Grisvard, P. (1991): Expansions in generalized eigenvectors of operators arising in the theory of elasticity. Diff. Int. Equat. *4*, 459–481. Zbl. 733.35082
Gohberg, I.C., Kreĭn, M.G. (1957): The basic propositions on defect numbers, root numbers, and indices of linear operators. Usp. Mat. Nauk *12*, No. 2, 43–118. Zbl. 88, 321. English transl.: Am. Math. Soc. Transl., Ser. II, *13*, 185–264 (1960).
Golubeva, Z.N. (1976): Some scalar diffraction problems and corresponding nonselfadjoint operators. Radiotekhn. Elektron. *21*, No. 2, 219–227. English transl.: Radio Eng. Electron. Phys. (2) *21* (1976), No. 7.
Greiner, P. (1971): An asymptotic expansion for the heat equation. Arch. Ration. Mech. Anal. *41*, No. 1, 163–218. Zbl. 238.35038
Grisvard, P. (1967): Caractérisation de quelques espaces d'interpolation. Arch. Ration. Mech. Anal. *25*, 40–63. Zbl. 187, 59
Grisvard, P. (1985): Elliptic Problems in Nonsmooth Domains. Pitman, London. Zbl. 695.35060
Grubb, G. (1968): A characterization of non-local boundary value problems associated with an elliptic operator. Ann. Sc. Norm. Sup. Pisa *22*, No. 3, 425–513. Zbl.182, 145
Grubb, G. (1971): On coerciveness and semiboundedness of general boundary problems. Isr. J. Math. *10*, 32–95. Zbl. 231.35027
Grubb, G. (1973): Weakly semibounded boundary problems and sesquilinear forms. Ann. Inst. Fourier Grenoble *23*, No. 2, 145–194. Zbl. 261.35011
Grubb, G. (1974): Properties of normal boundary problems for elliptic even–order systems. Ann. Sc. Norm. Sup. Pisa, Sci. Fis. Mat., Ser. IV, *1* , 1–61. Zbl. 309.35034
Grubb, G. (1977a): Boundary problems for systems of partial differential operators of mixed order. J. Funct. Anal. *26*, No. 2, 131–165. Zbl. 368.35030
Grubb, G. (1977b): Spectral asymptotics for Douglis–Nirenberg elliptic and pseudo–differential boundary problems. Commun. Partial Differ. Equations *2*, No. 11, 1071–1150. Zbl. 381.35034

Grubb, G. (1979): On coerciveness of Douglis–Nirenberg elliptic boundary value problems. Boll. Un. Mat. Ital., Ser. V, *16-B*, 1049–1080. Zbl. 456.35015

Grubb, G. (1986): Functional Calculus of Pseudo-Differential Boundary Problems. Birkhäuser, Boston Basel Stuttgart, second edition 1996. Zbl. 622.35001

Grubb, G. (1990): Pseudo–differential boundary problems in L_p-spaces. Commun. Partial Differ. Equations *15*, No. 3, 289–340. Zbl. 723.35091

Grubb, G., Geymonat, G. (1977): The essential spectrum of elliptic systems of mixed order. Math. Ann. *227*, 247–276. Zbl. 361.35050

Grubb, G., Geymonat, G. (1979): Eigenvalue asymptotics for selfadjoint elliptic mixed order systems with nonempty essential spectrum. Boll. Un. Mat. Ital., Ser. V, *16-B*, 1032–1048. Zbl. 423.35071

Hörmander, L. (1958): On the regularity of solutions of boundary problems. Acta Math. *99*, 225–264. Zbl. 83, 92

Hörmander, L. (1963): Linear Partial Differential Operators. Springer, Berlin Göttingen Heidelberg. Zbl. 108, 93

Hörmander, L. (1966): On the Riesz means of spectral functions and eigenvalue expansions for elliptic differential operators. Belfer Grad School Sci., Yeshiva Univ., Conf. Proceedings, *2*, 155–202.

Hörmander, L. (1983a,b, 1985a,b): The Analysis of Linear Partial Differential Operators. Vols. 1, 2, 3, 4. Springer, Berlin Heidelberg New York. Zbl. 521.35001, 521.35002, 601.35001, 612.35001

Il'in, V.A. (1991): Spectral Theory of Differential Operators. Nauka, Moscow. Zbl. 759.47026

Ivrii, V.Ya. (=Ivrij, V.Ya.) (1980): Second term of the spectral asymptotics for the Laplace–Beltrami operator on manifolds with boundary. Funkts. Anal. Prilozh. *14*, No. 2, 25–34. Zbl. 435.35059. English transl.: Funct. Anal. Appl. *14*, No. 2, 98–106 (1980).

Ivrii, V.Ya. (1984): Precise Spectral Asymptotics for Elliptic Operators. Lect. Notes Math. *1100*. Zbl. 565.35002

Ivrii, V.Ya. (1987): Precise spectral asymptotics for elliptic operators on manifolds with boundary. Sib. Mat. Zh. *28*, No. 1, 107–114. English transl.: Sib. Math. J. *28*, 80–86 (1987). Zbl. 651.35068

Ivrii, V.Ya. (1990–1992): Semiclassical microlocal analysis and precise spectral asymptotics. Preprints 1–9. Centre de Math. École Polytechnique.

Jerison, D., Kenig, C.E. (1995): The inhomogeneous Dirichlet problem in Lipschitz domains. J. Funct. Anal. *110*, No. 1, 161–219. Zbl. 832.35034

Kato, T. (1966): Perturbation Theory for Linear Operators. Springer, Berlin–Heidelberg–New York. Zbl. 148, 126

Kenig, C.E. (1994): Harmonic Analysis Techniques for Second Order Elliptic Boundary Values Problems. Am. Math. Soc. Zbl. 812.35001

Kondrat'ev, V.A. (1967): Boundary problems for elliptic equations in domains with conical or angular points. Tr. Mosk. Mat. O-va *16*, 209–292. Zbl. 162, 163. English transl.: Trans. Moscow Math. Soc. *16*, 227–313 (1967).

Kostometov, G.P. (1974): Asymptotic behavior of the spectrum of integral operators with a singularity on a diagonal. Mat. Sb. *94* (*136*), No. 3, 444–451. Zbl. 322.47028. English transl.: Math. USSR, Sb. *23* (1974), 417–424 (1975).

Kotko, L.A., Kreĭn, S.G. (1976): On completeness of the system of eigenfunctions and associated functions for boundary value problems with a parameter in boundary conditions. Dokl. Akad. Nauk SSSR *227*, No. 2, 288–290. English transl.: Sov. Math., Dokl. *17*, 390–393 (1976). Zbl. 338.35070

Kozhevnikov, A.N. (1973): Spectral problems for pseudodifferential systems elliptic in the Douglis–Nirenberg sense, and their applications. Mat. Sb. *92* (*134*), 68–88. Zbl. 278.35078. English transl.: Math. USSR, Sb. *21*, 63–90 (1973).

Kozhevnikov, A. (1993): The basic boundary value problems of static elasticity theory and their Cosserat spectrum. Math. Z. *213*, 241–274. Zbl. 793.73021

Kozhevnikov, A., Yakubov, S. (1995): On operators generated by elliptic boundary problems with a spectral parameter in boundary conditions. Integral Equations Oper. Theory *23*, No. 2, 205–231.

Kreĭn, S.G., Petunin, Yu.I., Semenov, E.M.(1978): Interpolation of Linear Operators. Nauka, Moscow. Zbl. 499.46044. English transl.: Am. Math. Soc., 1982.

Kumano-go, H. (1981): Pseudodifferential Operators. Massachusetts Institute of Technology. Zbl. 489.35003

Kupradze, V.D., ed. (1976): Three Dimensional Problems of the Mathematical Theory of Elasticity and Thermoelasticity. Moscow, Nauka. English transl.: North-Holland, New York, 1979. Zbl. 406.73001

Lapidus, M.L. (1991): Fractal drum, Inverse spectral problems for elliptic operator and a partial resolution of the Weyl–Berry conjecture. Trans. Am. Math. Soc. *325*, No. 2, 465–529. Zbl. 741.35048

Levendorskij, S.Z. (1989): Asymptotics of the spectrum of operators elliptic in the Douglis–Nirenberg sense. Tr. Mosk. Mat. O-va *52*, 34–57. Zbl. 707.35112. English transl.: Trans. Moscow Math. Soc., 33–57 (1991).

Levendorskij, S.Z. (1990): Asymptotic Distribution of Eigenvalues of Differential Operators. Kluwer Acad. Publishers, Dorderecht etc. Zbl.721.35049

Levendorskij, S.Z., Paneah, B.P. (1990): Degenerate elliptic equations and boundary problems. Itogi Nauki Tekh., Ser. Sovrem. Probl. Mat., Fundam. Napravleniya *63*, VINITI, Moscow, 131–200. Zbl. 739.35028. English transl.: Encycl. Math. Sci. *63*, Springer, Berlin Heidelberg New York, 131–201, 1994.

Levitin, M. (1993): Vibrations of viscous compressible fluid in bounded domains: spectral properties and asymptotics. Asymptotic Anal. *7*, 15–35. Zbl. 769.76064

Levitin, M., Vassiliev, D. (1996). Spectral asymptotics, renewal theorem, and the Berry conjecture for a class of fractals. Proc. London Math. Soc. *72*, 188–214.

Lidskij, V.B. (1962): The Fourier series expansion in terms of the principal functions of a non-selfadjoint elliptic operator. Mat. Sb. (N.S.) *57* (*99*), 137–150. Zbl. 206, 134

Lions, J.L., Magenes, E. (1968): Problèmes aux limites non homogènes et applications, Vol. 1, Dunod, Paris. Zbl. 165, 108

Lopatinskij, Ya.B. (1953): A method of reduction of boundary-value problems for systems of differential equations of elliptic type to a system of regular integral equations. Ukr. Mat. Zh. *5*, No. 2, 123–151. Zbl. 52,102. English transl.: Am. Math. Soc. Transl., Ser. II, *89* (1970), 149–183.

Lopatinskij, Ya.B. (1956): Decomposition of a polynomial matrix into a product. Nauchn. Zap. Polytechn. Inst., L'vov, Ser. Fiz. Mat. *2*, 3–7.

Markus, A.S. (1986): Introduction to the Spectral Theory of Polynomial Operator Pencils. Kishinev, Stiintsa. English transl.: Transl. Math. Monographs *71*, Am. Math. Soc., Providence, 1988. Zbl. 678.47006

Markus, A.S., Matsaev, V.I. (1981): Operators generated by sesquilinear forms and their spectral asymptotics. Mat. Issled., Kishinev *61*, 86–103. Zbl.476.47009

Markus, A.S., Matsaev, V.I. (1982): Comparison theorems for spectra of linear operators, and spectral asymptotics. Tr. Mosk. Mat. O-va *45*, 133–181. Zbl. 532.47012 English transl.: Trans. Moscow Math. Soc. *45*, 139–187 (1984).

Maruo, K. (1972): Asymptotic distribution of eigenvalues of nonsymmetric operators associated with strongly elliptic sesquilinear forms. Osaka J. Math. *9*, 547–560. Zbl.251.35077

Maruo, K., Tanabe, H. (1971): On the asymptotic distribution of eigenvalues associated with strongly elliptic sesquilinear forms. Osaka J. Math., *8*, 323–345. Zbl. 227.35073

Maz'ya, V.G. (1988): Boundary integral equations. Itogi Nauki Tekh., Ser. Sovrem. Probl. Mat., Fundam. Napravleniya *27*, VINITI, Moscow, 131–228. English transl.: Encycl. Math. Sci. *27*, Springer, Berlin Heidelberg New York, 127–222, 1991. Zbl. 780.45002

Melrose, R. (1980): Weyl's conjecture for manifolds with concave boundary. Geometry of the Laplace Operator. Honolulu/Hawaii 1979, Proc. Symp. Pure Math. *36*, 257–274. Am. Math. Soc., Providence, RI. Zbl. 436.58002

Melrose, R. (1984): The trace of the wave group. Microlocal Analysis. Proc. Conf., Boulder/Colo. 1983, Contemp. Math. *27*, 127–167. Zbl. 547.35095

Métivier, G. (1977): Valeurs propres de problèmes aux limites elliptiques irréguliers. Bull. Soc. Math. France *51-52*, 125–219. Zbl. 401.35088

Métivier, G. (1978): Valeurs propres d'opération définis par restriction de systèmes variationelles à des sous-espaces. J. Math. Pures Appl., Ser. IX, *57*, No. 2, 133–156. Zbl. 328.35029

Métivier, G. (1983): Estimation du reste en théorie spectrale. Rend. Sem. Mat. Univ. Torino, Fasc. Speciale. Convegno su Linear Partial and Pseudo-Differential Operators, Torino, 1982, 157–180. Zbl. 542.35055

Mikhailets, V.A. (1982): Asymptotics of the spectrum of elliptic operators and boundary conditions. Dokl. Akad. Nauk SSSR *266*, No. 5, 1059–1062. English transl.: Sov. Math., Dokl. 26, 464–468 (1982). Zbl. 513.35059

Mikhailets, V.A. (1989): A precise estimate of the remainder in the spectral asymptotics of general elliptic boundary problems. Funkts. Anal. Prilozh. *23*, No. 2, 65–66. English transl.: Funct. Anal. Appl. *23*, No. 2, 137–139 (1989). Zbl. 717.35063

Mikhailets, V.A. (1990): To the theory of general boundary problems for elliptic equations. In: Nonlinear Boundary Value Problems, Naukova Dumka, Kiev, *2*, 66–70.

Minakshisundaram, S., Pleijel, A. (1949): Some properties of the eigenfunctions of the Laplace operator on Riemannian manifolds. Can. J. Math. *1*, 242–256. Zbl. 41. 427

Mizohata, S. (1965): Sur les propriétés asymptotiques des valeurs propres pour les opérateurs elliptiques. J. Math. Kyoto Univ. *4-5*, 399–428. Zbl. 147, 96

Nečas, J. (1967): Les méthodes directes en théorie des équations elliptiques. Academia, Prague. Engl. Ed.: Teubner 1983. Zbl. 526.35003. Reprint: Wiley 1986.

Nirenberg, L. (1955): Remarks on strongly elliptic partial differential equations. Commun. Pure Appl. Math. *8*, 649–675. Zbl. 67, 76

Palais, R.S. (1965): Seminar on the Atiyah–Singer Index Theorem. Univ. Press, Princeton. Zbl. 137, 170

Panich, O.I. (1966): Elliptic boundary problems with a parameter only in boundary conditions. Dokl. Akad. Nauk SSSR *170*, No. 5, 1020–1023. English transl.: Sov. Math., Dokl. *7*, 1326–1329 (1967). Zbl. 171, 82

Panich, O.I. (1973): On boundary value problems with a parameter in boundary conditions. Mat. Fiz., Kiev *14*, 140–146. Zbl. 286.35029

Petrovskij, I.G. (1939): Sur l'analycité des solutions des systèmes d'équations différentielles. Mat. Sb. *5* (47), 3–68. Zbl. 22, 226

Pham The Lai (1976): Comportment asymptotique du noyau de la résolvante et des valeurs propres d'un opérateur elliptique non nécessairement autoadjoint. Isr. J. Math. *23*, No. 3–4, 221–250. Zbl. 339.35037

Pham The Lai, Robert, D. (1980): Sur une problème aux valeurs propres non linéaire. Isr. J. Math. *36*, No. 2, 169–187. Zbl. 444.35065

Pipher, J., Verchota, G.C. (1995): Dilation invariant estimates and the boundary Gårding inequality for higher order elliptic operators. Ann. Math. *142*, 1–38.

Plamenevskij, B.A. (1996): Elliptic boundary problems in domains with piecewise-smooth boundary. Itogi Nauki Tekh., Ser. Sovr. Probl. Mat., Fundam. Napravleniya *79*, VINITI, Moscow, in preparation. English transl.: in this volume.

Pliš, A. (1954): The problem of uniqueness for the solution of a system of partial differential equations. Bull. Acad. Pol. Sci. *2*, 55–57. Zbl. 55, 323

Rempel, S., Schulze, B.-W. (1982): Index Theory of Elliptic Boundary Problems. Akademie–Verlag, Berlin. Zbl. 504.35002

Roĭtberg, Ya.A. (1964): Elliptic boundary problems with nonhomogeneous boundary conditions and local increase of smoothness up to the boundary for generalized solutions. Dokl. Akad. Nauk SSSR *157*, No. 4, 798–801. English transl.: Sov. Math., Dokl. *5*, 1034–1038 (1964). Zbl. 173, 133

Roĭtberg, Ya.A. (1968): Homeomorphism theorems defined by elliptic operators. Dokl. Akad. Nauk SSSR *180*, No. 3, 542–545. English transl.: Sov. Math., Dokl. *9*, 656–660 (1968). Zbl. 182, 142

Roĭtberg, Ya.A. (1970): Homeomorphism theorems and a Green's formula for general elliptic boundary problems with nonnormal boundary conditions. Mat. Sb. *83* (125), No. 2, 181–213. Zbl. 216, 125. English transl.: Math. USSR, Sb. *12*, No. 2, 177–212 (1971).

Roĭtberg, Ya.A. (1971): On boundary values of generalized solutions of elliptic equations. Mat. Sb. *86*, No. 2, 248–267. Zbl. 221.35014. English transl.: Math. USSR, Sb. *15*, 241–260 (1972).

Roĭtberg, Ya.A. (1975): Theorem on a complete set of isomorphisms for systems elliptic in the Douglis–Nirenberg sense. Ukr. Mat. Zh. *27*, No. 4, 544–548. Zbl.313.35029. English transl.: Ukr. Math. J. *27*, 447–450 (1976).

Roĭtberg, Ya.A. (1991): Elliptic Boundary Value Problems in Generalized Functions. Preprints I–IV. Chernigov State Pedagogical Institute, Chernigov, Ukraine.

Roĭtberg, Ya.A., Sheftel', Z.G. (1967): Boundary value problems with a parameter in L_p for systems elliptic in the Douglis–Nirenberg sense. Ukr. Mat. Zh. *19*, 115–120. Zbl. 198, 440

Roĭtberg, Ya.A., Sheftel', Z.G. (1969): A theorem on homeomorphisms for elliptic systems and its applications. Mat. Sb. *78*, No. 3, 446–472. Zbl. 176, 409. English transl.: Math USSR, Sb. *7*, 439–465 (1969). Zbl. 191, 119

Roĭtberg, Ya.A., Sheftel', Z.G. (1972): Nonlocal boundary–value problems for elliptic equations and systems. Sib. Mat. Zh. *13*, No. 1, 165–181. English transl.: Sib. Math. J. *13*, 118–129 (1972). Zbl. 259.35031

Roĭtberg, Ya.A., Sheftel', Z.G. (1973): Theorems on isomorphisms for nonlocal elliptic boundary value problems and their applications. Ukr. Mat. Zh. *25*, No. 6, 761–771. English transl.: Ukr. Math. J. *25*, 630–638 (1974). Zbl. 286.35030

Rozenblum, G.V., Shubin, M.A., Solomyak, M.Z. (1989): Spectral theory of differential operators. Itogi Nauki Tekh., Ser. Sovrem. Probl. Matem., Fundam. Napravleniya *64*, VINITI, Moscow. English Transl.: Encycl. Math. Sci., *64*, Springer, Berlin Heidelberg New York, 1994. Zbl.805.35081

Safarov, Yu.G. (1988a): Exact asymptotics of the spectrum of a boundary problem, and periodic billiards. Izv. Akad. Nauk SSSR, Ser. Mat. *52*, 1230–1251. English transl.: Math. USSR, Izv. *33*, 553–573 (1989). Zbl. 682.35082

Safarov, Yu.G. (1988b): Asymptotics of the spectral function of a positive elliptic operator without the non–trapping condition. Funkt. Anal. Prilozh. *22*, No. 3, 53–65. English transl.: Funct. Anal. Appl. *22*, No. 3 (1988), 213–223. Zbl. 679.35074

Safarov, Yu.G., Vassiliev, D.G. (1988): Branching Hamiltonian billiards. Dokl. Akad. Nauk SSSR *301*, 271–275. English transl.: Sov. Math., Dokl. *38*, 64–68 (1989). Zbl. 671.58012

Safarov, Yu.G., Vassiliev, D.G. (1992): The asymptotic distribution of eigenvalues of differential operators. In: Spectral Theory of Operators, ed. by S. Gindikin, Am. Math. Soc. Transl., Ser. II, *150*, 55–111. Zbl. 751.58040

Safarov, Yu., Vassiliev, D. (1996): The Asymptotic Distribution of Eigenvalues of Partial Differential Operators. Am. Math. Soc. In preparation.

Samoĭlenko, I.S. (1972): Polynomial factorization of elliptic systems of differential equations and boundary value problems. Mat. Fiz., Kiev *11*, 129–133. Zbl. 245.35053

Schechter, M. (1959a): General boundary value problems for elliptic partial differential equations. Commun. Pure Appl. Math. *12*, 457–486. Zbl. 87, 302

Schechter, M. (1959b): Remarks on elliptic boundary value problems. Commun. Pure Appl. Math. *12*, 561–578. Zbl. 95, 79

Schechter, M. (1960a): A generalization of the problem of transmission. Ann. Sc. Norm. Sup. Pisa *14*, No. 3, 207–236. Zbl. 94, 296

Schechter, M. (1960b): Negative norms and boundary problems. Ann. Math. *72*, No. 3, 581–593. Zbl. 97, 84

Seeley, R.T. (1964): Extension of functions defined in a half-space. Proc. Am. Math. Soc. *15*, 625–626. Zbl. 127, 284

Seeley, R.T. (1965): Refinement of the functional calculus of Calderón and Zygmund. Konikl. Nederl. Acad. van Wetenschappen, Proc., Ser. A, *68*, No. 3, 521–531. Zbl. 141, 133

Seeley, R.T. (1966): Singular integrals and boundary value problems. Am. J. Math. *88*, 781–809. Zbl. 178, 176

Seeley, R.T. (1969a): The resolvent of an elliptic boundary problem. Am. J. Math. *91*, No. 4, 889–920. Zbl. 191, 118

Seeley, R.T. (1969b): Analytic extension of the trace associated with elliptic boundary problems. Am. J. Math. *91*, No. 4, 963–983. Zbl. 191, 119

Seeley, R.T. (1972): Interpolation in L^p with boundary conditions. Studia Math. *44*, 47–60. Zbl. 237.46041

Seeley, R.T. (1978): A sharp asymptotic remainder estimate for the eigenvalues of the Laplacian in a domain of $\mathbb{R}^3$. Adv. Math. *29*, 244-269. Zbl. 382.35043

Seeley, R.T. (1980): An estimate near the boundary for the spectral function of the Laplace operator. Am. J. Math. *102*, No. 5, 869–902. Zbl. 447.35029

Senator, K. (1967): Completely elliptic boundary conditions of Dirichlet type. Bull. Acad. Polon. Sci., Ser. Sci. Math. Astron. Phys., *15*, No. 8, 569–574. Zbl. 153, 427

Shapiro, Z.Ya. (1953): On general boundary problems for elliptic equations. Izv. Akad. Nauk SSSR, Ser. Mat., *17*, 539–562. Zbl. 51, 329

Shatalov, V.E. (1977): Some asymptotic expansions for Sobolev problems. Sib. Mat. Zh. *28*, No. 6, 1393–1410. Zbl. 395.35008. English transl.: Sib. Math. J. *18*, 987–1000 (1978).

Sheftel', Z.G. (1965): Energy inequalities and general boundary value problems for elliptic equations with discontinuous coefficients. Sib. Mat. Zh. *6*, No. 3, 636–668 (Russian). Zbl. 156, 343

Sheftel', Z.G. (1966): General theory of boundary problems for elliptic systems with discontinuous coefficients. Ukr. Mat. Zh. *18*, No. 3, 132–136. Zbl. 156, 344

Shkalikov, A.A. (1983): Boundary value problems for ordinary differential equations with a parameter in boundary conditions. Tr. Semin. Petrovskogo *9*, 190–229. Zbl. 553.34014. English transl.: J. Sov. Math. *33*, 1311–1342 (1986).

Shkred, A.V. (1989): On a linearization of spectral problems with a parameter in a boundary condition and properties of Keldysh's generated chains. Mat. Zametki *46*, No. 4, 99–109. English transl.: Math. Notes *46*, No. 4, 820–827 (1989). Zbl. 708.35060

Shubin, M. (1992): Spectral theory of elliptic operators on non-compact manifolds. Astérisque *207*, 35–108.

Skubachevskij, A.L. (1983): Nonlocal elliptic problems with a parameter. Mat. Sb. *121* (163), No. 2, 201–210. English transl.: Math. USSR, Sb. *49*, 197–206 (1984). Zbl. 564.35043

Skubachevskij, A.L. (1985): Solvability of elliptic problems with boundary conditions of Bitsadze–Samarskij type. Differ. Uravn. *21*, No. 4, 701–706. Zbl. 569.35026 English transl.: Differ. Equations *21*, 478–482 (1985). Zbl. 585.35027

Skubachevskij, A.L. (1986): Elliptic problems with nonlocal conditions near the boundary. Mat. Sb. *129* (171), No. 2, 279–302. English transl.: Math. USSR, Sb. *57*, 293–316 (1987). Zbl. 615.35028

Skubachevskij, A.L. (1991): On the stability of index of nonlocal elliptic problems. J. Math. Anal. Appl. *160*, No. 2, 323–341. Zbl. 780.47036

Slobodetskij, L.N. (1958): Generalized Sobolev spaces and their applications to boundary problems for partial differential equations. Uch. Zap. Leningr. Gos. Ped. Inst. *197*, 54–112. Zbl. 192, 228. English transl.: Am. Math. Soc. Transl., Ser. II, *57*, 207–275 (1966).

Slobodetskij, L.N. (1960): L_2-estimates for solutions of elliptic and parabolic systems. Vestn. Leningr. Univ. *15*, No. 7, 28–47. Zbl. 94, 296

Sobolev, S.L. (1950): Applications of Functional Analysis in Mathematical Physics. Leningrad University, Leningrad. English transl.: Am. Math. Soc., Providence, 1963. Zbl. 123, 90

Solomyak, M.Z. (1963): On linear elliptic first order systems. Dokl. Akad. Nauk SSSR *150*, No. 1, 48–51. English transl.: Sov. Math., Dokl. *4*, 604–607 (1963). Zbl. 163, 346

Solonnikov, V.A. (1964, 1966): On general boundary value problems for systems elliptic in the sense of Douglis–Nirenberg. I, II. Izv. Akad. Nauk SSSR, Ser. Mat., *28*, 665–706 (1964), and Tr. Mat. Inst. Steklova *92*, 233–297 (1966). Zbl. 128, 330 and 167, 149. English transl.: Am. Math. Soc. Transl., Ser. II, *56*, 193–232 (1964) and Proc. Steklov Inst. Math. *92*, 3–32 (1968).

Solonnikov, V.A. (1967): Estimates in L_p of solutions of elliptic and parabolic systems. Tr. Mat. Inst. Steklova *102*, 137–160. English transl.: Proc. Steklov Math. Inst. *102*, 157–185 (1970). Zbl. 204, 421

Sternin, B.Yu. (1966): Elliptic and parabolic equations on a manifold with boundary consisting of components of different dimensions. Tr. Mosk. Mat. O-va *15*, 346–382. Zbl. 161, 85. English transl.: Trans. Mosc. Math. Soc. *15*, 387–429 (1966).

Sternin, B.Yu. (1976): Relative elliptic theory and Sobolev's problem. Dokl. Akad. Nauk SSSR *230*, No. 2, 287–290. English transl.: Sov. Math., Dokl. *17* (1976), 1306–1309 (1977). Zbl. 365.58017

Taylor, M.E. (1981): Pseudodifferential Operators. Princeton Univ. Press, Princeton, N.J. Zbl. 453.47026

Trèves, F. (1980): Introduction to Pseudo-Differential and Fourier Integral Operators. Vols. 1, 2. Plenum Press, New York London. Zbl. 453.47027

Triebel, H. (1978): Interpolation. Function Spaces. Differential Operators. North-Holland, Amsterdam. Zbl. 387.46033

Vainberg, B.R. (1982): Asymptotic Methods in Equations of Mathematical Physics. Moscow University, Moscow (Russian). Zbl. 518.35002

Vainberg, B.R., Grushin, V.V. (1967): Uniformly nonelliptic problems, I, II. Mat. Sb. *72* (114), 602–636 and *73* (115), 126–154. English transl.: Math. USSR, Sb. *1*, 543–568 (1967) and *2*, 111–133 (1967). Zbl. 179, 433

Vassiliev, D.G. (1984): Two-term asymptotics of the spectrum of a boundary value problem under an interior reflection of general form. Funkts. Anal. Prilozh. *18*, No. 4, 1–13. English transl.: Funct. Anal. Appl. *18* (1984), No. 3, 267–277. Zbl. 574.35032

Vassiliev, D.G. (1986a): Asymptotics of the spectrum of a boundary value problem. Tr. Mosk. Matem. O-va *49*, 167–237. Zbl. 623.58024. English transl.: Trans. Moscow Math. Soc. *49*, 173–245 (1987). Zbl. 632.58036

Vassiliev, D.G. (1986b): Two–term asymptotic formula for a boundary value problem in the case of a piecewise–smooth boundary. Dokl. Akad. Nauk SSSR *286*, No. 5, 1043–1046. English transl.: Sov. Math., Dokl. *33*, 227–230 (1986). Zbl. 629.58021

Verchota, G.C. (1984): Layer potentials and regularity for the Dirichlet problem for Laplace equation in Lipschitz domains. J. Funct. Analysis *59*, 572–611. Zbl. 589.31005

Verchota, G.C. (1990): The Dirichlet problem for the polyharmonic equation in Lipschitz domain. Indiana Univ. Math. J. *39*, 671–702. Zbl. 724.31005

Vishik, M.I. (1950): On strongly elliptic systems of differential equations. Dokl. Akad. Nauk SSSR *74*, 881–884. Zbl. 41, 217

Vishik, M.I. (1951): On strongly elliptic systems of differential equations. Mat. Sb. *29*, 617–676. Zbl. 44, 95

Vishik, M.I. (1952): On general boundary value problems for elliptic differential equations. Tr. Mosk. Mat. O-va *1*, 187–246. Zbl. 47, 95. English transl.: Am. Math. Soc. Transl., Ser. II, *24*, 107–172 (1963) .

Vishik, M.I., Eskin, G.I. (1964): Convolution equations in a bounded domain. Usp. Mat. Nauk *20*, No. 3, 89–152. English transl.: Russ. Math. Surv. *20*, No. 3, 85–151 (1964). Zbl. 152, 342

Vishik, M.I., Eskin, G.I. (1967): Elliptic convolution equations in a bounded domain and its applications. Usp. Mat. Nauk *22*, No. 1, 15–76. English transl.: Russ. Math. Surv. *22*, No. 1, 13–75 (1967). Zbl. 167, 448

Voĭtovich, N.N., Katsenelenbaum, B.Z. , Sivov, A.N. (1977): Generalized Method of Eigen-Oscillations in the Diffraction Theory. Nauka, Moscow (Russian). Zbl. 384.35002

Volevich, L.R. (1965): Solvability of boundary value problems for general elliptic systems. Mat. Sb. *68* (110), No. 3, 373–416. Zbl. 141, 298. English transl.: Am. Math. Soc. Transl., Ser. II, *67*, 182–225 (1968).

Vol'pert, A.I. (1961): On the index and normal solvability of boundary problems for elliptic systems of differential equations on the plane. Tr. Mosk. Mat. O-va *10*, 41–87. Zbl. 123, 76

Yakubov, S.Ya. (1986): Multiple completeness of root vectors of unbounded polynomial operator pencil. Rev. Roum. Math. Pures Appl. *31*, No. 5, 423–438.

Yakubov, S.Ya.(1989): Multiple completeness for a system of operator pencils and elliptic boundary problems. Mat. Sb. *181*, No. 1, 95–113. English transl.: Math. USSR., Sb. *69*, No. 1, 99–119 (1991). Zbl. 716.47008

Yakubov, S. (1994): Completeness of Root Functions of Regular Differential Operators. Pitman Monographs and Surveys in Pure and Appl. Math. *71*, Longman.

II. Boundary Value Problems for Elliptic Pseudodifferential Operators

A. V. Brenner and E. M. Shargorodsky

Translated from the Russian
by A. Brenner

Contents

Introduction

The present article is intended to be a short introduction to the theory of boundary value problems for elliptic pseudodifferential operators. In this regard the traditional statement that a lack of space causes the omission of many interesting and important topics, and only the slight broaching of others, is an obvious one for us to make.

We have not attempted to put together a comprehensive survey of the existing literature since that would be beyond our forces, and the references are subjective and incomplete. The references to works about elliptic pseudodifferential operators on closed manifolds and about boundary value problems for elliptic differential operators are almost completely absent. An exposition of these theories and corresponding references can be found, for example, in (Agranovich M.S.(1990)) and the article of M.S.Agranovich in the present volume. We also use in this article some notions and facts from topology and functional analysis (manifolds, bundles, K-groups, Fredholm and nuclear operators, trace, etc.), assuming that these are familiar to the reader, who can make an acquaintance with them through standard textbooks or other articles of this series.

An important role in the article (and in the theory of pseudodifferential operators on manifolds with boundary in general) is played by the transmission property, which was considered by M.I.Vishik and G.I.Eskin for the first time in (Vishik M.I., Eskin G.I.(1965)). The necessity for the introduction of this property is forced on us by the following circumstance.

Let M be a smooth manifold with boundary $\Gamma = \partial M$ and interior Ω, and let M be embedded in a smooth manifold M_0 without boundary. Let E_0 and E_0' be smooth vector bundles over M_0, and write $E_0|_M = E$, $E_0'|_M = E'$. Finally, let $A : C_0^\infty(E_0) \to C^\infty(E_0')$ be a pseudodifferential operator and $u \in C_0^\infty(E) \equiv \pi_\Omega C_0^\infty(E_0)$, where π_Ω is the restriction operator from M_0 to Ω. (In the simplest case M is the closure of a domain Ω in the euclidean space $\mathbb{R}^n$, $M_0 = \mathbb{R}^n$, $E_0 = E_0' = \mathbb{R}^n \times \mathbb{C}$, $A : C_0^\infty(\mathbb{R}^n, \mathbb{C}) \to C^\infty(\mathbb{R}^n, \mathbb{C})$, $u \in C_0^\infty(M, \mathbb{C})$.) In order to define the action of A on u we have to extend the function u to M_0. We take the zero extension $e_\Omega u$ defined by $e_\Omega u = u$ on Ω

and $e_\Omega u = 0$ on $M_0 \backslash \Omega$. In general, the function $e_\Omega u$ is not in $C_0^\infty(E_0)$ since it may have a discontinuity along Γ. Thus the function $Ae_\Omega u$ may not be C^∞ smooth.

If we require the correlation $A_\Omega u \equiv \pi_\Omega A e \Omega u \in C^\infty(E')$ to hold, we have to narrow either the set of operators or the set of functions. In the first case we consider pseudodifferential operators with the transmission property (see §§ 2,3), in the second one we take the functions supported in M and vanishing on Γ (see § 4).

The article is devoted to analytic aspects of the theory of elliptic boundary value problems. There is almost no discussion of the topological point of view here. In particular, we omit an exposition of the index theory and its applications. Such subjectivity might be explained not only by the limitations in this volume, but also by the fact that the work of B.V.Fedosov devoted to the index theory has been already published in the present series.

The authors are grateful to M.S.Agranovich, R.V.Duduchava and M.A. Shubin for fruitful advice and to M.Eastham and G.Grubb for their kind help with the preparation of the English edition.

§1. Preliminary Ideas

1.1. Some Notation. We use the conventional notation in the following list.

$\mathbb{N}$ – the set of all positive integers.

$\mathbb{Z}$ – the set of all integers.

$\mathbb{Z}_+ = \mathbb{N} \cup \{0\}$.

$\mathbb{R}$ – the set of all real numbers.

$\mathbb{C}$ – the set of all complex numbers.

$\mathbb{R}^n$ – the standard real n-dimensional euclidean space.

$\mathbb{C}^n$ – the standard complex n-dimensional euclidean space.

$\mathbb{R}^n_\pm = \{x \in \mathbb{R}^n : x_n > \text{ (resp. } <) \; 0\}$; $\overline{\mathbb{R}^n_\pm}$ – the closures of the half-spaces $\mathbb{R}^n_\pm$ in $\mathbb{R}^n$; $\overline{\mathbb{R}^{2n}_{++}} = \overline{\mathbb{R}^n_+} \times \overline{\mathbb{R}^n_+}$. $\mathbb{Z}^n_+ = (\mathbb{Z}_+)^n$.

$\mathbb{C}_\pm = \{z \in \mathbb{C} : \operatorname{Im} z > \text{(resp. } <) \, 0\}$.

$\frac{\partial}{\partial x} = \left(\frac{\partial}{\partial x_1}, \dots, \frac{\partial}{\partial x_n}\right)$, where $(x_1, \dots, x_n) \in \mathbb{R}^n$.

$D = i^{-1}\frac{\partial}{\partial x}$, $D_j = i^{-1}\frac{\partial}{\partial x_j}$, where $i = \sqrt{-1} \in \mathbb{C}$.

$\partial^l_{x_j} = \frac{\partial^l}{\partial x^l_j}$, $l \in \mathbb{Z}_+$.

$D^\alpha = D_1^{\alpha_1} \dots D_n^{\alpha_n}$, where α is a multiindex, i.e. $\alpha = (\alpha_1, \dots, \alpha_n)$, $\alpha_j \in \mathbb{Z}_+$.

$\xi^\alpha = \xi_1^{\alpha_1} \dots \xi_n^{\alpha_n}$, where $\xi = (\xi_1, \dots, \xi_n) \in \mathbb{R}^n$.

$x \cdot \xi = x_1\xi_1 + \dots + x_n\xi_n$, where $x = (x_1, \dots, x_n) = (x', x_n) \in \mathbb{R}^n$ and $\xi = (\xi_1, \dots, \xi_n) = (\xi', \xi_n) \in \mathbb{R}^n$.

$|\alpha| = \alpha_1 + \dots + \alpha_n$, $\alpha! = \alpha_1! \dots \alpha_n!$, where α is a multiindex, $\alpha = (\alpha', \alpha_n)$.

$|x| = (x_1^2 + \dots + x_n^2)^{1/2}$, $\langle x \rangle = (1 + x_1^2 + \dots + x_n^2)^{1/2}$ for $x = (x_1, \dots, x_n) \in \mathbb{R}^n$.

$\Delta = \Delta_x = \frac{\partial^2}{\partial x_1^2} + \dots + \frac{\partial^2}{\partial x_n^2}$ – the Laplacian in $\mathbb{R}^n$.

1.2. Functional Spaces (Besov O.V., Il'in V.P., Nikolskij S.M. (1975), Kudryavtsev L.D., Nikolskij S.M. (1988), Nikolskij S.M. (1977), Triebel H. (1978), Triebel H. (1983)).

Here we continue our list of standard notation.

$\mathcal{E}(\Omega) = C^\infty(\Omega)$ – the space of all functions that are defined and are infinitely smooth in a domain $\Omega \subset \mathbb{R}^n$.

$\mathcal{D}(\Omega) = C_0^\infty(\Omega)$ – the space of C^∞-functions with compact supports in Ω.

$\mathcal{D}'(\Omega)$ – the space of all distributions (generalized functions) on Ω.

$\mathcal{E}'(\Omega)$ – the space of all distributions (generalized functions) with compact supports in Ω.

$L_p(\Omega)$, $1 \le p \le \infty$ – the space of all p–integrable functions on Ω,

$$\| u \|_{L_p(\Omega)} = \| u | L_p(\Omega) \| = \left(\int_\Omega |u(x)|^p \, \mathrm{d}x \right)^{1/p}$$

(with the usual change of the integral norm into the **sup** norm in the case $p = \infty$).

$S(\mathbb{R}^n)$ – the Schwartz space of C^∞ functions on $\mathbb{R}^n$, with derivatives vanishing faster than any power of $|x|$ as $|x| \to \infty$.

$S'(\mathbb{R}^n)$ – the corresponding dual space of all tempered distributions on $\mathbb{R}^n$.

$\mathbf{supp}\, u$ – the support of the (generalized) function u.

Also we denote by π_Ω the restriction operator of distributions from $\mathbb{R}^n$ to Ω, acting from $\mathcal{D}'(\mathbb{R}^n)$ into $\mathcal{D}'(\Omega)$, and by e_Ω the operator of zero extension, mapping a function u on Ω into the function $e_\Omega u$ which coinsides with u on Ω and is zero on $\mathbb{R}^n \setminus \Omega$. In the case $\Omega = \mathbb{R}^n_\pm$ we write $\pi_\pm$ and $e_\pm$ instead of π_Ω and e_Ω.

The direct and inverse Fourier transforms F, $F^{-1} : S(\mathbb{R}^n) \to S(\mathbb{R}^n)$ are defined by the formulae

$$\widehat{u}(\xi) \equiv F(u)(\xi) \equiv F_{x\to\xi}(u) = \int_{\mathbb{R}^n} \mathrm{e}^{-i\xi\cdot x} u(x) \, \mathrm{d}x \,,$$

$$\widetilde{f}(x) \equiv F^{-1}(f)(x) \equiv F^{-1}_{\xi\to x}(f) = (2\pi)^{-n} \int_{\mathbb{R}^n} \mathrm{e}^{ix\cdot\xi} f(\xi) \, \mathrm{d}\xi \,.$$

Using the duality we can extend the Fourier transforms to continuous operators $F, F' : S'(\mathbb{R}^n) \to S'(\mathbb{R}^n)$.

Let

$$M_0 = \left\{ \xi \in \mathbb{R}^n : |\xi| \le 2 \right\}$$

and

$$M_j = \left\{ \xi \in \mathbb{R}^n : 2^{j-1} \le |\xi| \le 2^{j+1} \right\}, \quad j \in \mathbb{N} \,.$$

Then for $s \in \mathbb{R}$, $1 \le p \le \infty$ and $1 \le q < \infty$ we define

$$B^s_{p,q}(\mathbb{R}^n) = \left\{ f \in S'(\mathbb{R}^n) : f \underset{S'}{=} \sum_{j=0}^{\infty} a_j \,, \quad \operatorname{supp} F a_j \subset M_j \,; \right.$$

$$\| \{a_j\} \|_{l_q^s(L_p)} = \left(\sum_{j=0}^{\infty} (2^{sj} \| a_j \|_{L_p})^q \right)^{1/q} < \infty \Bigg\} .$$

In the case $q = \infty$ we define

$$B_{p,\infty}^s(\mathbb{R}^n) = \Bigg\{ f \in S'(\mathbb{R}^n) : f \underset{S'}{=} \sum_{j=0}^{\infty} a_j , \quad \operatorname{supp} Fa_j \subset M_j ;$$

$$\| \{a_j\} \|_{l_\infty^s(L_p)} = \sup_{j \in \mathbb{Z}_+} 2^{sj} \| a_j \|_{L_p} < \infty \Bigg\} .$$

In the space $B_{p,q}^s(\mathbb{R}^n)$ we introduce the norm

$$\| f \|_{B_{p,q}^s(\mathbb{R}^n)} = \inf_{f = \sum a_j} \| \{a_j\} \|_{l_q^s(L_p)} .$$

For $s \in \mathbb{R}$ and $1 < p < \infty$, we also define the spaces

$$H_p^s(\mathbb{R}^n) = \{ f \in S'(\mathbb{R}^n) : \| f \|_{H_p^s(\mathbb{R}^n)} = \| F^{-1} \langle \xi \rangle^s F f \|_{L_p} < \infty \} .$$

The space $H_p^s(\mathbb{R}^n)$ is usually called the space of *Bessel potentials* (or the Liouville space, or the Lebesgue space) and the space $B_{p,q}^s(\mathbb{R}^n)$ is called the *Besov space.* Note that in contrast to the present work the symbol $H_p^s(\mathbb{R}^n)$ is often used for the Nikolskij spaces $B_{p,\infty}^s(\mathbb{R}^n)$, and the space of Bessel potentials is often denoted by $L_p^s(\mathbb{R}^n)$ (see, for example, Besov O.V., Il'in V.P., Nikolskij S.M. (1975), Kudryavtsev L.D., Nikolskij S.M. (1988), Nikolskij S.M. (1977)).

If $m \in \mathbb{Z}_+$, then $H_p^m(\mathbb{R}^n)$ coinsides with the *Sobolev space* $W_p^m(\mathbb{R}^n)$:

$$H_p^m(\mathbb{R}^n) = W_p^m(\mathbb{R}^n)$$

$$= \Bigg\{ f \in L_p(\mathbb{R}^n) : \|f\|_{W_p^m} = \sum_{|\alpha| \le m} \|D^\alpha f\|_{L_p} < \infty \Bigg\} .$$

Let us now write $s \in \mathbb{R}$ in the form $s = [s]^- + \{s\}^+$, where $[s]^- \in \mathbb{Z}$ and $0 < \{s\}^+ \le 1$, and introduce notation

$$(\Delta_h^1 f)(x) = f(x+h) - f(x)$$

and

$$(\Delta_h^2 f)(x) = f(x+2h) - 2f(x+h) + f(x) ,$$

where $h \in \mathbb{R}^n$. Then for $s > 0$, $1 \le p < \infty$ and $1 \le q < \infty$ an equivalent norm in $B_{p,q}^s(\mathbb{R}^n)$ is

$$\|f\|^*_{B_{p,q}^s(\mathbb{R}^n)} = \|f\|_{W_p^{[s]^-}} +$$

$$+ \sum_{|\alpha| = [s]^-} \left(\int_{\mathbb{R}^n} |h|^{-\{s\}^+ q} \|\Delta_h^2 D^\alpha f\|_{L_p}^q \frac{dh}{|h|^n} \right)^{1/q} \tag{1.1}$$

while, for $q = \infty$, an equivalent norm is

$$\|f\|^*_{B^s_{p,\infty}(\mathbb{R}^n)} = \|f\|_{W_p^{[s]^-}} + \sum_{|\alpha|=[s]^-} \sup_{0\neq h\in\mathbb{R}^n} |h|^{-\{s\}^+} \|\Delta_h^2 D^\alpha f\|_{L_p} .$$

Note also that for $s > 0$, the space $B^s_{\infty,\infty}(\mathbb{R}^n)$ coincides with the *Zygmund space* $Z^s(\mathbb{R}^n)$:

$$B^s_{\infty,\infty}(\mathbb{R}^n) = Z^s(\mathbb{R}^n) = \Bigg\{ f \in S'(\mathbb{R}^n) : \|f\|_{Z^s}$$

$$= \sum_{|\alpha|\leq[s]^-} \sup_{x\in\mathbb{R}^n} |D^\alpha f(x)| + \sum_{|\alpha|=[s]^-} \sup_{x\in\mathbb{R}^n,\, h\in\mathbb{R}^n\setminus\{0\}} \frac{|(\Delta_h^2 D^\alpha f)(x)|}{|h|^{\{s\}^+}} < \infty \Bigg\} . \tag{1.2}$$

Now let $s > 0$ and $s \notin \mathbb{N}$. Then $[s]^-$ is the integer part of s and $\{s\}^+$ is its fractional part: $[s]^- = [s]$, $\{s\}^+ = \{s\}$. In this case one can prove that $B^s_{p,p}(\mathbb{R}^n)$, $1 \leq p < \infty$, and $Z^s(\mathbb{R}^n)$ coincide with the Sobolev-Slobodeckij space $W_p^s(\mathbb{R}^n)$ and the Hölder space $C^s(\mathbb{R}^n)$ with the norms defined by the formulas (1.1) and (1.2) respectively and with Δ_h^2 replaced by Δ_h^1.

Note also that for all $s \in \mathbb{R}$ the equality $H_2^s(\mathbb{R}^n) = B^s_{2,2}(\mathbb{R}^n)$ is valid.

Now let $\Omega \subset \mathbb{R}^n$ be an arbitrary domain and let $X(\mathbb{R}^n)$ be one of the functional spaces defined above. We shall use the notation

$$X(\overline{\Omega}) = \pi_\Omega X(\mathbb{R}^n) ,$$

$$\widetilde{X}(\overline{\Omega}) = \{ f \in X(\mathbb{R}^n) : \operatorname{supp} f \subset \overline{\Omega} \} .$$

Now let M be a smooth compact n-dimensional manifold with boundary $\Gamma = \partial M$ embedded in a smooth closed manifold (i.e. a compact manifold without boundary) M_0. One can take $M_0 = 2M$ – a duplicate of M that may be obtained by pasting two copies of M along Γ (see Milnor J.W. (1965)). Thus we consider any smooth vector bundle E over M as a restriction to M of a smooth vector bundle E_0 over M_0 (see, for example, Palais R.S. (1965), Chapter X, §4, Theorem 5).

The spaces $X(E_0)$ (where $X(\mathbb{R}^n)$ is one of the spaces defined above) are defined in a standard way with a help of partition of unity.

Further on we denote by Ω the interior part of a manifold M and by π_Ω the restriction operator of functions from M_0 to Ω. As above, we define the spaces

$$X(\overline{E}) = \pi_\Omega X(E_0) ,$$

$$\widetilde{X}(\overline{E}) = \{ f \in X(E_0) : \operatorname{supp} f \subset M \} .$$

Also we define the standard trace operator $\gamma_j, j \in \mathbb{Z}_+$, by

$$\gamma_j : u \mapsto (\partial_{x_n}^j u)\big|_{x_n=0} .$$

It maps $S(\overline{\mathbb{R}^n_+})$ into $S(\mathbb{R}^{n-1})$, where $\mathbb{R}^{n-1}$ is identified with the boundary of the half-space $\mathbb{R}^n_+$. It can be proved that, for $s > 1/p$, and either $1 < p < \infty$ or $1 \leq p, q \leq \infty$, the mapping γ_0 may be extended to the continuous operator

$$\gamma_0 : H_p^s(\overline{\mathbb{R}_+^n}) \to B_{p,p}^{s-1/p}(\mathbb{R}^{n-1}) \ \big(\text{or } B_{p,q}^s(\overline{\mathbb{R}_+^n}) \to B_{p,q}^{s-1/p}(\mathbb{R}^{n-1})\big).$$

Moreover, the operator

$$\rho^{(m)} = \{\gamma_0, \dots, \gamma_{m-1}\} : H_p^s(\overline{\mathbb{R}_+^n}) \to \prod_{j=0}^{m-1} B_{p,p}^{s-j-1/p}(\mathbb{R}^{n-1})$$

$$\left(\text{or } B_{p,q}^s(\overline{\mathbb{R}_+^n}) \to \prod_{j=0}^{m-1} B_{p,q}^{s-j-1/p}(\mathbb{R}^{n-1})\right),$$

which is continuous for $s > m+1/p-1$, $m \in \mathbb{N}$, $1 < p < \infty$ (or $1 \le p, q \le \infty$), is also surjective. Analogous results are valid in the case of functional spaces defined on smooth compact manifolds with boundary.

1.3. Pseudodifferential Operators (Egorov Yu.V., Shubin M.A. (1988), Hörmander L. (1983, 1985), Pseudodifferential operators (1967), Shubin M.A. (1978), Taylor M.E. (1981), Trèves F. (1980))

Let $r \in \mathbb{R}$, $\rho, \delta \in [0,1]$ and $n_1, n_2 \in \mathbb{N}$ be arbitrary numbers. The set $\mathbf{S}_{\rho,\delta}^r(\mathbb{R}^{n_1} \times \mathbb{R}^{n_2})$ (or simply $\mathbf{S}_{\rho,\delta}^r$ if there is no misunderstanding) consists of functions $a \in C^\infty(\mathbb{R}^{n_1} \times \mathbb{R}^{n_2})$ possessing the estimates

$$\left|D_x^\beta D_\xi^\alpha a(x,\xi)\right| \le C_{\alpha\beta} \langle\xi\rangle^{r-\rho|\alpha|+\delta|\beta|}, \quad \forall (x,\xi) \in \mathbb{R}^{n_1} \times \mathbb{R}^{n_2}, \tag{1.3}$$

$\forall\, \alpha \in \mathbb{Z}_+^{n_2}$ and $\beta \in \mathbb{Z}_+^{n_1}$, where the constants $C_{\alpha\beta} = C_{\alpha\beta}(a)$ are independent of (x,ξ). Define also

$$\mathbf{S}_{\rho,\delta}^{-\infty} = \bigcap_{r\in\mathbb{R}} \mathbf{S}_{\rho,\delta}^r = \mathbf{S}_{1,0}^{-\infty}.$$

We denote by $\mathbf{S}^\mu(\mathbb{R}^{n_1} \times \mathbb{R}^{n_2})$, where $\mu \in \mathbb{C}$, the set of *polyhomogeneous* or *classical symbols* of order μ, i.e. those functions $a \in \mathbf{S}_{1,0}^{\mathbf{Re}\,\mu}$ such that

$$a \sim \sum_{j=0}^{\infty} a_{\mu-j}$$

for some sequence of functions $a_{\mu-j} \in \mathbf{S}_{1,0}^{\mathbf{Re}\,\mu-j}$ which are positively homogeneous[1] in ξ for $|\xi| \ge 1$ and of order $\mu - j$. Thus

$$a - \sum_{j=0}^{N-1} a_{\mu-j} \in \mathbf{S}_{1,0}^{\mathbf{Re}\,\mu-N}(\mathbb{R}^{n_1} \times \mathbb{R}^{n_2}), \quad \forall N \in \mathbb{N},$$

where

$$a_{\mu-j}(x, t\xi) = t^{\mu-j} a_{\mu-j}(x,\xi), \quad \text{for } |\xi| \ge 1,\ t \ge 1.$$

An operator $A : S(\mathbb{R}^n) \to S'(\mathbb{R}^n)$ is called the **pseudodifferential operator** of order r with symbol $a(x,\xi) \in \mathbf{S}_{\rho,\delta}^r(\mathbb{R}^n \times \mathbb{R}^n)$ if it acts as follows:

---

[1] In the following we sometimes omit the word "positively" and call the corresponding symbols homogeneous.

$$Au(x) \equiv a(x,D)u(x) \equiv \mathbf{OP}(a(x,\xi))u(x) = (2\pi)^{-n} \int_{\mathbb{R}^n} e^{ix\cdot\xi} a(x,\xi)(Fu)(\xi)\, d\xi\,.$$

We denote the set of all such operators by $\mathbf{OP}(\mathbf{S}^r_{\rho,\delta})$. The set $\mathbf{OP}(\mathbf{S}^\mu)$ is defined analogously. If $a(x,D) \in \mathbf{OP}(\mathbf{S}^\mu)$, the corresponding function $a^0 \equiv a_\mu$ (see the definition of $\mathbf{S}^\mu$) is called the *leading (or principal) symbol* of the operator.

For any $a \in \mathbf{S}^r_{1,0}$, $r, t \in \mathbb{R}$, $p \in (1,\infty)$ and $q \in [1,\infty]$, a pseudodifferential operator $a(x,D)$ admits an extension to a continuous operator

$$a(x,D) : H^t_p(\mathbb{R}^n) \to H^{t-r}_p(\mathbb{R}^n)\,, \quad B^t_{p,q}(\mathbb{R}^n) \to B^{t-r}_{p,q}(\mathbb{R}^n)\,.$$

When the symbol $a \in \mathbf{S}^0_{1,0}$ has compact support in x, the continuity of $a(x,D)$ in $L_p(\mathbb{R}^n)$ may be proved (see Taylor M.E. (1981)) by reduction to the Mikhlin–Hörmander theorem about Fourier L_p-multiplicators (see, for example, Hörmander L. (1983,1985), V.1). For arbitrary $a \in \mathbf{S}^0_{1,0}$ the statement may be proved (in the case of the space $L_p(\mathbb{R}^n)$) by using the argument from the proof of Theorem 3.5 in Hörmander's article (see Hörmander L. (1966)). In general, the continuity of the operator $a(x,D) : H^t_p(\mathbb{R}^n) \to H^{t-r}_p(\mathbb{R}^n)$ is proved by order reduction (see Triebel H. (1978), Triebel H. (1983)), and the continuity of $a(x,D) : B^t_{p,q}(\mathbb{R}^n) \to B^{t-r}_{p,q}(\mathbb{R}^n)$ by interpolation (Triebel H. (1978), Triebel H. (1983)).

Let M_0 be a smooth compact manifold (without boundary), and let E_0 and $E_0{}'$ be smooth vector bundles over M_0. The class of pseudodifferential operators $\mathbf{OP}(\mathbf{S}^r_{1,0})(E_0, E_0{}')$ (or $\mathbf{OP}(\mathbf{S}^\mu)(E_0, E_0{}')$) acting from $C^\infty(E_0)$ into $\mathcal{D}'(E_0{}')$ is defined in a standard way via partition of unity and local coordinates. Clearly, the pseudodifferential operators of the class $\mathbf{OP}(\mathbf{S}^r_{1,0})(E_0, E_0{}')$ are continuous in the corresponding Besov spaces and the spaces of Bessel potentials.

§2. Boundary Value Problems for Elliptic Pseudodifferential Operators with the Transmission Property (Boutet de Monvel Theory)

2.1. Motivation (Grubb G. (1989)). Let A be a pseudodifferential operator. We say that the operator A_Ω is the restriction of A to the domain $\Omega \subset \mathbb{R}^n$ if $A_\Omega = \pi_\Omega A e_\Omega$. Obviously A_Ω maps $C^\infty(\bar{\Omega})$ into itself if A is a differential operator, but this is not true for general pseudodifferential operators. The operator e_Ω generates singularities on $\partial\Omega$, and so $A_\Omega u$ may not belong to $C^\infty(\bar{\Omega})$ even if $u \in C_0^\infty(\bar{\Omega}) = \pi_\Omega C_0^\infty(\mathbb{R}^n)$. Thus, we need special restrictions on the symbol of A in order to preserve the property $A_\Omega : C_0^\infty(\bar{\Omega}) \to C^\infty(\bar{\Omega})$. First of all, we find these restrictions in the simplest model case where $n = 1$, $\Omega = \mathbb{R}_+$ and the symbol $a = a(\xi) \in \mathbf{S}^{-1}_{1,0}(\mathbb{R} \times \mathbb{R})$ is independent of x.

Let $\tilde{a}(x) = F^{-1}a(\xi)$. It follows from (1.3) that $a \in L_2(\mathbb{R})$, and so $\tilde{a} \in L_2(\mathbb{R})$. Moreover, $D_\xi a \in L_1(\mathbb{R})$ and hence $x\tilde{a} \in \dot{C}(\mathbb{R})$ (the space of continuous functions vanishing as $|x| \to \infty$). Similarly,

$$D_x^k(x^j\tilde{a}) = F^{-1}(\xi^k(-D_\xi)^j a) \in \dot{C}(\mathbb{R}) \quad \text{for } j-k \geq 1\,,$$

and so $\tilde{a}$ belongs to $C^\infty(\mathbb{R} \setminus \{0\})$ and is rapidly decreasing as $|x| \to \infty$.

Now for the operator $A = a(D) = \mathbf{OP}\big(a(\xi)\big) = F^{-1}a(\xi)F$ we obtain

$$A_\Omega u = \pi_+ a(D) e_+ u = \pi_+ F^{-1}(a(\xi)f(\xi))\,,$$

where $u \in S(\bar{\mathbb{R}}_+)$ and $f = F(e_+u)$. The function af belongs to $L_1(\mathbb{R})$ since $a, f \in L_2(\mathbb{R})$, and so $A_\Omega u \in \dot{C}(\bar{\mathbb{R}}_+)$. Since also

$$xA_\Omega u = \pi_+ F^{-1}[-D_\xi(a(\xi)f(\xi))] = -\mathbf{OP}\big(D_\xi a(\xi)\big)_\Omega(u) + \mathbf{OP}\big(a(\xi)\big)_\Omega(xu)$$

it follows that $xA_\Omega u \in \dot{C}(\bar{\mathbb{R}}_+)$. Similarly we obtain $x^N A_\Omega u \in \dot{C}(\bar{\mathbb{R}}_+)$, $\forall N \in \mathbb{N}$, and particularly $A_\Omega u(x) = o(x^{-N})$ when $x \to +\infty$.

On the other hand, the formula (a version of Green's formula)

$$\partial_x e_+ u = e_+ \partial_x u + u(0)\delta$$

implies that

$$\begin{aligned} \partial_x A_\Omega u &= \pi_+ \partial_x a(D) e_+ u \\ &= \pi_+ a(D) e_+ \partial_x u + (\pi_+ a(D)\delta)\cdot u(0) = A_\Omega \partial_x u + \pi_+ \tilde{a}(x)\cdot u(0)\,. \end{aligned}$$

Thus $\partial_x A_\Omega u \in \dot{C}(\bar{\mathbb{R}}_+)$ for an arbitrary function $u \in S(\bar{\mathbb{R}}_+)$ if and only if $\pi_+\tilde{a}(x) \in \dot{C}(\bar{\mathbb{R}}_+)$.

As above we obtain

$$\partial_x^j A_\Omega u = A_\Omega \partial_x^j u + \sum_{0 \leq l \leq j-1} \pi_+ \partial_x^l \tilde{a}(x) \cdot \gamma_{j-1-l} u\,,$$

where $\gamma_k u = \partial_x^k u(0)$. Now we can easily deduce the following

Lemma 2.1. *For any given symbol $a(\xi) \in \mathbf{S}_{1,0}^{-1}(\mathbb{R} \times \mathbb{R})$, the operator $a(D)_\Omega$ maps $S(\bar{\mathbb{R}}_+)$ into the space of functions $v \in C^\infty(\bar{\mathbb{R}}_+)$ such that $\partial_x^j v(x) = o(x^{-N})$, $\forall N \in \mathbb{N}$, as $x \to +\infty$. In this case $a(D)_\Omega$ maps $S(\bar{\mathbb{R}}_+)$ into itself if and only if $\pi_+\tilde{a}(x) \in S(\bar{\mathbb{R}}_+)$. This condition is also necessary and sufficient for the operator $a(D)_\Omega$ to map $C_0^\infty(\bar{\mathbb{R}}_+)$ into $C^\infty(\bar{\mathbb{R}}_+)$.*

It is easily seen from the proof of Lemma 2.1 that, for any given symbol $a(\xi) \in \mathbf{S}_{1,0}^{-1}(\mathbb{R} \times \mathbb{R})$, the condition $\pi_+\tilde{a}(x) \in S(\bar{\mathbb{R}}_+)$ is equivalent to the existence of the finite limits

$$\lim_{x \to 0+} \partial_x^k \tilde{a}(x) \qquad \forall k \in \mathbb{Z}_+\,.$$

Example 2.2. In the case of classical symbols $a(\xi) \in \mathbf{S}^{-1}(\mathbb{R} \times \mathbb{R})$ the condition under discussion may be expressed in terms of the function $a(\xi)$ itself without the inverse Fourier transform $\tilde{a}(x)$.

Now let $\sigma > 0$ and consider the symbols $a_1(\xi) = (\sigma + i\xi)^{-1}$ and $a_2(\xi) = (\sigma^2 + \xi^2)^{-1/2}$ from $\mathbf{S}^{-1}(\mathbb{R} \times \mathbb{R})$. Their leading homogeneous parts are $(i\xi)^{-1}$ and $|\xi|^{-1}$ respectively, and the following formulas hold for their inverse Fourier transforms : $\tilde{a}_1(x) = \chi_+(x)\exp(-\sigma x)$ (χ_+ is the characteristic function of the ray $\mathbb{R}_+$), $\tilde{a}_2(x) = \text{const} \cdot \ln x + O(1)$ as $x \to 0+$ (see Eskin (1973)). Thus, $a_1(D)_\Omega$ maps $S(\bar{\mathbb{R}}_+)$ into itself, but $\partial_x a_2(D)_\Omega u \notin \dot{C}(\bar{\mathbb{R}}_+)$, if $u(0) \neq 0$.

An arbitrary polyhomogeneous symbol independent of x and of order -1 may be written in the form

$$a(\xi) = c_1 a_1(\xi) + c_2 a_2(\xi) + a_3(\xi), \quad c_1, c_2 \in \mathbb{C},\ a_3 \in \mathbf{S}^{-2},$$

since its leading homogeneous part must be a linear combination of ξ^{-1} and $|\xi|^{-1}$. Clearly $\xi \cdot a_3 \in \mathbf{S}^{-1} \subset \mathbf{S}^{-1}_{1,0}$, and so $\partial_x a_3(D)_\Omega u \in \dot{C}(\bar{\mathbb{R}}_+)$. Then it follows from the afore-said that, for $u(0) \neq 0$, the derivative $\partial_x a(D)_\Omega u$ belongs to $\dot{C}(\bar{\mathbb{R}}_+)$ only if $c_2 = 0$ or, in other words, the leading homogeneous term in the symbol $a(\xi)$ has to be proportional to ξ^{-1}. Continuing this analysis, we easily find that the rest of the terms in the asymptotic development of $a(\xi)$ are proportional to ξ^{-2}, $\xi^{-3}, \ldots$ for $|\xi| \geq 1$, if the operator $a(D)_\Omega$ is to map $S(\bar{\mathbb{R}}_+)$ into itself. This conclusion leads to the definition of the spaces H_d, the main goal of the next section.

2.2. The Spaces H_d. (Boutet de Monvel L. (1971), Grubb G. (1986), Grubb G. (1989), Rempel S., Schulze B.-W. (1982a)).

Definition 2.3. For $d \in \mathbb{Z}$, we denote by H_d the space of all C^∞ functions f on $\mathbb{R}$ with the asymptotic behavior $f(t) \sim \sum_{j \leq d} s_j t^j$ as $|t| \to \infty$, in the sense that, for all $k, N \in \mathbb{Z}_+$ and $|t| \geq 1$

$$\Big| D_t^k [f(t) - \sum_{d-N<j\leq d} s_j t^j] \Big| \leq C_{kN} |t|^{d-k-N} . \tag{2.1}$$

Obviously the inequalities (2.1) are equivalent to the larger system

$$\Big| D_t^k t^m [f(t) - \sum_{d-N<j\leq d} s_j t^j] \Big| \leq C_{kmN} |t|^{d-k+m-N} \quad \text{for } |t| \geq 1 , \tag{2.2}$$

for all $k, m, N \in \mathbb{Z}_+$.

We also define $H = \bigcup_{d\in\mathbb{Z}} H_d$. The definition of H_d implies that for an arbitrary function $f \in H_d$ there is a unique decomposition

$$f(t) = \sum_{j=0}^{d} s_j t^j + f_{-1}(t) ,$$

where $f_{-1} \in H_{-1}$. Consequently $H = \mathbb{C}[t] \bigoplus H_{-1}$, where $\mathbb{C}[t]$ denotes the space of all polynomials with complex coefficients.

Lemma 2.4. *Let $\sigma > 0$, $d \in \mathbb{Z}$ and $f \in C^\infty(\mathbb{R})$. We introduce the functions*

$$q(\tau) = \tau^d f(\tau^{-1}) \quad \textit{for } \tau \in \mathbb{R} \setminus \{0\}$$

and

$$g(z) = (1+z)^d f\left(\frac{\sigma}{i}\frac{1-z}{1+z}\right) \quad \textit{for } z \in \Gamma_1 = \{z \in \mathbb{C} : |z| = 1\},\ z \neq -1 .$$

Then the following statements are equivalent:

(i) $f \in H_d$;
(ii) *q extends to a function from $C^\infty(\mathbb{R})$;*
(iii) *g extends to a function from $C^\infty(\Gamma_1)$.*

A function g belongs to $C^\infty(\Gamma_1)$ if and only if it may be expanded in a Fourier series $g(z) = \sum_{k\in\mathbb{Z}} b_k z^k$ with rapidly decreasing coefficients:

$$\sup_{k\in\mathbb{Z}} (1+|k|)^N |b_k| < \infty, \quad \forall N \in \mathbb{N}.$$

In this case $g(z) = g^+(z) + g^-(z)$, where $g^+(z) = \sum_{k\geq 0} b_k z^k$ and $g^-(z) = \sum_{k<0} b_k z^k$ admit analytic continuation to the sets $\{z \in \mathbb{C} : |z| < 1\}$ and $\{z \in \mathbb{C} : |z| > 1\}$ respectively.

So, for a given function $f \in H_{-1}$, we obtain the decomposition

$$f(t) = 2\sigma \sum_{k\in\mathbb{Z}} b_k \frac{(\sigma - it)^k}{(\sigma + it)^{k+1}} = f^+(t) + f^-(t),$$

$$f^+(t) = 2\sigma \sum_{k\geq 0} b_k \frac{(\sigma - it)^k}{(\sigma + it)^{k+1}}, \qquad f^-(t) = 2\sigma \sum_{k<0} b_k \frac{(\sigma - it)^k}{(\sigma + it)^{k+1}} \tag{2.3}$$

with rapidly decreasing coefficients. The functions $f^\pm$ admit analytic continuation on $\mathbb{C}_\mp$ with the expansions (2.3) being valid in $\bar{\mathbb{C}}_\mp$.

The functions $\sqrt{\sigma/\pi}(\sigma - it)^k(\sigma + it)^{-k-1}$ form an orthonormal basis in $L_2(\mathbb{R})$. Consider now the Laguerre functions

$$\varphi_k(x,\sigma) = \sqrt{2\sigma} F^{-1}_{t\to x} \frac{(\sigma - it)^k}{(\sigma + it)^{k+1}} .$$

It is easy to see that

$$\varphi_k(x,\sigma) = \begin{cases} \sqrt{2\sigma}\, \chi_+(x)(\sigma - \partial_x)^k (x^k e^{-\sigma x})/k!, & \text{for } k \in \mathbb{Z}_+, \\ \varphi_{-k-1}(-x,\sigma) & \text{for } k \in \mathbb{Z} \setminus \mathbb{Z}_+ . \end{cases}$$

The Laguerre functions form a complete orthogonal system of eigenfunctions of the Laguerre operator

$$\mathbf{L}_\sigma = -\sigma^{-1}\partial_x x \partial_x + \sigma x + 1 \quad \text{in} \quad L_2(\mathbb{R}_+)$$

with simple eigenvalues $2(k+1)$, $k \in \mathbb{Z}_+$.

Writing by $c_k = \sqrt{2\sigma} b_k$ and $\hat{\varphi}_k = F\varphi_k$, we obtain

$$f^+(t) = \sum_{k \geq 0} c_k \hat{\varphi}_k(t,\sigma), \quad f^-(t) = \sum_{k<0} c_k \hat{\varphi}_k(t,\sigma) \tag{2.4}$$

and consequently we have the decomposition $H_{-1} = H_{-1}^+ \bigoplus H_{-1}^-$, where $H_{-1}^\pm$ consists of functions $f^\pm$ as in (2.4) with rapidly decreasing coefficients. Note also that $H_{-1}^+ \perp H_{-1}^-$ in $L_2(\mathbb{R})$.

Lemma 2.5. *The following statements are equivalent:*

(i) $f^\pm \in H_{-1}^\pm$;

(ii) $f^\pm \in C^\infty(\mathbb{R})$ *admits an analytic continuation to* $\mathbb{C}_\mp$ *such that* $f^\pm \in C^\infty(\bar{\mathbb{C}}_\mp)$, *and the estimates* (2.2) *are valid in* $\bar{\mathbb{C}}_\mp$;

(iii) $u^\pm(x) = F_{t\to x}^{-1} f^\pm(t)$ *belongs to* $e_\pm S(\bar{\mathbb{R}}_\pm)$.

We also write $H^+ = H_{-1}^+$ and $H^- = H_{-1}^- \bigoplus \mathbb{C}[t]$, so that $H = H_{-1} \bigoplus \mathbb{C}[t] = H^+ \bigoplus H^-$.

Using (2.3) we can write an arbitrary function $f \in H_d$ in the form

$$f(t) = \sum_{j=0}^{d} s_j t^j + 2\sigma \sum_{k\in\mathbb{Z}} b_k \frac{(\sigma - it)^k}{(\sigma+it)^{k+1}} = \sum_{j=0}^{d} s_j t^j + \sum_{k\in\mathbb{Z}} c_k \hat{\varphi}_k(t,\sigma). \tag{2.5}$$

Consider now the projections $\Pi^\pm : H \to H^\pm$, $\Pi_{-1}^- : H \to H_{-1}^-$, $\Pi_{-1} : H \to H_{-1}$, defined by the following formulas:

$$\Pi^+ f(t) = \sum_{k\geq 0} c_k \hat{\varphi}_k(t,\sigma),$$

$$\Pi^- f(t) = \sum_{j=0}^{d} s_j t^j + \sum_{k<0} c_k \hat{\varphi}_k(t,\sigma),$$

$$\Pi_{-1}^- f(t) = \sum_{k<0} c_k \hat{\varphi}_k(t,\sigma),$$

$$\Pi_{-1} f(t) = \sum_{k\in\mathbb{Z}} c_k \hat{\varphi}_k(t,\sigma),$$

where $f \in H_d$ is a function of the form (2.5). Clearly

$$\Pi_{-1}\Pi^+ = \Pi^+\Pi_{-1} = \Pi^+, \qquad \Pi_{-1}\Pi^- = \Pi^-\Pi_{-1} = \Pi_{-1}^-.$$

Lemma 2.6. *For an arbitrary function* $f \in H_{-1}$,

$$\Pi^\pm f(t) = \frac{1}{2} f(t) \pm \frac{1}{2\pi i} \int_{\mathbb{R}} \frac{f(\tau)}{t-\tau} \mathrm{d}\tau,$$

where the integral is understood in the sense of the Cauchy principal value.

Note that the space $F^{-1}(\mathbb{C}[t])$ consists of "polynomials in δ'":

$$\sum_{j=0}^{d} s_j D_x^j \delta .$$

We denote this space by $\mathbb{C}[\delta']$. It then follows from Lemma 2.5 that

$$F^{-1}H = e_+ S(\bar{\mathbb{R}}_+) \oplus e_- S(\bar{\mathbb{R}}_-) \oplus \mathbb{C}[\delta'] ,$$

and hence

$$F^{-1}\Pi^+ F = e_+\pi_+ , \qquad F^{-1}\Pi^- F = I - e_+\pi_+ , \tag{2.6}$$

and moreover

$$(I - e_+\pi_+)v = e_-\pi_- v ,$$

for $v \in e_+S(\bar{\mathbb{R}}_+) \oplus e_-S(\bar{\mathbb{R}}_-)$. If now $u \in S(\bar{\mathbb{R}}_+)$ and $f = F(e_+u) \in H^+$, integration by parts gives

$$f(t) \sim \sum_{k=0}^{\infty} s_{-1-k}\, t^{-1-k} , \quad s_{-1-k} = (-i)^{k+1}\gamma_k u = -iD_x^k u(0) . \tag{2.7}$$

We next define linear functionals π' and Π' on $F^{-1}H$ and H respectively by

$$\pi' a = \begin{cases} \lim_{x\to 0+} \ a(x) , & \text{if } a \in e_+S(\bar{\mathbb{R}}_+) \oplus e_-S(\bar{\mathbb{R}}_-) , \\ 0 , & \text{if } a \in \mathbb{C}[\delta'] , \end{cases}$$

$$\Pi' = \pi' F^{-1} ,$$

and it is also possible to define the functional Π' without the inverse Fourier transform as follows. Suppose that the integral

$$\int^+ f(t)\mathrm{d}t = \int_{\mathbb{R}} f(t)\mathrm{d}t , \quad \text{if } f \in H \cap L_1(\mathbb{R}) = H_{-2} ,$$

and

$$\int^+ f(t)\mathrm{d}t = \int_{\Gamma} f(t)\mathrm{d}t , \quad \text{if } f \in H ,$$

admits an extension to a meromorphic function in $\mathbb{C}_+$ with $\Gamma \subset \mathbb{C}_+$ denoting a contour oriented counterclockwise around the poles. In particular,

$$\int^+ f(t)\mathrm{d}t = 0 \tag{2.8}$$

if $f \in H^-$ or $f \in H^+ \cap H_{-2}$. It turns out that

$$\Pi' f = \frac{1}{2\pi} \int^+ f(t)\mathrm{d}t \qquad \forall f \in H .$$

Now we can rewrite (2.7) in the form

$$s_{-1-k} = \frac{1}{2\pi i} \int^{+} t^k f(t) dt\,, \qquad f \in H^+\,,\; k \in \mathbb{Z}_+\,.$$

2.3. Pseudodifferential Operators with the Transmission Property (Boutet de Monvel L. (1971), Grubb G. (1986), Grubb G.(1989), Grubb G., Hörmander L. (1990), Hörmander L. (1983, 1985), Rempel S., Schulze B.-W. (1982a)).

Definition 2.7. We say that the symbol $a(x,\xi) \in \mathbf{S}^d_{1,0}(\mathbb{R}^n \times \mathbb{R}^n)$, $d \in \mathbb{R}$, possesses the H-property (or satisfies the H-condition) for $x_n = 0$ if, for all multiindices α and $\beta \in \mathbb{Z}^n_+$ and all $(x', \xi') \in \mathbb{R}^{n-1} \times \mathbb{R}^{n-1}$, the symbol $D^\beta_x D^\alpha_\xi a(x', 0, \xi', \xi_n)$ belongs to H as a function of ξ_n. We also say that $a(x,\xi)$ possesses the UH-property for $x_n = 0$ if the corresponding estimates of the type (2.2) are uniform in x', meaning that the constants in these estimates are independent of x'.

Suppose now that $f \in H$. Then by Lemma 2.5 $\pi_\pm F^{-1} f \in S(\bar{\mathbb{R}}_\pm)$, and so the operator $f(D_n)_{\mathbb{R}_\pm} = \pi_\pm f(D_n) e_\pm$ maps $S(\bar{\mathbb{R}}_\pm)$ into itself in the manner of Lemma 2.1. The next theorem appears as a generalization of this fact.

Theorem 2.8. *Let $d \in \mathbb{R}$ and $a \in \mathbf{S}^d_{1,0}(\mathbb{R}^n \times \mathbb{R}^n)$. Then the following statements are equivalent:*

(i) $a(x,D)_{\mathbb{R}^n_\pm} = \pi_\pm a(x,D) e_\pm$ *maps* $C^\infty_0(\overline{\mathbb{R}^n_\pm}) = \pi_\pm C^\infty_0(\mathbb{R}^n)$ *into* $C^\infty(\overline{\mathbb{R}^n_\pm}) = \pi_\pm C^\infty(\mathbb{R}^n)$;

(ii) *the symbol $a(x,\xi)$ satisfies the H-condition for $x_n = 0$;*

(iii) *for any given function $\chi \in C^\infty_0(\mathbb{R}^{n-1})$ the symbol $\chi(x')a(x,\xi)$ satisfies the UH-condition for $x_n = 0$;*

(iv) *for any given function $\chi \in C^\infty_0(\mathbb{R}^{n-1})$ there exist symbols*

$$s_{j,\alpha,\beta}(x',\xi') \in \mathbf{S}^{d-j-|\alpha|}_{1,0}(\mathbb{R}^{n-1} \times \mathbb{R}^{n-1})\,, \quad j \in \mathbb{Z},\, j \le d\,,$$

such that

$$\Big|\xi^l_n D^\beta_x D^\alpha_\xi \chi(x') a(x,\xi)\,|_{x_n=0} - \sum_{0 \le j+l \le d-|\alpha|+l} s_{j,\alpha,\beta}(x',\xi') \xi^{j+l}_n\Big| \le C_{l,\alpha,\beta} \langle \xi' \rangle^{d+1-|\alpha|+l} \langle \xi \rangle^{-1}\,, \qquad \forall\, \alpha, \beta \in \mathbb{Z}^n_+\,, \quad \forall\, l \in \mathbb{Z}_+\,; \tag{2.9}$$

(v) *for any given function $\chi \in C^\infty_0(\mathbb{R}^{n-1})$ and an arbitrary C^∞ function $\sigma(\xi') > 0$ on $\mathbb{R}^{n-1}$ which coincides with $|\xi'|$ when $|\xi'| \ge 1$ and $1/2$ when $|\xi'| \le 1/2$, there exist symbols*

$$b_{k,\alpha,\beta}(x',\xi') \in \mathbf{S}^{d+1/2-|\alpha|}_{1,0}(\mathbb{R}^{n-1} \times \mathbb{R}^{n-1})\,,$$

such that

$$D^\beta_x D^\alpha_\xi \chi(x') a(x,\xi)\,|_{x_n=0} = \sum_{0 \le j \le d-|\alpha|} s_{j,\alpha,\beta}(x',\xi')\, \xi^j_n + \sum_{k \in \mathbb{Z}} b_{k,\alpha,\beta}(x',\xi')\, \hat{\varphi}_k(\xi_n, \sigma(\xi'))\,, \qquad \forall\, \alpha, \beta \in \mathbb{Z}^n_+\,, \tag{2.10}$$

where $s_{j,\alpha,\beta}$ are the same as in (2.9), and $b_{k,\alpha,\beta}(x',\xi')$ form a rapidly decreasing sequence in the class $\mathbf{S}^{d+1/2-|\alpha|}_{1,0}(\mathbb{R}^{n-1}\times\mathbb{R}^{n-1})$, namely

$$\left(\sum_{k\in\mathbb{Z}}\left|(1+|k|)^N D^{\gamma}_{x'}D^{\theta}_{\xi'}b_{k,\alpha,\beta}(x',\xi')\right|^2\right)^{1/2}$$
$$\leq C_{\alpha,\beta,\gamma,\theta,N}\,\langle\xi'\rangle^{d+1/2-|\alpha|-|\theta|},\qquad \forall\gamma,\theta\in\mathbb{Z}^{n-1}_+,\quad \forall N\in\mathbb{Z}_+ .$$

Remark 2.9. One can easily deduce from the uniqueness of the functions $s_{j,\alpha,\beta}$ satisfying (2.9) and (2.10) that

$$D^{\beta'}_{x'}D^{\alpha}_{\xi}s_{j,0,(0,\beta_n)}(x',\xi')\,\xi_n^j$$
$$= s_{j,(\alpha',0),\beta}(x',\xi')\;D^{\alpha_n}_{\xi_n}\xi_n^j = s_{j-\alpha_n,\alpha,\beta}(x',\xi')\;\xi_n^{j-\alpha_n},$$

and $s_{j,\alpha,\beta}$ becomes zero after $d-j-|\alpha|+1$ differentiations with respect to ξ', and hence is a polynomial in ξ' of degree $\leq d-j-|\alpha|$.

In the view of Theorem 2.8, the fulfillment of the H-condition is necessary (and sufficient) for operators $a(x,D)_{\mathbb{R}^n_\pm}$ to map $C_0^\infty(\overline{\mathbb{R}^n_\pm})$ into $C^\infty(\overline{\mathbb{R}^n_\pm})$. A question arises: is the H-condition also necessary for an operator $a(x,D)_{\mathbb{R}^n_+}$ to map $C_0^\infty(\overline{\mathbb{R}^n_+})$ into $C^\infty(\overline{\mathbb{R}^n_+})$? For polyhomogeneous symbols the answer is positive.

Theorem 2.10. *Suppose that $d\in\mathbb{Z}$ and $a\in\mathbf{S}^d(\mathbb{R}^n\times\mathbb{R}^n)$, and let $a\sim\sum_{l=0}^{\infty}a_{d-l}$ be the corresponding asymptotic expansion of $a(x,\xi)$ into a sequence of functions homogeneous for $|\xi|\geq 1$. The following statements are equivalent:*

(i) *$a(x,D)_{\mathbb{R}^n_+}$ maps $C_0^\infty(\overline{\mathbb{R}^n_+})$ into $C^\infty(\overline{\mathbb{R}^n_+})$;*

(ii) *the symbol $a(x,\xi)$ satisfies the H-condition for $x_n=0$;*

(iii) *the symbol $a(x,\xi)$ satisfies the UH-condition for $x_n=0$;*

(iv) *$a-\sum_{l<N}a_{d-l}$ satisfies the H-condition for $x_n=0$ and any $N\in\mathbb{Z}_+$;*

(v) *$a-\sum_{l<N}a_{d-l}$ satisfies the UH-condition for $x_n=0$ and any $N\in\mathbb{Z}_+$;*

$$\text{(vi)}\qquad D^{\beta}_{x}D^{\alpha}_{\xi}a_{d-l}(x',0,0,-\xi_n)=(-1)^{d-|\alpha|-l}D^{\beta}_{x}D^{\alpha}_{\xi}a_{d-l}(x',0,0,\xi_n)$$
$$\textit{for}\quad |\xi_n|\geq 1,\quad \forall\alpha,\beta\in\mathbb{Z}^n_+,\ \forall l\in\mathbb{Z}_+ . \tag{2.11}$$

Thus, if $a\in\mathbf{S}^d(\mathbb{R}^n\times\mathbb{R}^n)$ and $d\in\mathbb{Z}$, then it follows from $a(x,D)_{\mathbb{R}^n_+}: C_0^\infty(\overline{\mathbb{R}^n_+})\to C^\infty(\overline{\mathbb{R}^n_+})$ that $a(x,D)_{\mathbb{R}^n_-}: C_0^\infty(\overline{\mathbb{R}^n_-})\to C^\infty(\overline{\mathbb{R}^n_-})$. This is not true in general even for polyhomogeneous symbols of noninteger order.

Theorem 2.11. *Let $d\in\mathbb{C}\setminus\mathbb{Z}$, $a\in\mathbf{S}^d(\mathbb{R}^n\times\mathbb{R}^n)$, and $a\sim\sum_{l=0}^{\infty}a_{d-l}$ be the corresponding decomposition into the sum of functions homogeneous for $|\xi|\geq 1$. Then the following statements are equivalent:*

(i) *$a(x,D)_{\mathbb{R}^n_+}$ maps $C_0^\infty(\overline{\mathbb{R}^n_+})$ into $C^\infty(\overline{\mathbb{R}^n_+})$;*

(ii) $D^{\beta}_{x}D^{\alpha}_{\xi}a_{d-l}(x',0,0,-1)=e^{\pi i(d-|\alpha|-l)}D^{\beta}_{x}D^{\alpha}_{\xi}a_{d-l}(x',0,0,1)$,

$$\forall\alpha,\beta\in\mathbb{Z}^n_+,\ \forall l\in\mathbb{Z}_+ . \tag{2.12}$$

From the above theorem, it follows that, for $a \in \mathbf{S}^d(\mathbb{R}^n \times \mathbb{R}^n)$ the operator $a(x,D)_{\mathbb{R}^n_-}$ maps $C_0^\infty(\overline{\mathbb{R}^n_-})$ into $C^\infty(\overline{\mathbb{R}^n_-})$ if and only if the Equations (2.12) are valid with $e^{-\pi i(d-|\alpha|-l)}$ instead of $e^{\pi i(d-|\alpha|-l)}$. Note that the equality $e^{-\pi i(d-|\alpha|-l)} = e^{\pi i(d-|\alpha|-l)}$ holds only for an integer d.

If the symbol $a(x,\xi) \in \mathbf{S}^d_{1,0}(\mathbb{R}^n \times \mathbb{R}^n)$, $d \in \mathbb{R}$, is not polyhomogeneous, the equivalent conditions for the operator $a(x,D)_{\mathbb{R}^n_+}$ to map $C_0^\infty(\overline{\mathbb{R}^n_+})$ into $C^\infty(\overline{\mathbb{R}^n_+})$ look more complicated than (2.12). We consider right away the more general case of the symbol class $\mathbf{S}^d_{\rho,\delta}$.

Definition 2.12. Let $d \in \mathbb{R}$ and $\rho, \delta \in [0,1]$. We denote by $\mathbf{S}^d_{\rho,\delta,\mathrm{tr}}(\mathbb{R}^{n-1} \times \mathbb{R})$ the set of symbols $a(x',\xi_n) \in \mathbf{S}^d_{\rho,\delta}(\mathbb{R}^{n-1} \times \mathbb{R})$ such that $\pi_+\tilde{a}(x',x_n) = \pi_+ F^{-1}_{\xi_n \to x_n} a(x',\xi_n)$ belongs to $C^\infty(\overline{\mathbb{R}^n_+})$.

Next theorem is a generalization of Lemma 2.1.

Theorem 2.13. *Let $d \in \mathbb{R}$ and $0 \le \delta < \rho \le 1$. The following conditions on $a \in \mathbf{S}^d_{\rho,\delta}(\mathbb{R}^n \times \mathbb{R}^n)$ are equivalent:*

(i) *$a(x,D)_{\mathbb{R}^n_+}$ maps $C_0^\infty(\overline{\mathbb{R}^n_+})$ into $C^\infty(\overline{\mathbb{R}^n_+})$;*

$$\text{(ii)}\ \ D_x^\beta D_\xi^\alpha a(x',0,0,\xi_n) \in \mathbf{S}^{d-\rho|\alpha|+\delta|\beta|}_{\rho,\delta,\mathrm{tr}}(\mathbb{R}^{n-1} \times \mathbb{R}), \quad \forall \alpha, \beta \in \mathbb{Z}^n_+; \tag{2.13}$$

(iii) *(2.13) is valid for $\beta' = (\beta_1, \dots, \beta_{n-1}) = 0$, $\alpha_n = 0$;*

(iv) *for arbitrary $j \in \mathbb{Z}_+$, $\alpha' \in \mathbb{Z}^{n-1}_+$ and $\xi' \in \mathbb{R}^{n-1}$*

$$D^j_{x_n} D^{\alpha'}_{\xi'} a(x',0,\xi',\xi_n) \in \mathbf{S}^{d+\delta j-\rho|\alpha'|}_{\rho,\delta,\mathrm{tr}}(\mathbb{R}^{n-1} \times \mathbb{R})$$

in the (x',ξ_n)-variables.

Corollary 2.14. *The operators $a(x,D)_{\mathbb{R}^n_\pm}$ map $C_0^\infty(\overline{\mathbb{R}^n_\pm})$ into $C^\infty(\overline{\mathbb{R}^n_\pm})$ if and only if the symbols $a(x,\xi)$ and $a(x,-\xi)$ (or, equivalently, $a(x,\xi)$ and $\overline{a(x,\xi)}$) satisfy the conditions of the Theorem* 2.13.

Now some words about the terminology. In some works (Boutet de Monvel L. (1971), Grubb G. (1986), Rempel S., Schulze B.-W. (1982a)) the transmission property was defined as the fulfilment of one of the equivalent conditions (ii)–(v) in Theorem 2.8. In others (Grubb G., Hörmander L. (1990), Hörmander L. (1983)) it was defined as the property of an operator $a(x,D)_{\mathbb{R}^n_+}$ to map $C_0^\infty(\overline{\mathbb{R}^n_+})$ into $C^\infty(\overline{\mathbb{R}^n_+})$. We shall consider boundary value problems in the case of pseudodifferential operators with polyhomogeneous symbols of integer order, so that both definitions of the transmission property coincide in view of Theorem 2.10. We also denote by $\mathbf{S}^d_{\mathrm{tr}}(\mathbb{R}^n \times \mathbb{R}^n)$ the subset of the class $\mathbf{S}^d(\mathbb{R}^n \times \mathbb{R}^n)$, $d \in \mathbb{Z}$, which contains symbols satisfying conditions (i)–(vi) in Theorem 2.10.

Suppose now that $a_j \in \mathbf{S}^{r_j}_{\rho,\delta}(\mathbb{R}^n \times \mathbb{R}^n)$, $j \in \mathbb{Z}_+$, where $r_j \to -\infty$ as $j \to \infty$ and $\bar{r}_k = \max_{j \ge k} r_j$. Then (see Shubin M.A. (1978), Hörmander L. (1983),

Taylor M.E. (1981)) there exists a symbol $a \in \mathbf{S}^{\bar{r}_o}_{\rho,\delta}(\mathbb{R}^n \times \mathbb{R}^n)$ such that $a \sim \sum_{j=0}^{\infty} a_j$, i.e. $a - \sum_{j<k} a_j \in \mathbf{S}^{\bar{r}_k}_{\rho,\delta}(\mathbb{R}^n \times \mathbb{R}^n)$.

If $0 \le \delta < \rho \le 1$ and the symbols a_j meet the requirements of Theorem 2.13 (Corollary 2.14), then the symbol a also satisfies them (see Grubb G., Hörmander L. (1990)). In particular, if $\delta = 0$, $\rho = 1$, and the a_j satisfy the H-condition for $x_n = 0$, then a also possesses the H-property (see Theorem 2.8). The analogous property is also valid for Theorems 2.10, 2.11 and the UH-condition.

It is easy to see from these remarks and the composition formulae (see Shubin M.A. (1978), Hörmander L. (1983, 1985), Taylor M.E. (1985)), that, if both $a(x,\xi)$ and $b(x,\xi)$ possess any one of the properties considered in Theorems 2.8, 2.10, 2.11, 2.13 and Corollary 2.14, then the symbol of the composition $a(x,D)b(x,D)$ possesses that property also. Similarly one proves that, if the symbol $a(x,\xi)$ satisfies the conditions of Theorem 2.13, then $a^*(x,D)_{\mathbb{R}^n_-}$ and ${}^t a(x,D)_{\mathbb{R}^n_-}$ map $C_0^\infty(\overline{\mathbb{R}^n_-})$ into $C^\infty(\overline{\mathbb{R}^n_-})$, where $a^*(x,D)$ and ${}^t a(x,D)$ are respectively adjoint and transposed operators of $a(x,D)$ (see Shubin M.A. (1978), Hörmander L. (1983)).

Taking into consideration that a pseudodifferential operator given by an *amplitude*

$$(Au)(x) = (2\pi)^{-n} \int_{\mathbb{R}^n} \int_{\mathbb{R}^n} e^{i(x-y)\cdot\xi} p(x,y,\xi) u(y) \, dy \, d\xi ,$$

may under certain conditions be written in terms of symbols (see Agranovich M.S. (1990), Shubin M.A. (1978), Taylor M.E. (1981)), we can easily reformulate Theorems 2.8, 2.10, 2.11, 2.13 and Corollary 2.14 in terms of an amplitude $p(x,y,\xi)$. We note only that the derivatives

$$D_x^\beta D_y^\gamma D_\xi^\alpha p(x,y,\xi) \,|_{x'=y', x_n=y_n=0} , \quad \alpha, \beta, \gamma \in \mathbb{Z}^n_+ ,$$

appear in the corresponding statements instead of $D_x^\beta D_\xi^\alpha a(x',0,\xi)$.

Using the change of variable formula in pseudodifferential operators (see Agranovich M.S. (1990), Shubin M.A. (1978), Taylor M.E. (1981)) we see that, for $1-\rho \le \delta < \rho$, the properties discussed in Theorems 2.8, 2.10, 2.11, 2.13 and Corollary 2.14 are invariant under diffeomorphisms preserving the boundary $\partial\mathbb{R}^n_\pm$. This fact permits us to consider pseudodifferential operators with the transmission property (in either sense) on a smooth manifold with boundary.

2.4 Trace and Potential Operators, Singular Green Operators (Boutet de Monvel L. (1971), Grubb G. (1986), Grubb G. (1989), Rempel S., Schulze B.-W. (1982)).

We consider the Neumann problem in the half-space $\mathbb{R}^n_+$ as the simplest model:

$$(1-\Delta)u = f , \quad \gamma_1 u = \phi .$$

The solution to the problem is given by the well known formula

$$u = Rf + K(\phi - \gamma_1 Rf)\,,$$

where $R = \mathbf{OP}(\langle\xi\rangle^{-2})_{\mathbb{R}^n_+}$ is the *volume potential* and

$$K\psi = \text{const}\cdot\pi_+\,\mathbf{OP}(\langle\xi\rangle^{-2})(\psi(x')\otimes\delta(x_n))$$

is the *single layer potential.*

Thus, if we wish to construct a calculus which contains both boundary problems and their inverses, we have to include in addition to operators of the form $a(x,D)_{\mathbb{R}^n_+}$:

trace operators (such as $\gamma_1 = \gamma_0\partial_{x_n}$ and in general $\gamma_0 b(x,D)$);
potential operators (such as K and in general $\pi_+ b(x,D)(\psi(x')\otimes\delta(x_n))$)
singular Green operators (such as $K\gamma_1 R$ and in general compositions of trace and potential operators).

If, in addition, we wish to multiply the operators in our calculus, we have to consider also singular Green operators of the form

$$G(A,B) = (AB)_{\mathbb{R}^n_+} - A_{\mathbb{R}^n_+}B_{\mathbb{R}^n_+}\,,$$

where A and B are pseudodifferential operators.

Before giving the exact definitions of trace and potential operators we fix our attention on their one-dimensional prototypes. Let

$$a(\xi_n)\in H\,,\ \ \Pi^\pm a = a^\pm \in H^\pm\,,\ \ u\in S(\bar{\mathbb{R}}_+)\,,\ \ f = F(e_+u)\in H^+$$

(see Lemma 2.5). Then, applying (2.6), the equality $\gamma_0 F^{-1} = \Pi' = (2\pi)^{-1}\int^+$ (see Section 2.2 for the definition of the plus-integral) and (2.8), we obtain

$$\gamma_0\mathbf{OP}_n\big(a(\xi_n)\big)_{\mathbb{R}^n_+}u = \gamma_0\pi_+F^{-1}\big[a(\xi_n)f(\xi_n)\big] = \gamma_0F^{-1}[\Pi^+(af)]$$

$$= \frac{1}{2\pi}\int^+ \Pi^+(a^+f + a^-f)\,\mathrm{d}\xi_n = \frac{1}{2\pi}\int^+ \Pi^+(a^-f)\,\mathrm{d}\xi_n = \frac{1}{2\pi}\int^+ a^-f\,\mathrm{d}\xi_n\,.$$

On the other hand, by (2.6) we obtain

$$\pi_+\mathbf{OP}_n\big(a(\xi_n)\big)\big(c\cdot\delta(x_n)\big) = \pi_+F^{-1}\big(a(\xi_n)c\big) = F^{-1}a^+c$$

on $\mathbb{R}_+$, where c is a constant. Now we introduce the **trace** and **potential operators** T and K by means of the following formulae:

$$(Tu)(x') \equiv \mathbf{OP}T\big(t(x',\xi)\big)u$$

$$= (2\pi)^{-n}\int\limits_{\mathbb{R}^{n-1}} \mathrm{e}^{ix'\cdot\xi'}\int^+ t(x',\xi)\widehat{e_+u}(\xi)\,\mathrm{d}\xi_n\mathrm{d}\xi'\,,\quad u\in C_0^\infty(\overline{\mathbb{R}^n_+})\,, \tag{2.14}$$

and

$$(K\phi)(x) \equiv \mathbf{OP}K\big(k(x',\xi)\big)\phi$$

$$= (2\pi)^{-n} \int\limits_{\mathbb{R}^n} e^{ix\cdot\xi} k(x',\xi)\hat{\phi}(\xi')\, d\xi\,, \quad x_n > 0, \ \ \phi \in C_0^\infty(\mathbb{R}^{n-1})\,, \tag{2.15}$$

where $t \in H^-$ and $k \in H^+$ with respect to ξ_n. More exactly

$$t(x',\xi) = \sum_{j=0}^{r-1} s_j(x',\xi')\xi_n^j + t'(x',\xi)\,, \tag{2.16}$$

where t' belongs to H^-_{-1} as a function in ξ_n, $r \geq 0$ is independent of (x',ξ'), and $s_j \in \mathbf{S}^{d-j}_{1,0}(\mathbb{R}^{n-1} \times \mathbb{R}^{n-1})$. The operator corresponding to the first term on the right hand of (2.16) is the sum of standard trace operators of the form

$$(2\pi)^{-n} \int\limits_{\mathbb{R}^{n-1}} e^{ix'\cdot\xi'} \int^+ s_j(x',\xi')\xi_n^j \widehat{e_+ u}(\xi)\, d\xi_n d\xi' = (-i)^j s_j(x',D')\gamma_j u\,.$$

Note that, in contrast to (2.9) and (2.10), the symbols $s_j(x',\xi')$ are not, generally speaking, polynomials in ξ'.

The "plus-integral" $\int^+$ in the trace operator T' corresponding to the symbol $t'(x',\xi)$ (see (2.16)) is an ordinary integral over $\mathbb{R}$ if $u \in S(\overline{\mathbb{R}^n_+})$.

The symbols t' and k are taken from the spaces $\mathbf{S}^d_{1,0}(\mathbb{R}^{n-1} \times \mathbb{R}^{n-1},\ H^-_{-1})$ and $\mathbf{S}^d_{1,0}(\mathbb{R}^{n-1} \times \mathbb{R}^{n-1},\ H^+_{-1}) = \mathbf{S}^d_{1,0}(\mathbb{R}^{n-1} \times \mathbb{R}^{n-1},\ H^+)$ respectively.

Definition 2.15. We define the symbol space $\mathbf{S}^d_{1,0}(\mathbb{R}^{n-1} \times \mathbb{R}^{n-1},\ H^\pm_{-1})$, $d \in \mathbb{R}$, as the set of functions $f(x',\xi',\xi_n)$ lying in $H^\pm_{-1}$ with respect to ξ_n and satisfying the estimates

$$\| \Pi^\pm_{-1} D^{\beta'}_{x'} D^{\alpha'}_{\xi'} D^m_{\xi_n} \xi^l_n f(x',\xi) \|_{L_{2,\xi_n}} \leq C_{\alpha',\beta',m,l} \langle \xi' \rangle^{d+1/2-|\alpha'|-m+l}\,, \tag{2.17}$$

$\forall\, (x',\xi') \in \mathbb{R}^{n-1} \times \mathbb{R}^{n-1}$, $\forall\, \alpha',\beta' \in \mathbb{Z}^{n-1}_+$, $\forall\, m,l \in \mathbb{Z}_+$, where $L_{2,\xi_n} = L_2(\mathbb{R})$, $\xi_n \in \mathbb{R}$.

The space $\mathbf{S}^d_{1,0}(\mathbb{R}^{n-1} \times \mathbb{R}^{n-1}, H^-_{r-1})$, $r \in \mathbb{N}$, consists of symbols of the type (2.16), where $s_j \in \mathbf{S}^{d-j}_{1,0}(\mathbb{R}^{n-1} \times \mathbb{R}^{n-1})$, $t' \in \mathbf{S}^d_{1,0}(\mathbb{R}^{n-1} \times \mathbb{R}^{n-1}, H^-_{-1})$,

$$\mathbf{S}^d_{1,0}(\mathbb{R}^{n-1} \times \mathbb{R}^{n-1}, H^-) = \bigcup_{r \in \mathbb{Z}_+} \mathbf{S}^d_{1,0}(\mathbb{R}^{n-1} \times \mathbb{R}^{n-1}, H^-_{r-1})\,.$$

We say that $f(x',\xi) \in \mathbf{S}^d(\mathbb{R}^{n-1} \times \mathbb{R}^{n-1}, H^\pm)$ if

$$f \in \mathbf{S}^d_{1,0}(\mathbb{R}^{n-1} \times \mathbb{R}^{n-1}, H^\pm)\,, \quad f \sim \sum_{l \in \mathbb{Z}_+} f_{d-l}\,,$$

where

$$f - \sum_{l<N} f_{d-l} \in \mathbf{S}^{d-N}_{1,0}(\mathbb{R}^{n-1} \times \mathbb{R}^{n-1},\ H^\pm)\,,$$

$\forall N \in \mathbb{N}$, and f_{d-l} are homogeneous functions in ξ for $|\xi'| \ge 1$ of order $d-l$. Symbols from $\mathbf{S}^d(\mathbb{R}^{n-1} \times \mathbb{R}^{n-1}, H^{\pm})$, are called *polyhomogeneous* or *classical potential (trace) symbols.*

Definition 2.16. An operator of the form (2.14) is called a **trace operator** of order d and class r ($d \in \mathbb{R}$, $r \in \mathbb{Z}_+$) if $t \in \mathbf{S}^d_{1,0}(\mathbb{R}^{n-1} \times \mathbb{R}^{n-1}, H^-_{r-1})$.

An operator of the form (2.15) is called a **potential operator** (or **coboundary operator**, or **Poisson operator**) of order $d+1$ if $k \in \mathbf{S}^d_{1,0}(\mathbb{R}^{n-1} \times \mathbb{R}^{n-1}, H^+)$.

Using "symbol-kernels"

$$\tilde{t}'(x', x_n, \xi') = \frac{1}{2\pi} F_{\xi_n \to x_n} t'(x', \xi', \xi_n), \quad \tilde{k}(x', x_n, \xi') = F^{-1}_{\xi_n \to x_n} k(x', \xi', \xi_n)$$

we may represent the trace operator T' of the class 0 and the potential operator K in the forms

$$(T'u)(x') \equiv \mathbf{OP}T(\tilde{t}'(x, \xi'))u$$
$$= (2\pi)^{1-n} \int\limits_{\mathbb{R}^{n-1}} e^{ix' \cdot \xi'} \int\limits_{\mathbb{R}_+} \tilde{t}'(x', x_n, \xi') F_{y' \to \xi'} u(y', x_n)\, dx_n d\xi',$$

and

$$(K\phi)(x) \equiv \mathbf{OP}K(\tilde{k}(x, \xi'))\phi$$
$$= (2\pi)^{1-n} \int\limits_{\mathbb{R}^{n-1}} e^{ix' \cdot \xi'} \tilde{k}(x', x_n, \xi') \widehat{\phi}(\xi')\, d\xi', \quad x_n > 0.$$

The symbol-kernels $\tilde{t}'$ and $\tilde{k}$ belong to the space $\mathbf{S}^d_{1,0}(\mathbb{R}^{n-1} \times \mathbb{R}^{n-1}, S(\bar{\mathbb{R}}_+))$ which consists of functions $\tilde{f}$ satisfying the estimates

$$\|D^{\beta'}_{x'} x_n^m D^l_{x_n} D^{\alpha'}_{\xi'} \tilde{f}(x, \xi')\|_{L_{2,x_n}(\mathbb{R}_+)} \le C_{\alpha', \beta', m, l} \langle \xi' \rangle^{d+1/2-m+l-|\alpha'|},$$
$$\forall (x', \xi') \in \mathbb{R}^{n-1} \times \mathbb{R}^{n-1}, \ \forall \alpha', \beta' \in \mathbb{Z}^{n-1}_+, \ \forall m, l \in \mathbb{Z}_+. \tag{2.18}$$

With the help of the elementary inequalities

$$\sup_t |p(t)| \le \sqrt{2} \|p\|^{1/2}_{L_2} \|D_t p\|^{1/2}_{L_2}, \quad \|p\|_{L_2} \le \frac{1}{\sqrt{\sigma}} \sup_t |(1+\sigma t) p(t)|,$$

$$\|p\|_{L_1} \le \frac{1}{\sqrt{\sigma}} \|(1+\sigma t)p(t)\|_{L_2}, \quad \sup_t |p(t)| \le \|D_t p\|_{L_1}, \tag{2.19}$$

which are valid for any function $p \in S(\bar{\mathbb{R}}_+)$, it is easy to replace (2.18) by equivalent estimates where L_1- or L_∞-norms appear instead of L_2-norms. In this case we have to take $\sigma = \langle \xi' \rangle$ in (2.19).

Theorem 2.17. *For an arbitrary potential operator K of order $d+1$ there exists a symbol $a(x, \xi) \in \mathbf{S}^d_{1,0}(\mathbb{R}^n \times \mathbb{R}^n)$ possessing the UH-property for $x_n = 0$ such that*

$$K\phi = \pi_+ a(x, D)(\phi(x') \otimes \delta(x_n)), \quad \forall \phi \in C_0^\infty(\mathbb{R}^{n-1}). \tag{2.20}$$

Analogously, for an arbitrary trace operator T' of order d and class 0 there exists a symbol $b(x,\xi) \in \mathbf{S}^d_{1,0}(\mathbb{R}^n \times \mathbb{R}^n)$ possessing the UH-property for $x_n = 0$ such that

$$T'u = \gamma_0 b(x,D)_{\mathbb{R}^n_+} u\,, \quad \forall u \in C_0^\infty(\overline{\mathbb{R}^n_+})\,. \tag{2.21}$$

Conversely, if the symbols $a(x,\xi)$ and $b(x,\xi) \in \mathbf{S}^d_{1,0}(\mathbb{R}^n \times \mathbb{R}^n)$ possess the UH-property for $x_n = 0$ then the operators defined by the formulae (2.20) and (2.21) are the potential operator of order $d+1$ and the trace operator of order d and class 0 respectively, and also

$$k(x',\xi) \sim \sum_{j\in\mathbb{Z}_+} \frac{(-1)^j}{j!} \Pi^+_{\xi_n} D^j_{x_n} \partial^j_{\xi_n} a(x',0,\xi)\,,$$

$$t'(x,\xi) = \Pi^-_{\xi_n} b(x',0,\xi)\,.$$

Now we pass to singular Green operators.

Definition 2.18. The space of symbols $\mathbf{S}^{d-1}_{1,0}(\mathbb{R}^{n-1} \times \mathbb{R}^{n-1}, H^+\widehat{\otimes} H^-_{-1})$, $d \in \mathbb{R}$, consists of functions $g'(x',\xi',\xi_n,\eta_n)$ which belong to H^+ with respect to ξ_n, to H^-_{-1} with respect to η_n, and satisfy the estimates

$$\begin{aligned}
&\|\Pi^+_{\xi_n} \Pi^-_{-1,\eta_n} D^{\beta'}_{x'} D^{\alpha'}_{\xi'} D^m_{\xi_n} \xi_n^{m'} D^l_{\eta_n} \eta_n^{l'} g'(x',\xi,\eta_n)\|_{L_{2,\xi_n,\eta_n}} \\
&\quad \le C_{\alpha',\beta',m,m',l,l'} \langle \xi' \rangle^{d-m+m'-l+l'-|\alpha'|}\,, \\
&\quad \forall \alpha', \beta' \in \mathbb{Z}^{n-1}_+\,, \ \forall m,m',l,l' \in \mathbb{Z}_+\,.
\end{aligned} \tag{2.22}$$

The space $\mathbf{S}^{d-1}_{1,0}(\mathbb{R}^{n-1} \times \mathbb{R}^{n-1}, H^+\widehat{\otimes} H^-_{r-1})$, $r \in \mathbb{N}$, consists of symbols of the form

$$g(x',\xi',\xi_n,\eta_n) = \sum_{j=0}^{r-1} k_j(x',\xi',\xi_n)\eta_n^j + g'(x',\xi',\xi_n,\eta_n)\,, \tag{2.23}$$

where

$$k_j \in \mathbf{S}^{d-j-1}_{1,0}(\mathbb{R}^n \times \mathbb{R}^n, H^+)\,, \quad g' \in \mathbf{S}^{d-1}_{1,0}(\mathbb{R}^{n-1} \times \mathbb{R}^{n-1}, H^+\widehat{\otimes} H^-_{-1})\,.$$

Further

$$\mathbf{S}^{d-1}_{1,0}(\mathbb{R}^{n-1} \times \mathbb{R}^{n-1}, H^+\widehat{\otimes} H^-) = \bigcup_{r\in\mathbb{Z}_+} \mathbf{S}^{d-1}_{1,0}(\mathbb{R}^{n-1} \times \mathbb{R}^{n-1}, H^+\widehat{\otimes} H^-_{r-1})\,.$$

Also we say that $g \in \mathbf{S}^{d-1}(\mathbb{R}^{n-1} \times \mathbb{R}^{n-1}, H^+\widehat{\otimes} H^-)$ if

$$g \in \mathbf{S}^{d-1}_{1,0}(\mathbb{R}^{n-1} \times \mathbb{R}^{n-1}, H^+\widehat{\otimes} H^-) \quad \text{and} \quad g \sim \sum_{l\in\mathbb{Z}_+} g_{d-l-1}\,,$$

where

$$g - \sum_{l<N} g_{d-l-1} \in \mathbf{S}^{d-1-N}_{1,0}(\mathbb{R}^{n-1} \times \mathbb{R}^{n-1}, H^+ \widehat{\otimes} H^-), \quad \forall N \in \mathbb{N},$$

and g_{d-l-1} is a homogeneous function in (ξ', ξ_n, η_n) for $|\xi'| \geq 1$ of order $d-1-l$. Symbols in $\mathbf{S}^{d-1}(\mathbb{R}^{n-1} \times \mathbb{R}^{n-1}, H^+ \widehat{\otimes} H^-)$ are called *polyhomogeneous* or *classical Green symbols.*

Definition 2.19. An operator G of the form

$$(Gu)(x) \equiv \mathbf{OP}G(g(x', \xi', \xi_n, \eta_n))u$$

$$= (2\pi)^{-n-1} \int_{\mathbb{R}^n} e^{ix\cdot\xi} \int^+ g(x', \xi', \xi_n, \eta_n)\, \widehat{e_+ u}(\xi', \eta_n)\, \mathrm{d}\eta_n\, \mathrm{d}\xi, \tag{2.24}$$

where $u \in C_0^\infty(\overline{\mathbb{R}^n_+})$, is called a **singular Green operator** of order d and of class r ($d \in \mathbb{R}$, $r \in \mathbb{Z}_+$) if the symbol g belongs to $\mathbf{S}^{d-1}_{1,0}(\mathbb{R}^{n-1} \times \mathbb{R}^{n-1}, H^+ \widehat{\otimes} H^-_{r-1})$.

According to (2.23) we have

$$G = \sum_{j=0}^{r-1} (-i)^j\, K_j \gamma_j + G',$$

where K_j is a potential operator of order $d-j$ with the symbol $k_j(x', \xi', \xi_n)$ and G' is a singular Green operator of order d and of class 0 with the symbol g':

$$(G'u)(x) \equiv \mathbf{OP}G(g'(x', \xi', \xi_n, \eta_n))u$$

$$= (2\pi)^{-n-1} \int_{\mathbb{R}^{n+1}} e^{ix\cdot\xi}\, g'(x', \xi', \xi_n, \eta_n)\, \widehat{e_+ u}(\xi', \eta_n)\, \mathrm{d}\eta_n\, \mathrm{d}\xi, \tag{2.25}$$

where $u \in C_0^\infty(\overline{\mathbb{R}^n_+})$.

Note that in (2.25) we have changed the "plus-integral" $\int^+$ into an ordinary integral along $\mathbb{R}$ as in the case of trace operators of zero class.

By analogy with the trace and the potential operators we can represent the singular Green operator of class 0 in terms of the symbol-kernel $\tilde{g}'(x', x_n, y_n, \xi') = \frac{1}{2\pi} F^{-1}_{\xi_n \to x_n} F_{\eta_n \to y_n}\, g'(x', \xi', \xi_n, \eta_n)$:

$$(G'u)(x) \equiv \mathbf{OP}G(\tilde{g}'(x', x_n, y_n, \xi'))u =$$

$$= (2\pi)^{1-n} \int_{\mathbb{R}^{n-1}} e^{ix'\cdot\xi'} \int_{\mathbb{R}_+} \tilde{g}'(x', x_n, y_n, \xi')\, F_{y' \to \xi'} u(y', y_n)\, \mathrm{d}y_n\, \mathrm{d}\xi'. \tag{2.26}$$

The symbol-kernel $\tilde{g}'$ belongs to the space $\mathbf{S}^{d-1}_{1,0}(\mathbb{R}^{n-1} \times \mathbb{R}^{n-1}, S(\overline{\mathbb{R}^2_{++}}))$, which consists of functions q satisfying the estimates

$$\| D^{\beta'}_{x'} x_n^m D^{m'}_{x_n} y_n^l D^{l'}_{y_n} D^{\alpha'}_{\xi'} q(x, y_n, \xi') \|_{L_{2,x_n,y_n}(\mathbb{R}^2_{++})}$$

$$\leq C_{\alpha', \beta', m, m', l, l'} \langle \xi' \rangle^{d-m+m'-l+l'-|\alpha'|},$$

$$\forall\, \alpha', \beta' \in \mathbb{Z}_+^{n-1}, \; \forall\, m, m', l, l' \in \mathbb{Z}_+ . \tag{2.27}$$

It follows from the definition of the spaces $H_{-1}^{\pm}$ that $\mathbf{S}_{1,0}^d(\mathbb{R}^{n-1} \times \mathbb{R}^{n-1}, H_{-1}^-)$ is the set of complex conjugate functions to the symbols in $\mathbf{S}_{1,0}^d(\mathbb{R}^{n-1} \times \mathbb{R}^{n-1}, H^+)$. It is possible to write down the decompositions of symbols

$$k \in \mathbf{S}_{1,0}^d(\mathbb{R}^{n-1} \times \mathbb{R}^{n-1}, H^+) \quad \text{and} \quad t' \in \mathbf{S}_{1,0}^d(\mathbb{R}^{n-1} \times \mathbb{R}^{n-1}, H_{-1}^-)$$

into Laguerre series:

$$k(x', \xi', \xi_n) = \sum_{j \in \mathbb{Z}_+} a_j(x', \xi')\, \hat{\varphi}_j(\xi_n, \sigma(\xi')),$$

$$t'(x', \xi', \xi_n) = \sum_{j \in \mathbb{Z}_+} b_j(x', \xi')\, \overline{\hat{\varphi}}_j(\xi_n, \sigma(\xi')), \tag{2.28}$$

where $\sigma \in C^\infty(\mathbb{R}^{n-1})$, $\sigma > 0$, $\sigma(\xi')$ equals $|\xi'|$ for $|\xi'| \geq 1$, and $1/2$ for $|\xi'| \leq 1/2$; and $\{a_j\}_{j \in \mathbb{Z}_+}$, $\{b_j\}_{j \in \mathbb{Z}_+}$ form rapidly decreasing sequences in $\mathbf{S}_{1,0}^{d+1/2}(\mathbb{R}^{n-1} \times \mathbb{R}^{n-1})$ in the sense that

$$\Big(\sum_{j \in \mathbb{Z}_+} \big| (1+j)^N D_{x'}^{\beta'} D_{\xi'}^{\alpha'} a_j(x', \xi') \big|^2 \Big)^{1/2} \leq C_{\alpha', \beta', N} \, \langle \xi' \rangle^{d+1/2-|\alpha'|}, \quad \forall\, \alpha', \beta' \in \mathbb{Z}_+^{n-1}, \; \forall\, N \in \mathbb{Z}_+, \tag{2.29}$$

with similar inequalities for $\{b_j\}_{j \in \mathbb{Z}_+}$.

For the corresponding symbol-kernels $\tilde{k}(x, \xi')$ and $\tilde{t}'(x, \xi')$ we have the series

$$\tilde{k}(x', x_n, \xi') = \sum_{j \in \mathbb{Z}_+} a_j(x', \xi')\, \varphi_j(x_n, \sigma(\xi'))$$

and

$$\tilde{t}'(x', x_n, \xi') = \sum_{j \in \mathbb{Z}_+} b_j(x', \xi')\, \varphi_j(x_n, \sigma(\xi')).$$

Now the symbol

$$g'(x', \xi', \xi_n, \eta_n) \in \mathbf{S}_{1,0}^{d-1}(\mathbb{R}^{n-1} \times \mathbb{R}^{n-1}, H^+ \widehat{\otimes} H_{-1}^-)$$

and the corresponding symbol-kernel

$$\tilde{g}'(x', x_n, y_n, \xi') \in \mathbf{S}_{1,0}^{d-1}(\mathbb{R}^{n-1} \times \mathbb{R}^{n-1}, S(\overline{\mathbb{R}_{++}^2}))$$

may be represented in the form

$$g'(x', \xi', \xi_n, \eta_n) = \sum_{j,l \in \mathbb{Z}_+} c_{jl}(x', \xi')\, \hat{\varphi}_j(\xi_n, \sigma(\xi'))\, \overline{\hat{\varphi}}_l(\eta_n, \sigma(\xi')),$$

$$\tilde{g}'(x', x_n, y_n, \xi') = \sum_{j,l \in \mathbb{Z}_+} c_{jl}(x', \xi')\, \varphi_j(x_n, \sigma(\xi'))\, \varphi_l(y_n, \sigma(\xi')), \tag{2.30}$$

where $\{c_{jl}\}_{j,l\in\mathbb{Z}_+}$ forms a rapidly decreasing sequence in $\mathbf{S}^d_{1,0}(\mathbb{R}^{n-1}\times\mathbb{R}^{n-1})$:

$$\sum_{j,l\in\mathbb{Z}_+} \left|(1+j)^N(1+l)^J D^{\beta'}_{x'} D^{\alpha'}_{\xi'} c_{jl}(x',\xi')\right|$$

$$\leq C_{\alpha',\beta',N,J}\,\langle\xi'\rangle^{d-|\alpha'|}, \quad \forall\alpha',\beta'\in\mathbb{Z}^{n-1}_+,\ \forall N,J\in\mathbb{Z}_+. \tag{2.31}$$

The above sets of symbols and symbol-kernels are Frechét spaces relatively to the topologies given by the semi-norms defined by the minimal values of the constants in the right-hand sides of the corresponding inequalities (2.17), (2.18), (2.22), (2.27).

It follows from (2.30) that

$$G' = \sum_{j,l\in\mathbb{Z}_+} K_{jl}\,T_l\,, \tag{2.32}$$

where

$$K_{jl} = \mathbf{OP}K\left(c_{jl}(x',\xi')\,\frac{\varphi_j(x_n,\sigma(\xi'))}{\sqrt{\sigma(\xi')}}\right),$$

$$T_l = \mathbf{OP}T\left(\sqrt{\sigma(\xi')}\,\varphi_l(y_n,\sigma(\xi'))\right).$$

Thus we have the next theorem.

Theorem 2.20. *An arbitrary singular Green operator of order d and class 0 may be represented in the form* (2.32), *where $K_{jl} = \mathbf{OP}K(k_{jl}(x',\xi))$ is a potential operator of order d, and $T_l = \mathbf{OP}T(t_l(x',\xi))$ is a trace operator of order and class 0, where $\{k_{jl}\}_{j,l\in\mathbb{Z}_+}$ forms a rapidly decreasing sequence in $\mathbf{S}^d_{1,0}(\mathbb{R}^{n-1}\times\mathbb{R}^{n-1},H^+)$ and the series*

$$\sum_{j,l\in\mathbb{Z}_+} k_{jl}(x',\xi',\xi_n)\,t_l(x',\xi',\eta_n)$$

converges in $\mathbf{S}^{d-1}_{1,0}(\mathbb{R}^{n-1}\times\mathbb{R}^{n-1},H^+\widehat{\otimes}H^-_{-1})$.

If we take m or l in (2.27) sufficiently large, then the exponent of the power on the right hand side of (2.27) may be made negative with arbitrary large modulus. So, for a given function $\chi\in C^\infty_0(\mathbb{R}^n_+)$, the operators $\chi\mathbf{OP}G(g')$ and $\mathbf{OP}G(g')\chi I$ map $S'(\overline{\mathbb{R}^n_+})$ into $C^\infty(\overline{\mathbb{R}^n_+})$. Analogously, it follows from (2.18) that the operator $\chi\mathbf{OP}K(k)$ maps $S'(\mathbb{R}^{n-1})$ into $C^\infty(\overline{\mathbb{R}^n_+})$. Operators of this kind are called **smoothing operators**.

Thus, a singular Green operator of class 0 maps $S'(\overline{\mathbb{R}^n_+})$ into $C^\infty(\mathbb{R}^n_+)$, and a potential operator maps $S'(\mathbb{R}^{n-1})$ into $C^\infty(\mathbb{R}^n_+)$. Singularities may occur only when $x_n = 0$.

Note in conclusion that it is possible to consider trace, potential and singular Green operators defined by (x',y')-dependent amplitudes.

2.5. Continuity. (Franke J. (1985), Franke J. (1986), Grubb G. (1990), Grubb G., Hörmander L. (1990), Rempel S., Schulze B.-W. (1982a)).

Theorem 2.21. *Let A be a pseudodifferential operator of order d possessing the UH-property for $x_n = 0$, let K be a potential operator of order d, and let G and T be respectively a singular Green operator and a trace operator of order d and class r, $d \in \mathbb{R}$, $r \in \mathbb{Z}_+$, $1 < p < \infty$, $1 \le q \le \infty$. Then the following mappings are continuous:*

$$A_{\mathbb{R}^n_+} : H^s_p(\overline{\mathbb{R}^n_+}) \to H^{s-d}_p(\overline{\mathbb{R}^n_+})$$

$$\big(or\ B^s_{p,q}(\overline{\mathbb{R}^n_+}) \to B^{s-d}_{p,q}(\overline{\mathbb{R}^n_+})\big)$$

for $s > 1/p - 1$;

$$G : H^s_p(\overline{\mathbb{R}^n_+}) \bigcup B^s_{p,p}(\overline{\mathbb{R}^n_+}) \to H^{s-d}_p(\overline{\mathbb{R}^n_+}) \bigcap B^{s-d}_{p,p}(\overline{\mathbb{R}^n_+})$$

$$\big(or\ B^s_{p,q}(\overline{\mathbb{R}^n_+}) \to B^{s-d}_{p,q}(\overline{\mathbb{R}^n_+})\big)$$

for $s > r + 1/p - 1$;

$$T : H^s_p(\overline{\mathbb{R}^n_+}) \bigcup B^s_{p,p}(\overline{\mathbb{R}^n_+}) \to B^{s-d-1/p}_{p,p}(\mathbb{R}^{n-1})$$

$$\big(or\ B^s_{p,q}(\overline{\mathbb{R}^n_+}) \to B^{s-d-1/p}_{p,q}(\mathbb{R}^{n-1})\big)$$

for $s > r + 1/p - 1$;

$$K : B^{s-1/p}_{p,p}(\mathbb{R}^{n-1}) \to H^{s-d}_p(\overline{\mathbb{R}^n_+}) \bigcap B^{s-d}_{p,p}(\overline{\mathbb{R}^n_+})$$

$$\big(or\ B^{s-1/p}_{p,q}(\mathbb{R}^{n-1}) \to B^{s-d}_{p,q}(\overline{\mathbb{R}^n_+})\big),$$

$\forall s \in \mathbb{R}$.

The analogues of Theorem 2.21 for Hölder spaces are obtained in (Rempel S., Schulze B.-W. (1982a)) and for a wide class of Besov-Triebel-Lizorkin spaces in (Franke J. (1985) and Franke J. (1986)).

For a pseudodifferential operator A possessing the UH-property for $x_n = 0$, and operators G and T of class zero we obtain the following relations after integrating by parts:

$$\begin{aligned} A_{\mathbb{R}^n_+} D^j_{x_n} &= \sum_{0\le l\le j-1} K^{(l)}_A \gamma_{j-1-l} + \big(AD^j_{x_n}\big)_{\mathbb{R}^n_+}, \\ GD^j_{x_n} &= \sum_{0\le l\le j-1} K^{(l)}_G \gamma_{j-1-l} + G^{(j)}, \\ TD^j_{x_n} &= \sum_{0\le l\le j-1} S^{(l)}_T \gamma_{j-1-l} + T^{(j)}, \end{aligned} \tag{2.33}$$

where

$$A = a(x, D), \quad G = \mathbf{OP}G(g(x', \xi', \xi_n, \eta_n)), \quad T = \mathbf{OP}T(t(x', \xi)),$$

$$K^{(l)}_A v = -(-i)^{j-l} \pi_+ A\big(v(x') \otimes D^l_{x_n} \delta(x_n)\big)$$

$$= -(-i)^{j-l}\pi_+ AD^l_{x_n}\big(v(x')\otimes\delta(x_n)\big) = \mathbf{OP}K\big(f_l(x',\xi)\big)v\,,$$

$$f_l(x',\xi) \sim -(-i)^{j-l}\sum_{k\in\mathbb{Z}_+}\frac{(-1)^k}{k!}\Pi^+_{\xi_n}\left[D^k_{x_n}\partial^k_{\xi_n}\xi^l_n\; a(x',0,\xi)\right],$$

$$K^{(l)}_G = -(-i)^{j-l}\mathbf{OP}K\big((-D_{y_n})^l\tilde{g}(x',x_n,0,\xi')\big)\,, \tag{2.34}$$

$$G^{(j)} = \mathbf{OP}G\big((-D_{y_n})^j\tilde{g}(x',x_n,y_n,\xi')\big)$$

$$= \mathbf{OP}G\Big(\Pi^-_{-1,\eta_n}\left[\eta^j_n g(x',\xi,\eta_n)\right]\Big)\,,$$

$$S^{(l)}_T = s_l(x',D')\,,\quad s_l(x',\xi') = -(-i)^{j-l}(-D_{x_n})^l\tilde{t}(x',0,\xi')\,,$$

$$T^{(j)} = \mathbf{OP}T\big((-D_{x_n})^j\tilde{t}(x',x_n,\xi')\big)$$

$$= \mathbf{OP}T\Big(\Pi^-_{-1,\xi_n}\left[\xi^j_n t(x',\xi)\right]\Big)\,.$$

Definition 2.22. Let A be a pseudodifferential operator possessing the UH-property for $x_n = 0$, let G and T be respectively a singular Green operator and a trace operator of class zero, and let $K^{(l)}_A$, $K^{(l)}_G$, $S^{(l)}_T$ be defined by the formulae (2.34). Then, for $m\in\mathbb{N}$, we say that the operator $A_{\mathbb{R}^n_+}+G$ is of class $-m$ if $K^{(l)}_A + K^{(l)}_G = 0$ for $0\le l\le m-1$. Analogously T is an operator of the class $-m$ if $S^{(l)}_T = 0$ for $0\le l\le m-1$.

Taking into consideration that

$$u\in H^s_p(\overline{\mathbb{R}^n_+}) \iff \exists v_\alpha\in H^{s+m}_p(\overline{\mathbb{R}^n_+}) : u = \sum_{|\alpha|\le m} D^\alpha v_\alpha\,,$$

and

$$u\in B^s_{p,q}(\overline{\mathbb{R}^n_+}) \iff \exists v_\alpha\in B^{s+m}_{p,q}(\overline{\mathbb{R}^n_+}) : u = \sum_{|\alpha|\le m} D^\alpha v_\alpha\,,$$

it is easy to prove the following statement using (2.33) and Theorem 2.21.

Theorem 2.23. *Let A be a pseudodifferential operator possessing the UH-property for $x_n = 0$, and let $A_{\mathbb{R}^n_+}+G$ and T be operators of order d and class r, $d\in\mathbb{R}$, $r\in\mathbb{Z}$. Then the following mappings are continuous for $s > r+1/p-1$:*

$$A_{\mathbb{R}^n_+}+G : H^s_p(\overline{\mathbb{R}^n_+}) \to H^{s-d}_p(\overline{\mathbb{R}^n_+})\,,$$

$$B^s_{p,q}(\overline{\mathbb{R}^n_+}) \to B^{s-d}_{p,q}(\overline{\mathbb{R}^n_+})\,,$$

$$T : H^s_p(\overline{\mathbb{R}^n_+})\bigcup B^s_{p,p}(\overline{\mathbb{R}^n_+}) \to B^{s-d-1/p}_{p,p}(\mathbb{R}^{n-1})\,,$$

$$B^s_{p,q}(\overline{\mathbb{R}^n_+}) \to B^{s-d-1/p}_{p,q}(\mathbb{R}^{n-1})\,.$$

This statement is precise in the sense that, if continuity takes place for smaller s than indicated, then the class of operators is smaller than indicated.

Now we pass to pseudodifferential operators with symbols in $\mathbf{S}^d_{\rho,\delta}(\mathbb{R}^n \times \mathbb{R}^n)$.

Definition 2.24. We say that the symbol $a(x',\xi_n) \in \mathbf{S}^d_{\rho,\delta}(\mathbb{R}^{n-1} \times \mathbb{R})$, where $d \in \mathbb{R}$ and $\rho, \delta \in [0,1]$, belongs to the space $\mathbf{S}^d_{\rho,\delta,\text{utr}}(\mathbb{R}^{n-1} \times \mathbb{R})$, if the function $x_n^N D_x^\beta \pi_+ \tilde{a}(x',x_n)$ is bounded in $\mathbb{R}^n_+$ for all $N \in \mathbb{Z}_+$ and $\beta \in \mathbb{Z}^n_+$, where $\tilde{a}(x',x_n) = F^{-1}_{\xi_n \to x_n} a(x',\xi_n)$.

Clearly $\mathbf{S}^d_{\rho,\delta,\text{utr}} \subset \mathbf{S}^d_{\rho,\delta,\text{tr}}$ (see Definition 2.12). The function

$$a(x',\xi_n) = \langle \xi_n \rangle \chi(\xi_n / \langle x' \rangle)\,,$$

where $\chi \in C_0^\infty(\mathbb{R})$ and $\chi(0) \neq 0$, is an example of a symbol in $\mathbf{S}^1_{1,0,\text{tr}} \setminus \mathbf{S}^1_{1,0,\text{utr}}$.

Theorem 2.25. *Let $d \in \mathbb{R}$, $0 \le \delta < \rho \le 1$, and let $a(x,\xi) \in \mathbf{S}^d_{\rho,\delta}(\mathbb{R}^n \times \mathbb{R}^n)$ with*

$$D^j_{x_n} D^{\alpha'}_{\xi'} a(x',0,0,\xi_n) \in \mathbf{S}^{d+\delta j - \rho|\alpha'|}_{\rho,\delta,\text{utr}}(\mathbb{R}^{n-1} \times \mathbb{R})\,, \quad \forall j \in \mathbb{Z}_+\,,\ \forall \alpha' \in \mathbb{Z}^{n-1}_+\,.$$

Also, let

$$K^+_a v = \pi_+ a(x,D)\big(v(x') \otimes \delta(x_n)\big)\,.$$

Then the operator

$$K^+_a : H^s_2(\mathbb{R}^{n-1}) \to H^t_2(\overline{\mathbb{R}^n_+})$$

is continuous for $t = \rho s - d - 1/2$ if $s > 0$, and for $t = s - d - 1/2$ if $s < 0$. In the case $s = 0$ the continuity takes place for $t = -d - 1/2$ if $\rho = 1$ and for $t < -d - 1/2$ if $\rho < 1$. The operator

$$a(x,D)_{\mathbb{R}^n_+} : H^s_2(\mathbb{R}^n_+) \to H^t_2(\mathbb{R}^n_+)$$

is continuous for $t = s - d - (1-\rho)(s - 1/2) = 1/2 - d + \rho(s - 1/2)$ if $s > 1/2$, and for $t = s - d$ if $-1/2 < s < 1/2$. In the case $s = 1/2$ the continuity takes place for $t = s - d$ if $\rho = 1$ and for $t < s - d$ if $\rho < 1$.

The example $a(x,\xi) = \chi(\xi_n / \langle \xi' \rangle^k)$, where $\chi \in C_0^\infty(\mathbb{R})$ with $\chi = 1$ in a neighbourhood of 0 and $k = 1/\rho > 1$ (in this case $d = 0$, $\delta = 0$), shows that Theorem 2.25 is exact in the sense that the exponent t in general may not be taken larger than stated.

Remark. 2.26. We considered above pseudodifferential operators in $\mathbb{R}^n_+$, i.e in a domain with a smooth boundary. Take now a model domain with nonsmooth boundary:

$$\mathbb{R}^n_{+,m} = \{x \in \mathbb{R}^n : x_m > 0, \dots, x_n > 0\}\,, \quad m < n\,.$$

Using the proofs of Theorems 2.21 and 2.25 adduced in Grubb G. (1990) and Grubb G., Hörmander L. (1990) it is not difficult to see that the statements of these theorems concerning the continuity of pseudodifferential operators remain valid if we change $\mathbb{R}^n_+$ into $\mathbb{R}^n_{+,m}$ and require the symbols to satisfy the conditions of the theorems relative to each of the variables $x_m, \ldots, x_n$ and not only x_n.

2.6. Composition Formulas. (Boutet de Monvel L. (1971), Grubb, G. (1986), Rempel S., Schulze B.-W. (1982a)).

Up to now we have considered only scalar operators, but all the above theory may be carried over to matrix operators.

Let now $A = a(x, D)$ be a pseudodifferential operator of order $d \in \mathbb{R}$ possessing the UH-property for $x_n = 0$, let K be a potential operator of order d, and let G and T be respectively a singular Green operator and a trace operator of order d and class $r \in \mathbb{Z}_+$. Let also $L = l(x', D') \in \mathbf{OP}(\mathbf{S}^d_{1,0})$ be a pseudodifferential operator of order d on $\partial\mathbb{R}^n_+$. Now we construct from these operators the so-called **Green operator** of order d and class r:

$$P = \begin{pmatrix} A_{\mathbb{R}^n_+} + G & K \\ T & L \end{pmatrix} : \begin{matrix} C_0^\infty(\overline{\mathbb{R}^n_+}, \mathbb{C}^N) \\ \bigoplus \\ C_0^\infty(\mathbb{R}^{n-1}, \mathbb{C}^J) \end{matrix} \longrightarrow \begin{matrix} C^\infty(\overline{\mathbb{R}^n_+}, \mathbb{C}^{N'}) \\ \bigoplus \\ C^\infty(\mathbb{R}^{n-1}, \mathbb{C}^{J'}) \end{matrix}, \tag{2.35}$$

where $C_0^\infty(\overline{\mathbb{R}^n_+}, \mathbb{C}^N) = C_0^\infty(\overline{\mathbb{R}^n_+}) \otimes \mathbb{C}^N$ etc., and $N, N', J, J' \in \mathbb{Z}_+$. We know from the previous section that P admits an extension to a continuous operator in the corresponding Besov spaces and spaces of Bessel potentials (and also in more general Besov-Triebel-Lizorkin spaces). All this gives the opportunity to consider compositions of operators of the type (2.35). The central fact of Boutet de Monvel's theory is that the composition of operators of the type (2.35) is again an operator of the same type. This fact may be proved in three steps (see Grubb G. (1986)): first, operators on the half-axis with symbols independent of x_n (and depending on x', ξ' as parameters) are considered, and then the case of x_n-dependent symbols. Finally, we state the composition formulae for Green operators of the type (2.35), considering these operators as pseudodifferential operators defined on $\mathbb{R}^{n-1}$ (relatively to x') with operator-valued symbols. In this way we may treat such operators by means of the standard calculus of pseudodifferential operators using the fact that the composition formulae for symbols are already obtained in the second step of the proof.

The function $a(x, \xi)$ is called the *interior symbol* of the operator P in (2.35), and the operator

$$\sigma_P(x', \xi', D_n) = \begin{pmatrix} a(x', 0, \xi', D_n)_{\mathbb{R}_+} + g(x', \xi', D_n) & k(x', \xi', D_n) \\ t(x', \xi', D_n) & l(x', \xi') \end{pmatrix} :$$

$$\begin{matrix} S(\bar{\mathbb{R}}_+) \otimes \mathbb{C}^N \\ \bigoplus \\ \mathbb{C}^J \end{matrix} \longrightarrow \begin{matrix} S(\bar{\mathbb{R}}_+) \otimes \mathbb{C}^{N'} \\ \bigoplus \\ \mathbb{C}^{J'} \end{matrix} \tag{2.36}$$

is called the *boundary symbol* of P. The leading (or principal) interior and boundary symbols for polyhomogeneous operators are defined in the usual way.

Applying the Fourier transform to the operator (2.36) we obtain

$$\hat{\sigma}_P(x',\xi') = \begin{pmatrix} \Pi^+_{\xi_n} a(x',0,\xi',\xi_n) + \Pi'_{\eta_n} g(x',\xi',\xi_n,\eta_n) & k(x',\xi',\xi_n) \\ \Pi'_{\xi_n} t(x',\xi',\xi_n) & l(x',\xi') \end{pmatrix} :$$

$$\begin{matrix} H^+\otimes\mathbb{C}^N \\ \bigoplus \\ \mathbb{C}^J \end{matrix} \longrightarrow \begin{matrix} H^+\otimes\mathbb{C}^{N'} \\ \bigoplus \\ \mathbb{C}^{J'} \end{matrix} . \qquad (2.37)$$

The operator $\hat{\sigma}_P$ is also called the *boundary symbol* of the operator P. The connection between (2.36) and (2.37) is the same as between the Wiener-Hopf and Toeplitz operators (see, for example, Prössdorf S. (1988)).

In the following we shall use the notation "$\circ$" for composition of operators and also in the appropriate formulae for symbols . We shall also use the notation "$\circ_n$" for operators and the corresponding symbols acting only in x_n.

Let P_j $(j=1,2)$ be the Green operator of order $d_j \in \mathbb{R}$ and class $r_j \in \mathbb{Z}_+$ of the type (2.35) given by

$$P_j = \begin{pmatrix} \pi_+ a_j(x,D)e_+ + \mathbf{OP}G\big(g_j(x',\xi,\eta_n)\big) & \mathbf{OP}K\big(k_j(x',\xi)\big) \\ \mathbf{OP}T\big(t_j(x',\xi)\big) & L_j(x',D') \end{pmatrix} .$$

Lemma 2.27. *If the symbols $a_j(x,\xi)$ are independent of x_n, then the following formulae are valid:*

1) $(a_1)_{\mathbb{R}^n_+} \circ_n k_2 = \Pi^+_{\xi_n}\big[a_1(x',\xi)\, k_2(x',\xi)\big] \in \mathbf{S}^{d-1}_{1,0}(H^+)$,

2) $g_1 \circ_n k_2 = (2\pi)^{-1} \int^+ g_1(x',\xi,\eta_n)\; k_2(x',\xi',\eta_n)\, \mathrm{d}\eta_n \in \mathbf{S}^{d-1}_{1,0}(H^+)$,

3) $k_1 \circ_n l_2 = k_1(x',\xi)\, l_2(x',\xi') \in \mathbf{S}^{d-1}_{1,0}(H^+)$,

4) $t_1 \circ_n (a_2)_{\mathbb{R}^n_+} = \Pi^-_{\xi_n}\big[t_1(x',\xi)\, a_2(x',\xi)\big] \in \mathbf{S}^{d}_{1,0}(H^-_{r-1})$,

5) $t_1 \circ_n g_2 = (2\pi)^{-1} \int^+ t_1(x',\xi)\; g_2(x',\xi,\eta_n)\, \mathrm{d}\xi_n \in \mathbf{S}^{d}_{1,0}(H^-_{r_2-1})$,

6) $l_1 \circ_n t_2 = l_1(x',\xi')\, t_2(x',\xi) \in \mathbf{S}^{d}_{1,0}(H^-_{r_2-1})$,

7) $t_1 \circ_n k_2 = (2\pi)^{-1} \int^+ t_1(x',\xi)\, k_2(x',\xi)\, \mathrm{d}\xi_n \in \mathbf{S}^{d}_{1,0}(\mathbb{R}^{n-1}\times\mathbb{R}^{n-1})$,

8) $l_1 \circ_n l_2 = l_1(x',\xi')\, l_2(x',\xi') \in \mathbf{S}^{d}_{1,0}(\mathbb{R}^{n-1}\times\mathbb{R}^{n-1})$,

9) $(a_1)_{\mathbb{R}^n_+} \circ_n g_2 = \Pi^+_{\xi_n}\big[a_1(x',\xi)\, g_2(x',\xi,\eta_n)\big] \in \mathbf{S}^{d-1}_{1,0}(H^+\hat{\otimes}H^-_{r_2-1})$,

10) $g_1 \circ_n (a_2)_{\mathbb{R}^n_+} = \Pi^-_{\eta_n}\big[g_1(x',\xi,\eta_n)\, a_2(x',\xi',\eta_n)\big] \in \mathbf{S}^{d-1}_{1,0}(H^+\hat{\otimes}H^-_{r-1})$,

11) $g_1 \circ_n g_2 = (2\pi)^{-1} \int^+ g_1(x',\xi,\xi_n,\zeta_n)\, g_2(x',\xi',\zeta_n,\eta_n)\, \mathrm{d}\zeta_n$

$$\in \mathbf{S}^{d-1}_{1,0}(H^+\hat{\otimes}H^-_{r_2-1}),$$

12) $k_1 \circ_n t_2 = k_1(x', \xi', \xi_n)\, t_2(x', \xi', \eta_n) \in \mathbf{S}^{d-1}_{1,0}(H^+ \hat{\otimes} H^-_{r_2-1})$,

where $\mathbf{S}^{d-1}_{1,0}(H^+) = \mathbf{S}^{d-1}_{1,0}(\mathbb{R}^{n-1} \times \mathbb{R}^{n-1}, H^+)$, *etc.*, $d = d_1 + d_2$, $r = \max\{r_1 + d_2, 0\}$.

Lemma 2.28. *If the symbols* $a_j(x, \xi)$ *are independent of* x_n, *then*

13) $g(a_1, a_2) \equiv (a_1 a_2)_{\mathbb{R}^n_+} - (a_1)_{\mathbb{R}^n_+} \circ_n (a_2)_{\mathbb{R}^n_+} =$

$$\Pi^+_{\xi_n} \Pi^-_{\eta_n} \left[\frac{\left(a_1^+(x', \xi', \xi_n) - a_1^+(x', \xi', \eta_n)\right) \left(a_2^-(x', \xi', \xi_n) - a_2^-(x', \xi', \eta_n)\right)}{i(\xi_n - \eta_n)} \right]$$
$$\in \mathbf{S}^{d-1}_{1,0}(H^+ \hat{\otimes} H^-_{d_2^+ - 1}),$$

where

$$a_1^+(x', \xi) = \Pi^+_{\xi_n} a_1(x', \xi', \xi_n),$$
$$a_2^-(x', \xi) = \Pi^-_{\xi_n} a_2(x', \xi', \xi_n), \quad d_2^+ = \max\{d_2, 0\}.$$

Another formula for $g(a_1, a_2)$ was obtained in § 2.6 of the monograph (Grubb, G. (1986)). The relations 1)–13) give rise to the composition rule for the boundary symbols σ_{P_j}.

If the symbols $a_j(x, \xi)$ are x_n−dependent, then the formulae 1), 4), 9), 10), 13) from the previous lemmas have to be modified. In this case the Taylor decompositions

$$a_j(x, \xi) = \sum_{m<M} \frac{x_n^m}{m!} \partial^m_{x_n} a_j(x', 0, \xi) + x_n^M r_{j,M}(x, \xi).$$

are applied (see § 2.7 in Grubb, G . (1986)).

Lemma 2.29. *The following relations hold:*

1′) $(a_1)_{\mathbb{R}^n_+} \circ_n k_2 = k \in \mathbf{S}^{d-1}_{1,0}(H^+)$,

$$k(x', \xi) \sim \sum_{m \in \mathbb{Z}_+} \frac{1}{m!} \Pi^+_{\xi_n} \Big((-D_{\xi_n})^m \left[\partial^m_{x_n} a_1(x', 0, \xi)\, k_2(x', \xi) \right] \Big);$$

4′) $t_1 \circ_n (a_2)_{\mathbb{R}^n_+} = t \in \mathbf{S}^d_{1,0}(H^-_{r-1})$,

$$t(x', \xi) \sim \sum_{m \in \mathbb{Z}_+} \frac{1}{m!} \Pi^-_{\xi_n} \Big(\left[D^m_{\xi_n} t_1(x', \xi) \right] \partial^m_{x_n} a_2(x', 0, \xi) \Big);$$

9′) $(a_1)_{\mathbb{R}^n_+} \circ_n g_2 = g' \in \mathbf{S}^{d-1}_{1,0}(H^+ \hat{\otimes} H^-_{r_2-1})$,

$$g'(x', \xi, \eta_n) \sim \sum_{m \in \mathbb{Z}_+} \frac{1}{m!} \Pi^+_{\xi_n} \Big((-D_{\xi_n})^m \left[\partial^m_{x_n} a_1(x', 0, \xi)\, g_2(x', \xi, \eta_n) \right] \Big);$$

10′) $g_1 \circ_n (a_2)_{\mathbb{R}^n_+} = g'' \in \mathbf{S}^{d-1}_{1,0}(H^+ \hat{\otimes} H^-_{r-1})$,

$$g''(x',\xi,\eta_n) \sim \sum_{m\in\mathbb{Z}_+} \frac{1}{m!}\,\Pi^-_{\eta_n}\left(\left[D^m_{\eta_n} g_1(x',\xi,\eta_n)\right]\partial^m_{x_n} a_2(x',0,\xi',\eta_n)\right);$$

13′) $(a_1a_2)_{\mathbb{R}^n_+} - (a_1)_{\mathbb{R}^n_+} \circ_n (a_2)_{\mathbb{R}^n_+} = g(a_1,a_2) \in \mathbf{S}^{d-1}_{1,0}(H^+\hat{\otimes}H^-_{d_2^+-1})$,

$$g(a_1,a_2)\;(x',\xi,\eta_n)$$

$$\sim \sum_{m,l,k\in\mathbb{Z}_+} \frac{(-1)^{m+l}}{m!l!k!}\,D^m_{\xi_n}D^k_{\eta_n} g\big(\partial^m_{x_n} a_1(x',0,\xi),\,D^l_{\eta_n}\partial^{l+k}_{x_n} a_2(x',0,\xi',\eta_n)\big),$$

where $d = d_1 + d_2$, $r = \max\{r_1 + d_2, 0\}$, $d_2^+ = \max\{d_2, 0\}$, *and the symbols in the asymptotic decomposition in the formula* 13′) *are computed with the use of* 13).

The composition formula for the operators

$$P_j(x',\xi',D_n) = \begin{pmatrix} a_j(x',\xi',D_n)_{\mathbb{R}_+} + g_j(x',\xi',D_n) & k_j(x',\xi',D_n) \\ t_j(x',\xi',D_n) & l_j(x',\xi') \end{pmatrix}:$$

$$\begin{matrix} C_0^\infty(\bar{\mathbb{R}}_+)\otimes\mathbb{C}^{N_j} \\ \bigoplus \\ \mathbb{C}^{J_j} \end{matrix} \longrightarrow \begin{matrix} C^\infty(\bar{\mathbb{R}}_+)\otimes\mathbb{C}^{N'_j} \\ \bigoplus \\ \mathbb{C}^{J'_j} \end{matrix}, \quad N_1 = N'_2,\; J_1 = J'_2 \qquad (2.38)$$

follows from the previous lemmas.

Now we are ready to formulate the composition rule for the operators P_j.

Theorem 2.30. *Let* P_j $(j = 1, 2)$ *be the Green operators of order* $d_j \in \mathbb{R}$ *and class* $r_j \in \mathbb{Z}_+$. *Then their composition* $P = P_1P_2$ *is the Green operator of order* $d_1 + d_2$ *and class* $\max\{r_1 + d_2, r_2\}$, *and moreover*

$$P(x',\xi',D_n) \sim \sum_{\alpha\in\mathbb{Z}_+^{n-1}} \frac{1}{\alpha!}\,D^\alpha_{\xi'} P_1(x',\xi',D_n)\circ D^\alpha_{x'} P_2(x',\xi',D_n).$$

Remark. 2.31. So far, we have considered Green operators of nonnegative class. In the case where $A_{\mathbb{R}^n_+} + G$ and T are operators of class $r \in \mathbb{Z}$, P is called (see (2.35)) a Green operator of order d and class r. It is easy to see, with the help of Theorem 2.23, that Theorem 2.30 remains valid for Green operators of negative class, i.e. we may consider there $r_j \in \mathbb{Z}$.

It is also possible to consider Green operators of the type (2.35), where $A_{\mathbb{R}^n_+} + G$ is of order $d \in \mathbb{R}$, and K, T and L have orders $s \in \mathbb{R}$, $h \in \mathbb{R}$ and $s + h - d$ respectively. All the results of the section may be transferred to these operators. Moreover, these operators may be reduced to Green operators considered above, with all the components of the same order, with the help of the order reduction operators, but this will be discussed in detail in the next section.

2.7. Boundary Value Problems on Manifolds. (Boutet de Monvel L. (1971), Grubb, G. (1986), Grubb, G. (1989), Grubb, G. (1990), Rempel S., Schulze B.-W. (1982a)).

Let M be a smooth n–dimensional compact manifold with boundary $\Gamma = \partial M$ and interior $\Omega = M \setminus \partial M$. Let also E and E' be smooth vector bundles over M, and similarly for F and F' over Γ. Consider the operator

$$P = \begin{pmatrix} A_\Omega + G & K \\ T & L \end{pmatrix} : \begin{matrix} C^\infty(E) \\ \bigoplus \\ C^\infty(F) \end{matrix} \longrightarrow \begin{matrix} C^\infty(E') \\ \bigoplus \\ C^\infty(F') \end{matrix} . \tag{2.39}$$

Let W be an open (generally speaking disconnected) subset in M, with $W' = W \cap \Gamma$ (W' may be empty), such that the bundles $E\,|_W$, $E'\,|_W$, $F\,|_{W'}$ and $F'\,|_{W'}$ are trivial. Also we suppose that any connected component W_0 of W, intersecting with the boundary, is diffeomorphic to $W_0' \times [0,1)$, where $W_0' = W_0 \cap \Gamma$. We denote by

$$\chi_E : E\,|_W \to V \times \mathbb{C}^N , \quad \chi_{E'} : E'\,|_W \to V \times \mathbb{C}^{N'} ,$$

$$\chi_F : F\,|_{W'} \to V' \times \mathbb{C}^J , \quad \chi_{F'} : F'\,|_{W'} \to V' \times \mathbb{C}^{J'}$$

the trivializations of the corresponding bundles, where V is an open subset in $\overline{\mathbb{R}^n_+}$, $V' = V \cap \partial\overline{\mathbb{R}^n_+}$, and $N, N', J, J' \in \mathbb{Z}_+$ are the layer dimensions of the bundles. The operator P_W defined by the commutative diagram

$$\begin{array}{ccc} C^\infty(E) \bigoplus C^\infty(F) & \xrightarrow{P} & C^\infty(E') \bigoplus C^\infty(F') \\ \uparrow \chi_E^* \oplus \chi_F^* & & \downarrow (\chi_{E'})_* \oplus (\chi_{F'})_* \\ C_0^\infty(V, \mathbb{C}^N) \bigoplus C_0^\infty(V', \mathbb{C}^J) & \xrightarrow{P_W} & C^\infty(V, \mathbb{C}^{N'}) \bigoplus C^\infty(V', \mathbb{C}^{J'}) \end{array}$$

is called the *local representation* of P over W (relative to the given trivialization).

Let $\phi, \psi \in C_0^\infty(V)$. Then the operator P_W induces the operator

$$\psi P_W \phi I : \begin{matrix} C_0^\infty(\overline{\mathbb{R}^n_+}, \mathbb{C}^N) \\ \bigoplus \\ C_0^\infty(\mathbb{R}^{n-1}, \mathbb{C}^J) \end{matrix} \longrightarrow \begin{matrix} C^\infty(\overline{\mathbb{R}^n_+}, \mathbb{C}^{N'}) \\ \bigoplus \\ C^\infty(\mathbb{R}^{n-1}, \mathbb{C}^{J'}) \end{matrix} ,$$

where we denote by ϕI the multiplication operator of all the components by ϕ or $\phi\,|_{\partial\mathbb{R}^n_+}$ respectively.

Definition 2.32. We say that the operator P of the form (2.39) is a **Green operator** of order $d \in \mathbb{R}$ and class $r \in \mathbb{Z}$ if, for an arbitrary open set $W \subset M$ possessing the above properties and arbitrary functions $\phi, \psi \in C_0^\infty(V)$, the induced operator $\psi P_W \phi I$ is a Green operator of order d and class r.

The correctness of the definition follows from the invariance of the UH–property relative to diffeomorphisms preserving the boundary (see the end of Section 2.3) and Theorems 2.17, 2.20 and 2.30.

Using Theorems 2.21 and 2.23 we can prove in a standard way that, if P is a Green operator of order d and class r, then the mappings

$$P : H_1(s,p) = \begin{matrix} H_p^s(E) \\ \bigoplus \\ B_{p,p}^{s-1/p}(F) \end{matrix} \longrightarrow H_2(s,p) = \begin{matrix} H_p^{s-d}(E') \\ \bigoplus \\ B_{p,p}^{s-d-1/p}(F') \end{matrix}, \tag{2.40}$$

$$P : B_1(s,p,q) = \begin{matrix} B_{p,q}^s(E) \\ \bigoplus \\ B_{p,q}^{s-1/p}(F) \end{matrix} \longrightarrow B_2(s,p,q) = \begin{matrix} B_{p,q}^{s-d}(E') \\ \bigoplus \\ B_{p,q}^{s-d-1/p}(F') \end{matrix}$$

are continuous for $s > r + 1/p - 1$, where $s \in \mathbb{R}$, $p \in]1,\infty[$ and $q \in [1,\infty]$.

Using the partition of unity and passing to local coordinates, we can transfer the results of the last section to Green operators of the form (2.39).

In the rest of this section we shall deal only with **polyhomogeneous Green operators**, i.e. operators P of the form (2.39) such that, for arbitrary W, ϕ, ψ possessing the above properties, the operator $\psi P_W \phi I$ is polyhomogeneous.

We can consider a Green operator P (see (2.39)) as an **operator of a boundary value problem for a pseudodifferential operator** A, defined on an embracing manifold $M_0 \supset M$ (without boundary). Also we recollect that A is called an **elliptic pseudodifferential operator** if (see Agranovich M.S. (1990), Hörmander L. (1983, 1985), Rempel S., Schulze B.-W. (1982a), Trèves F. (1980)) its leading symbol $\sigma_A : \mathrm{pr}^* E_0 \to \mathrm{pr}^* {E_0}'$ is an isomorphism, where $\mathrm{pr} : T^*M_0 \setminus \{0\} \to M_0$ is the canonical projection, and T^*M_0 is the cotangent bundle ($E_0 \mid_M = E$, $E_0' \mid_M = E'$).

Clearly, if A is an elliptic pseudodifferential operator, then the fiber dimensions of the bundles E and E' coincide: $N = N'$. From Theorem 2.8 and the discussion after Theorem 2.11, it follows that an elliptic pseudodifferential operator possessing the $H-$property on Γ certainly has integer order $d \in \mathbb{Z}$.

The symbol $\sigma_A : \mathrm{pr}^* E \to \mathrm{pr}^* E'$ is called the *leading (or principal) interior symbol* of the Green operator P.

Now we return to the operator $\psi P_W \phi I$. Let the functions ψ and ϕ equal 1 in some neighbourhood of the point $(z', 0) \in V'$, and consider at this point the leading boundary symbol of the operator $\psi P_W \phi I$:

$$\sigma_{\psi P_W \phi I}(z', \xi', D_n) = \begin{pmatrix} a^0(z', 0, \xi', D_n)_{\mathbb{R}_+} + g^0(z', \xi', D_n) & k^0(z', \xi', D_n) \\ t^0(z', \xi', D_n) & l^0(z', \xi') \end{pmatrix} :$$

$$\begin{matrix} S(\bar{\mathbb{R}}_+) \otimes \mathbb{C}^N \\ \bigoplus \\ \mathbb{C}^J \end{matrix} \longrightarrow \begin{matrix} S(\bar{\mathbb{R}}_+) \otimes \mathbb{C}^{N'} \\ \bigoplus \\ \mathbb{C}^{J'} \end{matrix}. \tag{2.41}$$

It is easy to prove (see, for example, Rempel S., Schulze B.-W. (1982a)) that the operator (2.41) is the local representative of some bundle morphism

$$\sigma_\Gamma(P)\ :\ \begin{array}{c}\pi_{\mathrm{r}}^* E|_\Gamma\otimes S(\bar{\mathbb{R}}_+)\\ \bigoplus\\ \pi_{\mathrm{r}}^* F\end{array} \longrightarrow \begin{array}{c}\pi_{\mathrm{r}}^* E'|_\Gamma\otimes S(\bar{\mathbb{R}}_+)\\ \bigoplus\\ \pi_{\mathrm{r}}^* F'\end{array}, \tag{2.42}$$

where $\pi_{\mathrm{r}} : S^*\Gamma \to \Gamma$ is the cosphere bundle realized in the cotangent bundle $T^*\Gamma$ with the help of some Riemannian metric on Γ.

Definition 2.33. The morphism $\sigma_\Gamma(P)$ is called the *leading boundary symbol* of the Green operator P.

Analogously to $\sigma_\Gamma(P)$ we can define (see (2.37)) the morphism

$$\widehat{\sigma_\Gamma}(P)\ :\ \begin{array}{c}\pi_{\mathrm{r}}^* E|_\Gamma\otimes H^+\\ \bigoplus\\ \pi_{\mathrm{r}}^* F\end{array} \longrightarrow \begin{array}{c}\pi_{\mathrm{r}}^* E'|_\Gamma\otimes H^+\\ \bigoplus\\ \pi_{\mathrm{r}}^* F'\end{array}, \tag{2.43}$$

which is also called the *leading boundary symbol* of the Green operator P.

Definition 2.34. A Green operator is called **elliptic** if its leading (interior and boundary) symbols are isomorphisms.

Now we formulate without proof one of the main results of Boutet de Monvel's theory.

Theorem 2.35. *Let P be an* **elliptic Green operator** *of order $d \in \mathbb{Z}$ and class $r \in \mathbb{Z}$. Then there exists the* **parametrix** *Q of P, which is also an elliptic Green operator of order $-d$ and class $r-d$, such that $PQ - I$ and $QP - I$ are Green operators of order $-\infty$ and classes $r-d$ and r respectively. The operators*

$$P\ :\ H_1(s,p)\ \to\ H_2(s,p)\quad \text{and}\quad P\ :\ B_1(s,p,q)\ \to\ B_2(s,p,q)$$

(see (2.40)) are of Noether type[2] *(see, for example,* Prössdorf S. (1988)*) for $s > r + 1/p - 1$, where $s \in \mathbb{R}$, $p \in]1,\infty[$, $q \in [1,\infty]$. The finite dimensional kernel of these operators belongs to C^∞ and does not depend on s,p,q nor on the choice of spaces, and their images are defined by a finite number of orthogonality conditions to C^∞-functions, these conditions not depending on s,p,q nor on the choice of spaces.*

Analogous results for Hölder spaces were obtained in (Rempel S., Schulze B.-W. (1982a)) and for a wide class of Besov-Triebel-Lizorkin spaces in (Franke J. (1985), Franke J. (1986)). It was also proved in (Rempel S., Schulze B.-W. (1982a)) that if the Green operator P is of Noether type in the pair of spaces $H_1(s,2)$, $H_2(s,2)$ then it is elliptic.

Let

$$(\alpha_j)_{1\le j\le j_0}\,,\quad (\beta_k)_{1\le k\le k_0}\,,\quad (\rho_l)_{1\le l\le l_0}\,,\quad (\delta_m)_{1\le m\le m_0}$$

be given collections of integers and let (E_j), (E'_k) be smooth vector bundles over M, and similarly for (F_l), (F'_m) over Γ. We define the operators analogous

[2] i.e. are Fredholm operators.

to the corresponding components in (2.39): $A_{kj,\Omega} + G_{kj}$ of order $d + \beta_k + \alpha_j$ and class $r + \alpha_j$, K_{kl} of order $d + \beta_k + \rho_l$, T_{mj} of order $d + \delta_m + \alpha_j$ and class $r + \alpha_j$, L_{ml} of order $d + \delta_m + \rho_l$, $d, r \in \mathbb{Z}$. We combine them to form the operator, analogous to (2.39),

$$P = \begin{matrix} \oplus_{1\le j\le j_0} C^\infty(E_j) \\ \bigoplus \\ \oplus_{1\le l\le l_0} C^\infty(F_l) \end{matrix} \longrightarrow \begin{matrix} \oplus_{1\le k\le k_0} C^\infty(E'_k) \\ \bigoplus \\ \oplus_{1\le m\le m_0} C^\infty(F'_m) \end{matrix} . \tag{2.44}$$

This operator we shall also call a **Green operator**.

As in (2.40) the operators

$$P : H_1(s,p) = \begin{matrix} \oplus_{1\le j\le j_0} H_p^{s+\alpha_j}(E_j) \\ \bigoplus \\ \oplus_{1\le l\le l_0} B_{p,p}^{s+\rho_l-1/p}(F_l) \end{matrix} \longrightarrow$$

$$\longrightarrow H_2(s,p) = \begin{matrix} \oplus_{1\le k\le k_0} H_p^{s-d-\beta_k}(E'_k) \\ \bigoplus \\ \oplus_{1\le m\le m_0} B_{p,p}^{s-d-\delta_m-1/p}(F'_m) \end{matrix} ,$$

$$P : B_1(s,p,q) = \begin{matrix} \oplus_{1\le j\le j_0} B_{p,q}^{s+\alpha_j}(E_j) \\ \bigoplus \\ \oplus_{1\le l\le l_0} B_{p,q}^{s+\rho_l-1/p}(F_l) \end{matrix} \longrightarrow$$

$$\longrightarrow B_2(s,p,q) = \begin{matrix} \oplus_{1\le k\le k_0} B_{p,q}^{s-d-\beta_k}(E'_k) \\ \bigoplus \\ \oplus_{1\le m\le m_0} B_{p,q}^{s-d-\delta_m-1/p}(F'_m) \end{matrix} , \tag{2.45}$$

are continuous for $s > r + 1/p - 1$, where $s \in \mathbb{R}$, $p \in]1,\infty[$, and $q \in [1,\infty]$.

For the Green operator of the form (2.44) we may also define the leading interior symbol

$$\sigma_A : \mathrm{pr}^* E_0 \to \mathrm{pr}^* E_0{}' ,$$

where

$$E = \underset{1\le j\le j_0}{\oplus} E_j \,, \; E' = \underset{1\le k\le k_0}{\oplus} E'_k \,,$$

and the leading boundary symbol $\sigma_\Gamma(P)$ of the form (2.42) (or $\widehat{\sigma_\Gamma}(P)$ of the form (2.43)).

Definition 2.36. A Green operator of the type (2.44) is called **elliptic (injectively elliptic, surjectively elliptic) in the Douglis-Nirenberg sense** if its leading interior and boundary symbols are isomorphisms (injective morphisms, surjective morphisms) of the corresponding bundles.

The following result is a generalization of Theorem 2.35.

Theorem 2.37. *Let P be an elliptic (injectively elliptic, surjectively elliptic) Green operator in the Douglis-Nirenberg sense of the type* (2.44). *Then there* *exists the* **parametrix (left parametrix, right parametrix)** Q *of* P*, which is also an elliptic (injectively elliptic, surjectively elliptic) Green operator in the Douglis-Nirenberg sense. The operators*

$$P : H_1(s,p) \to H_2(s,p) \quad \textit{and} \quad P : B_1(s,p,q) \to B_2(s,p,q)$$

(see (2.45)*) are of Noether type (half-Noether type in the corresponding sense: are* Φ_+ *or* Φ_- *operators, i.e. in the first case they have finite dimensional kernels and closed images, in the second, images of finite codimension; see, for example,* Prössdorf S. (1988)*) for* $s > r + 1/p - 1$*, where* $s \in \mathbb{R}$*,* $p \in]1, \infty[$ *and* $q \in [1, \infty]$*. In the case of injective ellipticity the finite dimensional kernel of these operators belongs to* C^∞ *and does not depend on* s, p, q *nor on the choice of spaces. In this case the following smoothness property of solutions holds: if* $Pu = f$*, where* $f \in H_2(s,p)$ *(or* $B_2(s,p,q)$*), and* $u \in H_1(s',p')$ *(or* $B_1(s',p',q')$*), with* $s > r+1/p-1$ *and* $s' > r+1/p'-1$*,* $s, s' \in \mathbb{R}$*,* $p, p' \in]1, \infty[$*,* $q, q' \in [1, \infty]$*, then* $u \in H_1(s,p)$ *(or* $B_1(s,p,q)$*). In the case of surjective ellipticity the images of operators under consideration are defined by a finite number of orthogonality conditions to* C^∞ *functions and these conditions do not depend on* s, p, q *nor on the choice of spaces.*

The above theorem can also be generalized to operators acting in Besov-Triebel-Lizorkin spaces and, in particular, Hölder spaces (see (Franke J. (1985), Franke J. (1986), Rempel S., Schulze B.-W. (1982a)) in the same way as Theorem 2.35.

Green operators of the type (2.44) are reduced to operators of the form (2.39) by the order reduction procedure.

The operators with the symbols $(\langle \xi' \rangle \pm i\xi_n)^m$ were used as operators carrying out isomorphisms of different functional spaces over $\mathbb{R}^n_+$ (see, for example, Eskin G.I. (1973), Peetre J. (1959)). Unfortunately, these operators are not of the class $S^m_{1,0}(\mathbb{R}^{n-1} \times \mathbb{R}^{n-1})$ (for example, $D^2_{\xi_1}(\langle \xi' \rangle \pm i\xi_n)$ is $O(\langle \xi' \rangle^{-1})$, but not $O(\langle \xi \rangle^{-1})$. In order to construct similar operators with symbols in $S^m_{1,0}$, we need the following functions:

$$\chi_\pm \in S(\mathbb{R}), \quad \chi_\pm(0) = 1, \quad \operatorname{supp} F^{-1}\chi_\pm \subset \bar{\mathbb{R}}_\pm,$$

$$\sigma_1 \in C^\infty(\mathbb{R}), \quad \sigma_1 > 0, \quad \sigma_1(t) = |t| \text{ for } |t| \geq 1,$$

$$\sigma_1(t) = 1/2 \text{ for } |t| \leq 1/2, \quad \sigma(\xi', \mu) = \sigma_1(|(\xi', \mu)|),$$

where μ is a positive parameter to be fixed in the future.

Define

$$\lambda^m_\pm(\xi, \mu) = \left(\sigma(\xi', \mu)\, \chi_\pm\!\left(\frac{\xi_n}{a\sigma(\xi', \mu)}\right) \pm i\xi_n\right)^m, \quad m \in \mathbb{Z}, \tag{2.46}$$

where $a > 0$ is large enough for negative powers to be defined. More exactly

$$\left| \frac{\sigma\chi_\pm(\xi_n/a\sigma) \pm i\xi_n}{\sigma \pm i\xi_n} \right| = \left| 1 + \frac{(\chi_\pm(\xi_n/a\sigma) - 1)\sigma}{\sigma \pm i\xi_n} \right| \geq \frac{1}{2},$$

when

$$\sup_{\xi_n} \left| \frac{(\chi_\pm(\xi_n/a\sigma) - 1)\sigma}{\sigma \pm i\xi_n} \right| \equiv \sup_t \left| \frac{\chi_\pm(t) - 1}{1 \pm iat} \right| \leq \frac{1}{2}$$

with the last inequality being valid when

$$a \geq 2 \sup_t |\chi_\pm{}'(t)| \,.$$

It can be easily checked that $\lambda_\pm^m(\xi,\mu) \in \mathbf{S}^m(\mathbb{R}^{n+1} \times \mathbb{R}^{n+1})$ and the corresponding leading homogeneous symbols are equal to

$$\lambda_\pm^{m,0}(\xi,\mu) = \left(|(\xi',\mu)| \, \chi_\pm\Big(\frac{\xi_n}{a|(\xi',\mu)|}\Big) \pm i\xi_n \right)^m , \quad m \in \mathbb{Z}\,.$$

When the parameter $\mu > 0$ is fixed, $\lambda_\pm^m(\xi,\mu) \in \mathbf{S}^m(\mathbb{R}^n \times \mathbb{R}^n)$ and the corresponding leading homogeneous symbols are equal to $\lambda_\pm^{m,0}(\xi, 0)$ (when $|\xi| \geq 1$).

Define now the operators

$$\Lambda_{\pm,\mu}^m = \mathbf{OP}(\lambda_\pm^m) \in \mathbf{OP}\big((\mathbf{S}^m(\mathbb{R}^n \times \mathbb{R}^n)\big)\,.$$

Lemma 2.38. *Let $s \in \mathbb{R}$, $p \in]1,\infty[$, $q \in [1,\infty]$, $m \in \mathbb{Z}$, and let $a > 0$ be sufficiently large (see (2.46)). Then for arbitrary $\mu > 0$ the operators*

$$\Lambda_{\mp,\mu,\mathbb{R}_\pm^n}^m : H_p^s(\overline{\mathbb{R}_\pm^n}) \to H_p^{s-m}(\overline{\mathbb{R}_\pm^n}) \quad \Big(\text{or } B_{p,q}^s(\overline{\mathbb{R}_\pm^n}) \to B_{p,q}^{s-m}(\overline{\mathbb{R}_\pm^n})\Big),$$

$$\Lambda_{\pm,\mu}^m : \tilde{H}_p^s(\overline{\mathbb{R}_\pm^n}) \to \tilde{H}_p^{s-m}(\overline{\mathbb{R}_\pm^n}) \quad \Big(\text{or } \tilde{B}_{p,q}^s(\overline{\mathbb{R}_\pm^n}) \to \tilde{B}_{p,q}^{s-m}(\overline{\mathbb{R}_\pm^n})\Big)$$

are isomorphisms and

$$\Lambda_{\mp,\mu,\mathbb{R}_+^n}^j \, \Lambda_{\mp,\mu,\mathbb{R}_+^n}^k = \Lambda_{\mp,\mu,\mathbb{R}_+^n}^{j+k}, \quad \forall\, j,k \in \mathbb{Z}\,.$$

Clearly $\lambda_\pm^m \in \mathbf{S}_{\mathrm{tr}}^m(\mathbb{R}^n \times \mathbb{R}^n)$, $m \in \mathbb{Z}$, (see Section 2.3).

Now we move on to order reduction operators on manifolds. As we have done earlier, we assume that M is embedded in a compact manifold M_0 without boundary, and that E is the restriction to M of a smooth vector bundle E_0 over M_0. According to the theorem about the "collar" (see, for example, (Munkres R. (1966)) there exists a neighbourhood U of the boundary $\Gamma = \partial M$ in M_0 which can be identified with $\Sigma_2 = \Gamma\times]-2,2[$, and moreover the bundle $E_0\,|_U$ is identified with the lifting of $E\,|_\Gamma$ to U.

We denote by x_n the variable with the range $]-2,2[$. Also we set $\Sigma_c = \Gamma\times]-c,c[$ for $0 < c \leq 2$. We define the symbol $\lambda^m{}_{-,E}$ which coincides with

$$\Big(\alpha(x_n)\big[\sigma(\xi',\mu)\,\chi_-\big(\xi_n/a\sigma(\xi',\mu)\big) - i\xi_n\big] + \big(1-\alpha(x_n)\big)\sigma(\xi,\mu)\Big)^m I_{E_0}$$

on Σ_2, and with $\sigma(\xi,\mu)I_{E_0}$ on $M_0 \setminus \Sigma_2$. Here $\alpha \in C_0^\infty(\mathbb{R})$, $0 \leq \alpha \leq 1$, $\alpha(t) = 1$ when $|t| \leq 1$, and $\alpha(t) = 0$ when $|t| \geq 3/2$.

It may be seen easily that, for sufficiently large $a > 0$, the symbol $\lambda^m_{-,E} \in \mathbf{S}^m_{\mathrm{tr}}$ is elliptic. We denote by $\Lambda^{(m)}_{-,\mu,E}$ the corresponding pseudodifferential operator. As follows from Lemma 2.38, the operator $\Lambda^{(m)}_{-,\mu,E,\Omega}$ is elliptic with a parameter (see Grubb, G. (1986), Grubb, G. (1989) or §3 of the present work). Thus (see Grubb, G. (1986), Grubb, G. (1989) or §3 of the present work) there exists a family of parametrices $\mathcal{R}_\mu$ such that, for sufficiently large μ, the operator $\mathcal{R}_\mu$ is the inverse to $\Lambda^{(m)}_{-,\mu,E,\Omega}$ in the space $C^\infty(E)$. Taking into account Theorem 2.35 we obtain the following result.

Lemma 2.39. *Let $s \in \mathbb{R}$, $p \in]1,\infty[$, $q \in [1,\infty]$, $m \in \mathbb{Z}$, and let $a > 0$ and $\mu > 0$ be sufficiently large. Then the operators*

$$\Lambda^{(m)}_{-,\mu,E,\Omega} : H^s_p(E) \to H^{s-m}_p(E) \quad \left(\text{or } B^s_{p,q}(E) \to B^{s-m}_{p,q}(E)\right),$$

are isomorphisms.

Acting on (2.44) from the left with the operator

$$\underset{1\le k\le k_0}{\oplus} \Lambda^{(-\beta_k)}_{-,\mu,E'_k,\Omega} \bigoplus \underset{1\le m\le m_0}{\oplus} \Lambda^{(-\delta_m)}_{\Gamma,F'_m}$$

and from the right with the operator

$$\underset{1\le j\le j_0}{\oplus} \Lambda^{(-\alpha_j)}_{-,\mu,E_j,\Omega} \bigoplus \underset{1\le l\le l_0}{\oplus} \Lambda^{(-\rho_l)}_{\Gamma,F_l},$$

where $\Lambda^{(\rho)}_{\Gamma,F}$ is an elliptic pseudodifferential operator on Γ carrying out the isomorphisms $B^r_{p,q}(F) \to B^{r-\rho}_{p,q}(F)$ (see, for example, Agranovich M.S. (1990), Rempel S., Schulze B.-W. (1982a)), we obtain a Green operator of the type (2.39) (see Section 2.6).

Now we return to Theorem 2.35. We find from the composition formulae (see Section 2.6) that the interior and the boundary leading symbols of the operator P and its parametrix Q are mutually inverse. If Q and Q' are the two parametrices of P, then $Q - Q'$ is a Green operator of order $-\infty$ and class $r - d$. Conversely, if Q is a parametrix of P and C is a Green operator of order $-\infty$ and class $r - d$, then $Q + C$ is also a parametrix of P.

The elliptic boundary value problem of the type (2.39) does not exist for any given elliptic pseudodifferential operator $A \in \mathbf{OP}(\mathbf{S}^d_{\mathrm{tr}})$. Consider the morphism

$$\Pi^+\sigma_A : \pi \mathrm{r}^* E\,|_\Gamma \otimes H^+ \to \pi \mathrm{r}^* E'\,|_\Gamma \otimes H^+$$

(cf. (2.43)). It may be proved (see Rempel S., Schulze B.-W. (1982a)) that this is a Fredholm family with the index element $\mathrm{ind}_{S^*\Gamma}\Pi^+\sigma_A$ belonging to the group $K(S^*\Gamma)$, (the so-called Atiyah-Jänich index; see, for example, Rempel S., Schulze B.-W. (1982a)).

It turns out that a necessary and sufficient condition for the existence of a boundary value problem of the type (2.39), corresponding to the elliptic pseudodifferential operator A, is that

$$\operatorname{ind}_{S^*\Gamma} \Pi^+ \sigma_A \in \pi\mathrm{r}^* K(\Gamma),$$

where $\pi\mathrm{r}^* K(\Gamma)$ is the inverse image of the group $K(G)$ relative to the projection $\pi\mathrm{r} : S^*\Gamma \to \Gamma$ (see Boutet de Monvel L. (1971), Rempel S., Schulze B.-W. (1982a)).

According to Theorems 2.35 and 2.37 the **index of an elliptic Green operator** is independent of s, p, q and the choice of spaces. There is a deep theorem on the index of elliptic boundary value problems expressing the index in terms of topological invariants (see Fedosov B.V. (1991), Atiyah M.F., Bott R. (1964), Boutet de Monvel L. (1971), Rempel S., Schulze B.-W. (1982a)) analogous to the well known Atiyah-Singer Index Formula (see, for example, Palais R.S. (1965)).

Note that the foregoing theory can be generalized to the case of elliptic complexes (see Dynin A.S. (1972), Franke J. (1985), Pillat U., Schulze B.-W. (1980), Rempel S., Schulze B.-W. (1982a), Schulze B.-W. (1979)).

2.8. About Some Applications. The boundary value problem theory for elliptic pseudodifferential operators is a generalization of the same theory for elliptic differential operators. The discussion of the latter theory is presented in the article of M.S.Agranovich in the present volume (see also the literature cited there). We show here, without going into details, how classical elliptic boundary value problems can be embedded in the scheme of Boutet de Monvel. For simplicity we consider the case of a scalar elliptic differential operator.

Let M be a smooth compact n-dimensional manifold with boundary $\Gamma = \partial M$, and let $\Omega = M \setminus \partial M$ be the interior of the manifold (in particular, it is possible to regard M as the closure of a bounded domain $\Omega \subset \mathbb{R}^n$ with C^∞ boundary Γ). Let $A : C^\infty(M, \mathbb{C}) \to C^\infty(M, \mathbb{C})$ be an elliptic differential operator of order m with C^∞-smooth coefficients. The ellipticity means that at any point $x \in M$ in any local coordinate system the leading symbol of the operator A satisfies the condition:

$$\sigma_A(x,\xi) \equiv a^0(x,\xi) \neq 0 \qquad \forall \xi \in \mathbb{R}^n \setminus \{0\}. \tag{2.47}$$

The following boundary value problem is posed for the operator A:

$$Au = f \in C^\infty(M, \mathbb{C}), \tag{2.48}$$

$$\gamma_0 T_j u = g_j \ \in \ C^\infty(\Gamma, \mathbb{C}), \qquad j = 1, \ldots, l, \tag{2.49}$$

where T_j are differential operators with C^∞-smooth coefficients, and γ_0 is the trace operator restricting the function to Γ. Denote by $\gamma_0 T$ the vector formed by the operators $\gamma_0 T_j$, $j = 1, \ldots, l$. The following Green operator corresponds to the boundary value problem (2.48), (2.49):

$$P = \begin{pmatrix} A \\ \gamma_0 T \end{pmatrix} : C^\infty(M, \mathbb{C}) \longrightarrow \begin{matrix} C^\infty(M,\mathbb{C}) \\ \oplus \\ C^\infty(\Gamma,\mathbb{C}^l) \end{matrix}. \tag{2.50}$$

Now we take an arbitrary point on the boundary Γ and introduce local coordinates in its neighbourhood. According to Definition 2.36 the ellipticity of the operator (2.50) means that its leading boundary symbol (see (2.41))

$$\begin{pmatrix} a^0(x',0,\xi',D_n) \\ \gamma_0 t^0(x',0,\xi',D_n) \end{pmatrix} : S(\bar{\mathbb{R}}_+) \longrightarrow \begin{matrix} S(\bar{\mathbb{R}}_+) \\ \oplus \\ \mathbb{C}^l \end{matrix} \tag{2.51}$$

is an isomorphism for all ξ' in $\mathbf{S^{n-2}} = \{\zeta' \in \mathbb{R}^{n-1} : |\zeta'| = 1\}$ and for all $(x',0)$ from the neighbourhood under consideration. Here we denote by $t^0(x,\xi)$ the vector formed by the leading symbols $t_j^0(x,\xi)$ of the differential operators T_j $(j = 1,\dots,l)$, written in the local coordinates.

In other words, the boundary value problem (2.48), (2.49) is elliptic if the problem

$$a^0(x',0,\xi',D_n)v(x_n) = \varphi(x_n)\,, \quad x_n > 0\,, \tag{2.52}$$

$$t_j^0(x',0,\xi',D_n)v(x_n)\,|_{x_n=0} = \psi_j\,, \quad j = 1,\dots,l\,, \tag{2.53}$$

for an ordinary differential equation on the half-axis, has a unique solution $v \in S(\bar{\mathbb{R}}_+)$ for all $\varphi \in S(\bar{\mathbb{R}}_+)$ and $\psi_j \in \mathbb{C}$ (and for all x',ξ' in consideration). The last condition is called the **Shapiro-Lopatinskij condition.**

Let $\xi' \in \mathbb{R}^{n-1}$ with $|\xi'| = 1$, and let $\nu_1^+(x',\xi'),\dots,\nu_r^+(x',\xi')$ be the roots of the polynomial $a^0(x',0,\xi',\nu)$ in $\nu \in \mathbb{C}$ which lie in the upper half-plane $\mathbf{Im}\,\nu > 0$. It is easily seen that for $n > 2$ the order m of the operator A is even and $r = m/2$. Indeed, for $n > 2$ the unit sphere $\mathbf{S^{n-2}}$ is connected, and we can transfer the point ξ' into the point $-\xi'$ along a smooth curve lying in $\mathbf{S^{n-2}}$. During this process the number of roots lying in the upper (lower) complex half-space would not change, since it follows from (2.47) that our polynomial has no real roots. On the other hand, if $\nu(x',\xi')$ is a root of the polynomial $a^0(x',0,\xi',\nu)$ then $-\nu(x',\xi')$ is the root of the polynomial $a^0(x',0,-\xi',\nu)$. Thus, the number of roots of the polynomial $a^0(x',0,-\xi',\nu)$ in the upper half-plane is equal both to the number of roots of the polynomial $a^0(x',0,\xi',\nu)$ in the upper half-plane and to the number of its roots in the lower half-plane. Thus $r = m - r$ and hence $m = 2r$.

In the case $n = 2$ elliptic differential operators of odd order exist, for example the Cauchy-Riemann operator

$$\overline{\partial} = \frac{1}{2}\Big(\frac{\partial}{\partial x_1} + i\frac{\partial}{\partial x_2}\Big)\,.$$

Elliptic boundary value problems of the type (2.48), (2.49) for such operators do not exist since it follows from the Shapiro-Lopatinskij condition (as may be easily seen) that $r = m/2 = l$ (see also Section 4.1 below).

Suppose now that these equalities hold. Then we put

$$a^+(x',\xi',\nu) = \prod_{k=1}^{r}\big(\nu - \nu_k^+(x',\xi')\big)$$

and define the remainders from the division of $t_j^0(x',0,\xi',\nu)$ by $a^+(x',\xi',\nu)$:

$$t_j'(x',\xi',\nu) = t_j^0(x',0,\xi',\nu) \ \mathbf{mod}\, a^+(x',\xi',\nu)\,,$$

$$t_j'(x',\xi',\nu) = \sum_{k=1}^{r} t_{jk}(x',\xi')\nu^{k-1}\,, \quad j = 1,\ldots,l = r\,.$$

It can be proved (see, for example, Rempel S., Schulze B.-W. (1982a), Section 3.1.1.3) that an equivalent formulation of the Shapiro-Lopatinskij condition is

$$\mathbf{det}\,\|t_{jk}(x',\xi')\| \neq 0 \quad \text{for } \xi' \neq 0\,,$$

which in its turn is equivalent to linear independence of the polynomials $t_j^0(x',0,\xi',\nu)$ modulo $a^+(x',\xi',\nu)$ for $\xi' \neq 0$. The Shapiro-Lopatinskij condition is also called the **complementing condition** or the **covering condition**.

The passage from differential operators to pseudodifferential ones is natural for two reasons. The first is connected with the construction of a functional calculus. The inverse operator of an elliptic differential operator on a closed manifold is already pseudodifferential. The analogous situation occurs for manifolds with boundary: the parametrix of a classical elliptic differential boundary value problem is an operator in the Boutet de Monvel algebra which is neither a differential operator nor a differential boundary value problem. The framework of the theory of pseudodifferential operators gives greater freedom of action in various transformations of boundary value problems, for example in calculation of compositions of boundary value problems with operators solving auxiliary boundary value problems. Also the operators elliptic in the Douglis-Nirenberg sense are reduced to operators elliptic in the Petrovskij sense with the help of pseudodifferential order reduction operators. The second reason is connected with the Index theory, where in order to apply the K-theory we have to widen the class of possible deformations relative to which the index is stable.

The Boutet de Monvel algebra plays an important role in the spectral theory of differential operators (see, for example, Ivrii V.Ya. (1980), Ivrii V.Ya. (1984), where, in particular, the solution of the H. Weyl problem about the second term in the asymptotics for the fundamental frequencies of the oscillating membrane is contained).

Singular integral operators of the form

$$(S_{G,k}\varphi)(z) = \frac{1}{\pi}\iint_G \left(\frac{\overline{\zeta - z}}{\zeta - z}\right)^k \frac{1}{|\zeta - z|^2}\varphi(\zeta)\ \mathrm{d}G_\zeta\,, \tag{2.54}$$

where $z \in G \subset \mathbb{C}$, $k \in \mathbb{Z}\setminus\{0\}$, and G is a bounded domain, appear in a number of problems in the theory of generalized analytical functions, the theory of quasi-conformal mappings and Riemann surfaces, the theory of partial differential equations and others. These operators are pseudodifferential operators with the symbols

$$\text{const} \cdot \left(\frac{\xi_1 - i\xi_2}{\xi_1 + i\xi_2}\right)^k$$

(see, for example, Mikhlin S.G. (1962)) and thus possess the transmission property. Singular integral operators in $L_p(G)$, $1 < p < \infty$, containing operators of the form (2.54) and the corresponding Bergman operators were investigated by other methods in the works (Vekua I.N. (1988), Dzhangibekov G. (1989), Dzhuraev A.D. (1987), Komyak I.I. (1980) and others) and in Besov spaces in (Bliev N.K. (1985)). In the case of infinitely smooth coefficients and boundary ∂G, the Boutet de Monvel theory allows the results of the above works to be carried over to a wide class of Besov-Triebel-Lizorkin spaces and, in particular, to Hölder spaces (see in this case Duduchava R.V., Saginashvili A.I., Shargorodsky E.M. (1995)).

§3. Parameter-Dependent Boundary Value Problems

3.1. Parameter-Dependent Pseudodifferential Operators. (Agranovich M.S. (1990), Grubb, G. (1986), Grubb, G. (1989)). We have already met parameter – dependent pseudodifferential operators in the construction of order reduction operators. Later we will discuss other applications of pseudodifferential operators with a parameter and their boundary value problems. We only point out now that they are essential for the spectral theory of operators and the theory of parabolic boundary value problems.

It can be seen easily that the symbol of the differential operator $-\Delta_x + \mu^2 I$ is elliptic not only relative to the variable ξ dual to x for an arbitrary fixed $\mu \in \mathbb{R}$, but also relative to the variables $(\xi, \mu) \in \mathbb{R}^{n+1}$. For pseudodifferential operators the situation is more complicated. Take, for example, the elliptic pseudodifferential operator of the second order

$$A = \frac{D_{x_1}^4 + D_{x_2}^4}{D_{x_1}^2 + D_{x_2}^2 + 1} = \mathbf{OP}\left(\frac{\xi_1^4 + \xi_2^4}{\xi_1^2 + \xi_2^2 + 1}\right)$$

and add $\mu^2 I$ to it. We obtain the pseudodifferential operator with the symbol

$$q(\xi, \mu) = \frac{\xi_1^4 + \xi_2^4}{\xi_1^2 + \xi_2^2 + 1} + \mu^2,$$

which does not belong to $\mathbf{S}^2_{1,0}(\mathbb{R}^3 \times \mathbb{R}^3)$ since $D^3_{\xi_1} q(\xi, \mu)$ is $\mathbf{O}\Big(\langle(\xi_1, \xi_2)\rangle^{-1}\Big)$, but not $\mathbf{O}\Big(\langle(\xi_1, \xi_2, \mu)\rangle^{-1}\Big)$. Only first and second derivatives admit "good" estimates: they are $\mathbf{O}\Big(\langle(\xi_1, \xi_2, \mu)\rangle\Big)$ and $\mathbf{O}(1)$ respectively. Clearly, it is natural to define general classes which allow us to count the number of "good" derivatives with respect to the variables (ξ, μ). For brevity we use the notation

$$|\xi, \mu| = |(\xi, \mu)|, \quad \langle \xi, \mu \rangle = \langle (\xi, \mu) \rangle = \sqrt{|\xi|^2 + \mu^2 + 1}, \quad \rho(\xi, \mu) = \frac{\langle \xi \rangle}{\langle \xi, \mu \rangle}.$$

Definition 3.1. Let $\nu \in \mathbb{R}$, and $n_1, n_2 \in \mathbb{N}$.

1). The space of symbols $\mathbf{S}^{d,\nu}_{1,0}(\mathbb{R}^{n_1} \times \overline{\mathbb{R}^{n_2+1}_+})$ of order $d \in \mathbb{R}$ and regularity ν consists of functions $a(x,\xi,\mu) \in C^\infty(\mathbb{R}^{n_1} \times \mathbb{R}^{n_2} \times \bar{\mathbb{R}}_+)$ possessing the estimates

$$\left|D^\beta_x D^\alpha_\xi D^j_\mu a(x,\xi,\mu)\right| \le C_{\alpha,\beta,j}\Big(\langle\xi\rangle^{\nu-|\alpha|}\langle\xi,\mu\rangle^{d-\nu-j} + \langle\xi,\mu\rangle^{d-|\alpha|-j}\Big)$$

$$= C_{\alpha,\beta,j}\Big(\rho(\xi,\mu)^{\nu-|\alpha|}+1\Big)\langle\xi,\mu\rangle^{d-|\alpha|-j},$$

$$\forall (x,\xi,\mu) \in \mathbb{R}^{n_1} \times \mathbb{R}^{n_2} \times \bar{\mathbb{R}}_+,\ \forall \alpha \in \mathbb{Z}^{n_2}_+,\ \forall \beta \in \mathbb{Z}^{n_1}_+,\ \forall j \in \mathbb{Z}_+,$$

where the constants $C_{\alpha,\beta,j} = C_{\alpha,\beta,j}(a)$ are independent of (x,ξ,μ).

2). The space of polyhomogeneous symbols $\mathbf{S}^{d,\nu}(\mathbb{R}^{n_1} \times \overline{\mathbb{R}^{n_2+1}_+})$ of order $d \in \mathbb{C}$ and regularity ν consists of functions $a \in \mathbf{S}^{\mathbf{Re}\, d,\,\nu}(\mathbb{R}^{n_1} \times \overline{\mathbb{R}^{n_2+1}_+})$, with the asymptotic decomposition $a \sim \sum_{l\in\mathbb{Z}_+} a_{d-l}$, in the sense that

$$a - \sum_{l<N} a_{d-l} \in \mathbf{S}^{\mathbf{Re}\, d-N,\,\nu-N}_{1,0}(\mathbb{R}^{n_1} \times \overline{\mathbb{R}^{n_2+1}_+})$$

for all $N \in \mathbb{N}$, where $a_{d-l}(x,\xi,\mu)$ is homogeneous in (ξ,μ) of order $d-l$ for $|\xi| \ge 1$.

We now define

$$\bigcap_{N\in\mathbb{R}} \mathbf{S}^{d,N}_{1,0} \equiv \mathbf{S}^{d,\infty}_{1,0}, \qquad \bigcap_{N\in\mathbb{R}} \mathbf{S}^{d,N} \equiv \mathbf{S}^{d,\infty},$$

$$\bigcap_{N\in\mathbb{R}} \mathbf{S}^{d-N,\,\nu-N}_{1,0} = \bigcap_{N\in\mathbb{R}} \mathbf{S}^{-N,\,\nu'-N}_{1,0} \equiv \mathbf{S}^{-\infty,\,\nu'-\infty}_{1,0}$$

$$\equiv \mathbf{S}^{-\infty,\,\nu'-\infty}, \qquad \nu' = \nu - d.$$

It is clear that for a fixed μ a symbol from $\mathbf{S}^{d,\nu}_{1,0}(\mathbb{R}^{n_1} \times \overline{\mathbb{R}^{n_2+1}_+})$ belongs to $\mathbf{S}^d_{1,0}(\mathbb{R}^{n_1} \times \mathbb{R}^{n_2})$. Hence we can carry over all the main constructions and formulae of the standard theory of pseudodifferential operators to μ-dependent pseudodifferential operators with symbols (amplitudes) in $\mathbf{S}^{d,\nu}_{1,0}(\mathbb{R}^n \times \overline{\mathbb{R}^{n+1}_+})$ $\left(\mathbf{S}^{d,\nu}_{1,0}(\mathbb{R}^{2n} \times \overline{\mathbb{R}^{n+1}_+})\right)$, provided that we have the regularity control of the corresponding symbols (amplitudes). For example, the composition of two pseudodifferential operators

$$\mathbf{OP}\big(a_j(x,\xi,\mu)\big)u(x) = (2\pi)^{-n}\int_{\mathbb{R}^n} e^{ix\cdot\xi}(a_j(x,\xi,\mu)(Fu)(\xi)\, d\xi$$

with symbols in $\mathbf{S}^{d_j,\nu_j}_{1,0}(\mathbb{R}^n \times \overline{\mathbb{R}^{n+1}_+})$, $j = 1,2$, is a pseudodifferential operator from $\mathbf{OP}(\mathbf{S}^{d,\nu}_{1,0})$, where $d = d_1 + d_2$ and $\nu = m(\nu_1,\nu_2) = \min\{\nu_1,\nu_2,\nu_1+\nu_2\}$. The appearance of the quantity $m(\nu_1,\nu_2)$ is caused by the inequality

$$(\rho^{\nu_1}+1)(\rho^{\nu_2}+1) = \rho^{\nu_1}+\rho^{\nu_2}+\rho^{\nu_1+\nu_2}+1 \leq 3\rho^{m(\nu_1,\nu_2)}+1$$

for $0 < \rho \leq 1$, which arises in the estimation of the terms in the asymptotic decomposition of the symbol of the composite pseudodifferential operator. During the passage to the adjoint operator, or from a pseudodifferential operator defined by an amplitude to the pseudodifferential operator defined by its symbol, or again in the case of a change of variables, the regularity does not alter.

Let $a(x,\xi,\mu)$ be a polyhomogeneous symbol of order d with leading part $a^0(x,\xi,\mu) \equiv a_d(x,\xi,\mu)$. Denote by $a^h(x,\xi,\mu)$ the function homogeneous in (ξ,μ) for $|\xi| > 0$ and coinciding with $a^0(x,\xi,\mu)$ for $|\xi| \geq 1$. If possible, the symbol a^h is extended by continuity to $\xi = 0$ (in the case $n = 1$, $\mathbb{R} \setminus \{0\}$ falls into two parts $\mathbb{R}_\pm$ and the extensions a^h to $\bar{\mathbb{R}}_\pm$ must be considered independently).

Definition 3.2. The symbol

$$a(x,\xi,\mu) \in \mathbf{S}^{d,\nu}(\mathbb{R}^n \times \overline{\mathbb{R}^{n+1}_+}) \otimes \mathbb{C}^{N\times N}, \quad N \in \mathbb{N},$$

is called *parameter-elliptic (or elliptic with parameter)* if a_0 can be chosen such that, for some constants $C, C_0 > 0$, the symbol $a^0(x,\xi,\mu)$ is invertible for $|\xi,\mu| \geq C_0$ and

$$|\, a^0(x,\xi,\mu)^{-1} \,| \leq C\,\langle\xi,\mu\rangle^{-\mathbf{Re}\, d} \quad \text{for } |\xi,\mu| \geq C_0\,.$$

For $\nu > 0$ the parameter-ellipticity is equivalent to the purely algebraic condition that the matrix $a^h(x,\xi,\mu)$ is invertible for all $(\xi,\mu) \neq 0$.

Theorem 3.3. *Let $\nu \geq 0$, and let $a \in \mathbf{S}^{d,\nu}(\mathbb{R}^n \times \overline{\mathbb{R}^{n+1}_+}) \otimes \mathbb{C}^{N\times N}$ be an elliptic symbol. Then there exists a* **parametrix**

$$\mathbf{OP}\big(b(x,\xi,\mu)\big) \in \mathbf{OP}\Big(\mathbf{S}^{-d,\nu}(\mathbb{R}^n \times \overline{\mathbb{R}^{n+1}_+}) \otimes \mathbb{C}^{N\times N}\Big)$$

with the leading symbol $b^0(x,\xi,\mu) = a^0(x,\xi,\mu)^{-1}$ (for sufficiently large $|\xi,\mu|$), and with the following main property:

$$\mathbf{OP}(a)\,\mathbf{OP}(b) - I\,, \quad \mathbf{OP}(b)\,\mathbf{OP}(a) - I \in \mathbf{OP}\big(\mathbf{S}^{-\infty,\,\nu-\infty} \otimes \mathbb{C}^{N\times N}\big)\,.$$

As usual the parametrix is uniquely defined up to smoothing operators from $\mathbf{OP}\big(\mathbf{S}^{-\infty,\,\nu+d-\infty} \otimes \mathbb{C}^{N\times N}\big)$.

We now define the spaces

$$H^{s,(\mu)}(\mathbb{R}^n) = \big\{\, u \in S'(\mathbb{R}^n) : \langle\xi,\mu\rangle^s\, \hat{u}(\xi) \in L_2(\mathbb{R}^n)\big\}\,,$$

equipped with the norms

$$\|u\|_{H^{s,(\mu)}(\mathbb{R}^n)} \equiv \|u\|_{s,(\mu)} = (2\pi)^{-n} \big\|\, \langle\xi,\mu\rangle^s\, \hat{u}(\xi)\,\big\|_{L_2(\mathbb{R}^n)}\,, \quad s \in \mathbb{R}\,.$$

These spaces coincide with the ordinary spaces $H_2^s(\mathbb{R}^n)$ as far as their members are concerned, but their norms are μ-dependent. As above, the corresponding spaces $H^{s,(\mu)}(\bar{\Omega})$ and $H^{s,(\mu)}(E)$ can also be defined. It is easily seen that, for $s \geq 0$, we have the norm equivalence

$$\|u\|_{s,(\mu)}^2 \simeq \|u\|_{H_2^s(\mathbb{R}^n)}^2 + \langle\mu\rangle^{2s} \|u\|_{L_2(\mathbb{R}^n)}, \tag{3.1}$$

which is uniform with respect to $\mu \geq 0$. We shall discuss in this paragraph only operators in the spaces $H^{s,(\mu)}$, that is, we restrict ourselves to the framework of L_2-theory.

Theorem 3.4. *Let* $a(x,\xi,\mu) \in \mathbf{S}_{1,0}^{d,\nu}\big(\mathbb{R}^n \times \overline{\mathbb{R}_+^{n+1}}\big)$ *and* $A_\mu = \mathbf{OP}\big(a(x,\xi,\mu)\big)$. *Then*

$$\left\|A_\mu u\right\|_{s-d,(\mu)} \leq C_s\left(\langle\mu\rangle^{-\nu} + 1\right)\|u\|_{s,(\mu)}, \quad \forall\, s \in \mathbb{R}, \ \forall\, \mu \in \bar{\mathbb{R}}_+ .$$

The proof is based on the reduction of A_μ, with the help of operators of the form $\mathbf{OP}\big(\langle\xi,\mu\rangle^m\big) \in \mathbf{OP}\big(\mathbf{S}^{m,\infty}\big)$, to a pseudodifferential operator from $\mathbf{OP}\big(\mathbf{S}_{1,0}^{0,\nu}\big)$ acting in $L_2(\mathbb{R}^n)$ $\big($note that $m(\nu,+\infty) = \nu\big)$ and the application of standard norm estimates of pseudodifferential operators in $L_2(\mathbb{R}^n)$ through the finite number of symbol semi-norms.

3.2. Operators in the Half-Space (Grubb G. (1986), Grubb G. (1989)). We say that the symbol $a(x,\xi,\mu) \in \mathbf{S}_{1,0}^{d,\nu}\big(\mathbb{R}^n \times \overline{\mathbb{R}_+^{n+1}}\big)$ possesses the H-property (or satisfies the H-condition) for $x_n = 0$ if, for all multiindices $\alpha, \beta \in \mathbb{Z}_+^n$ and for all $l, j \in \mathbb{Z}_+$, the inequalities

$$\begin{aligned}&\Big|\xi_n^l D_x^\beta D_\xi^\alpha D_\mu^j a(x',0,\xi,\mu) - \sum_{0\leq k+l\leq d+l-|\alpha|-j} s_{k,\alpha,\beta,j}(x',\xi',\mu)\xi_n^{k+l}\Big| \\ &\quad\leq C_{\alpha,\beta,j,l}(x')\left(\rho(\xi',\mu)^{\nu-|\alpha|+l} + 1\right)\langle\xi',\mu\rangle^{d+1+l-|\alpha|-j}\langle\xi,\mu\rangle^{-1}\end{aligned} \tag{3.2}$$

are valid, where $C_{\alpha,\beta,j,l}$ are continuous functions and $s_{k,\alpha,\beta,j}$ are uniquely defined polynomials in (ξ',μ) of order $\leq d-k-|\alpha|-j$ with C^∞ coefficients (cf. Remark 2.9). We also say that the symbol $a(x,\xi,\mu)$ possesses the UH-property for $x_n = 0$ if the estimates (3.2) are uniform in x', i.e. the $C_{\alpha,\beta,j,l}$ are independent of x'.

For symbols in $\mathbf{S}^{d,\nu}\big(\mathbb{R}^n \times \overline{\mathbb{R}_+^{n+1}}\big)$, $d \in \mathbb{Z}$, the H- and UH-properties are equivalent to the property that (cf. Theorem 2.10)

$$D_x^\beta D_\xi^\alpha D_\mu^j a_{d-l}(x',0,0,-\xi_n,0) = (-1)^{d-l-|\alpha|-j} D_x^\beta D_\xi^\alpha D_\mu^j a_{d-l}(x',0,0,\xi_n,0)$$
$$\text{for} \quad |\xi_n| \geq 1, \quad \forall \alpha,\beta \in \mathbb{Z}_+^n, \ \forall j,l \in \mathbb{Z}_+ .$$

The space $\mathbf{S}_{1,0}^{d-1,\nu}\big(\mathbb{R}^{n-1} \times \overline{\mathbb{R}_+^n}, S(\bar{\mathbb{R}}_+)\big)$ of *symbol-kernels* of potential operators (or coboundary operators, or Poisson operators) of order $d \in \mathbb{R}$ and regularity $\nu \in \mathbb{R}$ consists of functions $\tilde{f}$ satisfying the estimates (cf. (2.18))

$$\left\| D_{x'}^{\beta'} x_n^m D_{x_n}^l D_{\xi'}^{\alpha'} D_\mu^j \tilde{f}(x,\xi',\mu) \right\|_{L_{2,x_n}(\mathbb{R}_+)}$$
$$\leq C_{\alpha',\beta',m,l,j} \left(\rho(\xi',\mu)^{\nu-|\alpha'|-(m-l)_+} + 1\right) \langle \xi',\mu \rangle^{d-1/2-m+l-|\alpha'|-j},$$
$$\forall \alpha',\beta' \in \mathbb{Z}_+^{n-1},\ \forall j,m,l \in \mathbb{Z}_+, \text{ where } N_\pm = \max\{\pm N, 0\}. \tag{3.3}$$

This space is also the symbol-kernel space of trace operators of order $d-1$, class 0 and regularity ν. In general a trace operator of class $r \geq 0$ and regularity ν is the sum of an operator of class 0 with regularity ν and operators of the form $S_{j,\mu}\gamma_j$, where $S_{j,\mu}$ is a pseudodifferential operator on $\mathbb{R}^{n-1}$ of regularity ν and $j \leq r-1$.

For singular Green operators of order d, class 0 and regularity ν, the symbol-kernel space $\mathbf{S}_{1,0}^{d-1,\nu}(\mathbb{R}^{n-1} \times \overline{\mathbb{R}_+^n}, S(\overline{\mathbb{R}_{++}^2}))$, $d,\nu \in \mathbb{R}$, is the set of functions $\tilde{g}$ satisfying the estimates (cf. (2.27))

$$\left\| D_{x'}^{\beta'} x_n^k D_{x_n}^{k'} y_n^m D_{y_n}^{m'} D_{\xi'}^{\alpha'} D_\mu^j \tilde{g}(x,y_n,\xi',\mu) \right\|_{L_{2,x_n,y_n}(\mathbb{R}_{++}^2)}$$
$$\leq C_{\alpha',\beta',m,m',k,k',j} \left(\rho(\xi',\mu)^{\nu-|\alpha'|-(k-k')_+-(m-m')_+} + 1\right) \times$$
$$\times \langle \xi',\mu \rangle^{d-k+k'-m+m'-|\alpha'|-j},\quad \forall \alpha',\beta' \in \mathbb{Z}_+^{n-1},\ \forall j,k,k',m,m' \in \mathbb{Z}_+.$$

A singular Green operator of class $r \geq 0$ and regularity ν is the sum of a singular Green operator of class 0 with regularity ν and operators of the form $K_{j,\mu}\gamma_j$, where the $K_{j,\mu}$ are potential operators of regularity ν and $j \leq r-1$.

The corresponding (parameter-dependent) symbol spaces are denoted as follows:

$$\mathbf{S}_{1,0}^{d-1,\nu}(\mathbb{R}^{n-1} \times \overline{\mathbb{R}_+^n}, H^+)$$

for the potential operators,

$$\mathbf{S}_{1,0}^{d-1,\nu}(\mathbb{R}^{n-1} \times \overline{\mathbb{R}_+^n}, H_{r-1}^-)$$

for the trace operators, and

$$\mathbf{S}_{1,0}^{d-1,\nu}(\mathbb{R}^{n-1} \times \overline{\mathbb{R}_+^n}, H^+ \widehat{\otimes} H_{r-1}^-)$$

for singular Green operators.

As usual, the spaces of polyhomogeneous symbols $\mathbf{S}^{d-1,\nu}(\cdots)$ etc. consist of symbols having asymptotic expansions into series of positive homogeneous functions relative to (ξ,μ) (resp. (ξ,η_n,μ)) for $|\xi'| \geq 1$.

The (parameter-dependent) potential operators K_μ, trace operators T_μ and singular Green operators G_μ are defined in the same way as in Section 2.4. The results of Section 2.6 remain true for them and also for operators $A_{\mu,\mathbb{R}_+^n}$. Moreover, the regularity of almost all compositions of operators with regularity ν_j $(j=1,2)$, considered there, is $m(\nu_1,\nu_2)$. The single exception is the operator $(A_{1,\mu}A_{2,\mu})_{\mathbb{R}_+^n} - A_{1,\mu,\mathbb{R}_+^n} A_{2,\mu,\mathbb{R}_+^n}$, with regularity $m(\nu_1 - 1/2, \nu_2 - 1/2)$. It is pointed out in the preprint (Grubb G. (1989)), that the author is preparing the work about boundary problems with a parameter in L_p-spaces where, in particular, the possibility of changing $m(\nu_1-1/2,\nu_2-1/2)$ into $m(\nu_1-\varepsilon,\nu_2-\varepsilon)$, $\forall\varepsilon>0$ is proved.

Finally, we note that it is possible to consider potential, trace and singular Green operators defined by (x',y')-dependent amplitudes.

3.3. Boundary Value Problems on Manifolds (Grubb G. (1986), Grubb G. (1989)) As in Section 2.7, we introduce (parameter-dependent) Green operators of order $d\in\mathbb{R}$, class $r\in\mathbb{Z}_+$ and regularity $\nu\in\mathbb{R}$:

$$P_\mu=\begin{pmatrix} A_{\mu,\Omega}+G_\mu & K_\mu \\ T_\mu & L_\mu\end{pmatrix} : \begin{matrix} C^\infty(E) \\ \bigoplus \\ C^\infty(F)\end{matrix} \longrightarrow \begin{matrix} C^\infty(E') \\ \bigoplus \\ C^\infty(F')\end{matrix} . \tag{3.4}$$

The operator P_μ acts continuously from

$$H_1(s,\mu)=H^{s,(\mu)}(E)\oplus H^{s-1/2,(\mu)}(F)$$

into

$$H_2(s,\mu)=H^{s-d,(\mu)}(E')\oplus H^{s-d-1/2,(\mu)}(F')$$

for $s>r-1/2$, and its norm is estimated by the value $C_s\left(\langle\mu\rangle^{-\nu}+1\right)$, $\forall\mu\in\bar{\mathbb{R}}_+$. As usual, the limitation $s>r-1/2$ arises because of the operators G_μ and T_μ. The continuity and the norm estimate for the operator $A_{\mu,\Omega}$ holds for $s>-1/2$, and for all $s\in\mathbb{R}$ in the case of K_μ and L_μ.

In the following we shall not consider operators of negative classes ($r<0$), since that would involve technical difficulties, and it will also be convenient to distinguish between the regularity ν_1 of the operator A_μ and the regularity ν_2 of the operators G_μ, K_μ, T_μ, L_μ. In the rest of this paragraph we shall deal only with polyhomogeneous Green operators.

We denote by $a^0(x,\xi,\mu)$ the leading interior symbol of the operator P_μ (i.e. the leading symbol of the pseudodifferential operator A_μ) and by

$$\sigma_P(x',\xi',\mu,D_n)$$
$$=\begin{pmatrix} a^0(x',0,\xi',\mu,D_n)_{\mathbb{R}_+}+g^0(x',\xi',\mu,D_n) & k^0(x',\xi',\mu,D_n) \\ t^0(x',\xi',\mu,D_n) & l^0(x',\xi',\mu)\end{pmatrix} \tag{3.5}$$

its leading boundary symbol (here all the symbols are written in local coordinates, and for brevity we deviate from the more explicit notation of (2.41)). As in the case of a^0 in Section 3.1, we now construct for g^0, k^0, t^0, l^0 the functions g^h, k^h, t^h, l^h which coincide with them for $|\xi'|\geq 1$ and are homogeneous relative to (ξ,μ) (resp. (ξ,η_n,μ)) for $\xi'\neq 0$. We denote by $\sigma_P^h(x',\xi',\mu,D_n)$ the operator which is given by the formula (3.5) when a^0, g^0, k^0, t^0, l^0 are replaced by a^h, g^h, k^h, t^h, l^h.

Definition 3.5. Let $d\in\mathbb{R}$, $r\in[0,d_+]$ (see (3.3)), $\nu_1\geq 1$, $\nu_2>0$. The operator P_μ of the type (3.4) is called *parameter-elliptic* if the following three conditions are fulfilled:

(I) $a^h(x,\xi,\mu) : \mathbb{C}^N \to \mathbb{C}^{N'}$ is a bijection for $(\xi,\mu) \neq 0$;

$$\text{(II)}\ \sigma_P(x',\xi',\mu,D_n)\ :\ \begin{matrix} H_2^{d_+}(\bar{\mathbb{R}}_+)\otimes\mathbb{C}^N \\ \bigoplus \\ \mathbb{C}^J \end{matrix} \longrightarrow \begin{matrix} H_2^{d_-}(\bar{\mathbb{R}}_+)\otimes\mathbb{C}^{N'} \\ \bigoplus \\ \mathbb{C}^{J'} \end{matrix}$$

is a bijection for $|\xi'| \geq 1$ and $\mu \geq 0$;

(III) for each $\mu \geq 0$ the limit of $\sigma_P^h(x',\xi',\mu,D_n)$ exists as $\xi' \to 0$ and is a bijection from $H_2^{d_+}(\bar{\mathbb{R}}_+) \otimes \mathbb{C}^N \oplus \mathbb{C}^J$ to $H_2^{d_-}(\bar{\mathbb{R}}_+) \otimes \mathbb{C}^{N'} \oplus \mathbb{C}^{J'}$ (in the case $n = 2$ we suppose for simplicity, that

$$\sigma_P^h(x',0+,\mu,D_n) = \sigma_P^h(x',0-,\mu,D_n)) .$$

Note that, as in Section 2.7, the fulfillment of condition (I) and the poly-homogenity of A_μ involve $N = N'$ and $d \in \mathbb{Z}$.

Theorem 3.6. *Let $d \in \mathbb{Z}$, $r \in [0,d_+]$, $\nu_1 \geq 1$, $\nu_2 > 0$, and let P_μ be a parameter-elliptic Green operator of order d, class r and regularity ν_2 (ν_1 for the pseudodifferential part). Then the parametrix Q_μ of P_μ exists, and is also a parameter-elliptic Green operator of order $-d$, class d_-, regularity $\nu_3 = \min\{\nu_1 - 1/2, \nu_2\} \geq 0$ (ν_1 for the pseudodifferential part), such that $P_\mu Q_\mu - I$ and $Q_\mu P_\mu - I$ are Green operators of order $-\infty$, classes d_- and d_+ respectively and regularity ν_3 (ν_1 for the pseudodifferential part). Moreover, the operator Q_μ is the inverse to P_μ for $\mu \geq \mu_0$ with sufficiently large μ_0.*

The results expounded in the present section allow the resolvent of the Green operator (2.39) to be investigated, and hence the construction of the functional calculus. We consider below an important particular case.

3.4. Realizations and their Resolvents (Grubb G. (1986), (1989)).

Consider a (parameter-independent) Green operator of the special kind:

$$P = \begin{pmatrix} A_\Omega + G \\ T \end{pmatrix}\ :\ C^\infty(E) \longrightarrow \begin{matrix} C^\infty(E) \\ \bigoplus \\ C^\infty(F) \end{matrix}\ , \tag{3.6}$$

where A and G are operators of order $d \in \mathbb{Z}_+$, G and T are of class $r \in [0,d]$, and

$$T = \bigoplus_{0\leq j\leq d-1} T_j\,, \quad T_j : C^\infty(E) \to C^\infty(F_j)$$

is a trace operator of order j with

$$\dim F_j \geq 0\,, \qquad \bigoplus_{0\leq j\leq d-1} F_j = F$$

(if $d = 0$ the operator T is absent).

Definition 3.7. An operator $B = (A+G)_T$, acting in the space $L_2(E)$ with the domain

$$D(B) = \{u \in H_2^d(E) : Tu = 0\},$$

and which is the restriction of $A_\Omega + G$ to $D(B)$, is called the *H_2^d-realization* (or simply *realization*) of the operator A, defined by the system $\{A_\Omega + G, T\}$. A realization is called *elliptic* if the operator (3.6) is elliptic.

Elliptic realizations possess properties which are typical for elliptic operators. Let $B = (A+G)_T$ be an elliptic realization of order d. Then its parametrix $\mathcal{R} = (A^{-1})_\Omega + G_0$ exists, where A^{-1} is a parametrix for A on E_0, and G_0 is a singular Green operator of order $-d$ and class 0. Then $\mathcal{R}$ maps $L_2(E)$ into $D(B)$ and $B\mathcal{R} = I + S$, where S is an integral operator with a C^∞ smooth kernel and finite-dimensional image. The operator $\mathcal{R}B - I$ also has a finite-dimensional image in $D(B)$ lying in $C^\infty(E)$.

The composition of two realizations

$$B_1 = (A_1 + G_1)_{T^{(1)}} \quad \text{and} \quad B_2 = (A_2 + G_2)_{T^{(2)}},$$

of orders $d_1 \in \mathbb{Z}_+$ and $d_2 \in \mathbb{Z}_+$ respectively, is the operator

$$C = \left(A_{1,\Omega} + G_1\right)\left(A_{2,\Omega} + G_2\right)$$

with the domain

$$D(C) = \Big\{u \in H_2^{d_2}(E) : T^{(2)}u = 0, \quad \left(A_{2,\Omega} + G_2\right)u \in H_2^{d_1}(E),\ T^{(1)}\left(A_{2,\Omega} + G_2\right)u = 0\Big\}$$

and C is the extension of the realization $B_3 = (A_3 + G_3)_{T^{(3)}}$, where (see Section 2.6)

$$\begin{aligned} A_3 &= A_1 A_2, \\ G_3 &= A_{1,\Omega}A_{2,\Omega} - (A_1A_2)_\Omega + A_{1,\Omega}G_2 + G_1A_{2,\Omega} + G_1G_2, \\ T_3 &= \begin{pmatrix} T^{(2)} \\ T^{(1)}\left(A_{2,\Omega} + G_2\right) \end{pmatrix}. \end{aligned}$$

If B_2 is an elliptic realization, then $B_2u \in H_2^{d_1}(E)$ implies $u \in H_2^{d_1+d_2}(E)$ and so $D(C) = D(B_3)$, i.e. $C = B_3$. If in addition B_1 is elliptic, then $B_3 = (A_3 + G_3)_{T^{(3)}}$ is also elliptic.

Consider now the operator

$$P_\lambda = \begin{pmatrix} A_\Omega + G - \lambda I \\ T \end{pmatrix}, \qquad \lambda = e^{i\theta}\mu^d,\ \theta \in [0, 2\pi[,\ \mu \in \bar{\mathbb{R}}_+, \tag{3.7}$$

of regularity $\nu_2 \in [1/2, d]$ relative to the parameter μ (for the pseudodifferential part $\nu_1 = d$). We are not able to apply the definition of parameter-ellipticity that we have already given (see Definition 3.5) since the components T_j of the operator T have different orders. So we have first to carry out the reduction of orders, leading to the operator

$$P'_\lambda = \begin{pmatrix} A_\Omega + G - \lambda I \\ \bigoplus_{0\le j\le d-1} \Lambda^{(d-j)}_{\mu,\Gamma,F_j}\, T_j \end{pmatrix}, \qquad \lambda = e^{i\theta}\,\mu^d\,, \tag{3.7'}$$

where $\Lambda^{(\rho)}_{\mu,\Gamma,F}$ is an elliptic pseudodifferential operator on Γ with the leading symbol $\langle \xi', \mu\rangle^\rho$ effecting the isomorphisms $H^{s,(\mu)}(F) \to H^{s-\rho,(\mu)}(F)$, $\forall\, s \in \mathbb{R}$ (uniformly in $\mu \ge 0$). We say that the operator (3.7) is *elliptic with parameter (or parameter-elliptic)* μ for some $\theta = \theta_0$ if the operator (3.7′) is covered by Definition 3.5.

Suppose now that the operator (3.7) is elliptic with the parameter μ for $\theta = \theta_0$. Then it will remain elliptic with the parameter for $\|\theta - \theta_0\| \le \varepsilon$ with sufficiently small $\varepsilon > 0$. Moreover, it can be proved, that the fulfillment of Condition III in Definition 3.5 implies the normality of T.

Definition 3.8. The system of trace operators $T = \{T_0, \dots, T_{d-1}\}$ is called *normal* if the operators $T_j : C^\infty(E) \to C^\infty(F_j)$ are of the form

$$T_j = \sum_{0\le k\le j} S_{jk}\,\gamma_k + T'_j\,,$$

where T'_j is a trace operator of order j and class 0, $S_{jk} : C^\infty(E\,|_\Gamma) \to C^\infty(F_j)$ is a pseudodifferential operator of order $j-k$ and, moreover, the leading coefficient S_{jj} is the operator induced by a surjective bundle morphism $E\,|_\Gamma \to F_j$ (locally it is the operator of multiplication by a matrix-valued function $s_{jj}(x')$, $x' \in \Gamma$).

The boundary problem $(A_\Omega + G)u = f$, $Tu = \phi$ and the realization $B = (A+G)_T$ are also called *normal* if T is a normal system of trace operators and the manifold Γ is non-characteristic for A.

Theorem 3.9. *Let* (3.7) *be a parameter-elliptic operator for* $\theta = \theta_0 \in [0, 2\pi[$. *Then there exist numbers* $\delta, \varepsilon > 0$ *such that, for*

$$\lambda \in W_{\delta,\varepsilon} = \{\zeta \in \mathbb{C} : |\zeta| \ge \delta\,,\ |\arg\zeta - \theta_0| \le \varepsilon\}\,,$$

the resolvent $\mathcal{R}_\lambda = (B - \lambda I)^{-1}$ *of the realization* $B = (A+G)_T$ *is defined and, moreover,* $\mathcal{R}_\lambda = A^{-1}_{\lambda,\Omega} + G_\lambda$, *where* A^{-1}_λ *is a parametrix for* $A - \lambda I$ *and* G_λ *is a singular Green operator of order* $-d$, *class* 0 *and regularity* $\nu_2 \in [1/2, d]$ *relative to the parameter* $\mu = |\lambda|^{1/d}$. *Also,*

$$\begin{aligned} \|\mathcal{R}_\lambda f\|_{H_2^{s+d}} + \langle\lambda\rangle^{1+s/d}\,\|\mathcal{R}_\lambda f\|_{L_2} &\le C_s\,\big(\|f\|_{H_2^s} + \\ &+ \langle\lambda\rangle^{s/d}\,\|f\|_{L_2}\big)\,,\quad \forall\, s \in \bar{\mathbb{R}}_+\,,\ \forall\,\lambda \in W_{\delta,\varepsilon}\,. \end{aligned} \tag{3.8}$$

In the case $d > n$ *the resolvent* $\mathcal{R}_\lambda$ *is a nuclear operator and*

$$\mathbf{tr}\, A^{-1}_{\lambda,\Omega} \sim \sum_{j\in\mathbb{Z}_+} b_j \lambda^{\frac{n-j}{d}-1} + \sum_{l\in\mathbb{N}} c_l \lambda^{-l-1}\ln\lambda\,,$$

$$\mathbf{tr}\, G_\lambda = r_{-d-1}\lambda^{\frac{n-d-1}{d}} + \ldots + r_{-d-\nu-n}\lambda^{-\frac{d+\nu}{d}} + \mathbf{O}\Big(\langle\lambda\rangle^{-1-\frac{\nu_2-1/4}{d}}\Big)\,,$$

where ν *is the largest integer strictly less then* ν_2.[3]

[3] One can prove that ν_2 is an integer or half-integer, and thus $\nu < \nu_2 - 1/4$.

Before we start the discussion of the functional calculus for elliptic realizations, we note that the operators of the type (3.7) play an important role in singular perturbation theory (see, for example, Frank L.S. (1979), Frank L.S., Wendt W.D. (1982), Frank L.S., Wendt W.D. (1984), Grubb G. (1986), Grubb G. (1989)).

3.5. The Exponent and Complex Powers of an Elliptic Realization (Grubb G. (1986), Grubb G. (1989)). Suppose that (3.7) is a parameter-elliptic operator for $\theta \in [\pi/2, 3\pi/2]$. Then, according to Theorem 3.9, the resolvent $\mathcal{R}_\lambda = (B - \lambda I)^{-1}$ of an elliptic realization $B = (A+G)_T$ exists for λ belonging to

$$W_{\delta,\pi/2+\varepsilon} = \{\zeta \in \mathbb{C} : |\zeta| \geq \delta\,,\ |\arg\zeta - \pi| \leq \pi/2 + \varepsilon\}\,,$$

where $\delta, \varepsilon > 0$. It follows from the estimate (3.8) that

$$\|\mathcal{R}_\lambda\|_{L_2 \to L_2} = \mathbf{O}(\langle\lambda\rangle^{-1})\,.$$

Hence the equation

$$\mathbb{U}(t) \equiv \exp(-tB) = \frac{i}{2\pi}\int\limits_\gamma \mathrm{e}^{-\lambda t}\mathcal{R}_\lambda\,\mathrm{d}\lambda\,, \tag{3.9}$$

where $t > 0$, $\gamma = \partial W_{\delta',\pi/2+\varepsilon'}$, $\delta' > \delta$ and $\varepsilon' \in]0,\varepsilon[$, defines a continuous operator which maps $L_2(E)$ into $H_2^s(E)$, $\forall s \in \bar{\mathbb{R}}_+$, and thus is nuclear. Its trace formula will be given a little later. Here we note that, as usual, the semi-group $\exp(-tB)$ is connected with the evolution problem

$$\partial_t u + A_\Omega u + Gu = f(x,t)\,, \quad x \in \Omega\,,\ t > 0\,, \tag{3.10}$$

$$Tu = 0 \qquad \text{on } \Gamma \times \mathbb{R}_+\,, \tag{3.11}$$

$$u\,|_{t=0} = u_0(x)\,, \quad x \in \Omega\,, \tag{3.12}$$

which is called **parabolic** if (3.7) is a parameter-elliptic operator for $\theta \in [\pi/2, 3\pi/2]$. Such a definition of parabolicity coincides with the classical one (see, for example, Agranovich M.S., Vishik M.I. (1964), Eidelman S.D. (1990)) in the case of differential boundary value problems. Problems of the type (3.10)–(3.12) (generally with non-homogeneous boundary conditions (3.11)) are considered in the control theory (see, for example, Lasiecka I., Triggiani R. (1983), Nambu T. (1979), Pedersen M. (1988), Triggiani R. (1979)) and in initial-boundary value problems for the Navier-Stokes system (see Grubb G., Solonnikov V.A. (1987), Grubb G., Solonnikov V.A. (1988)) and others.

After the addition of an appropriate constant ρ to A (which implies the substitution of $\mathbb{U}(t)$ by $\mathrm{e}^{-\rho t}\mathbb{U}(t)$) we may suppose that $\mathcal{R}_\lambda$ exists for

$$\lambda \in V_{\delta_0,\pi/2+\varepsilon_0} = \{\zeta \in \mathbb{C} : |\zeta| \leq \delta_0 \text{ or } |\arg\zeta - \pi| \leq \pi/2 + \varepsilon_0\}\,,$$

where $\delta_0, \varepsilon_0 > 0$, and the curve γ (see (3.9)) lies in the half-plane $\{\zeta \in \mathbb{C} : \mathbf{Re}\,\zeta > 0\}$; for example,

$$\gamma = \partial V_{\delta_1,\pi/2+\varepsilon_1}\,, \quad \text{where} \quad \delta_1 \in [0,\delta_0[\,,\ \varepsilon_1 \in]0,\varepsilon_0[\,.$$

Also we suppose that the operator $A - \lambda I$ is invertible for $\lambda \in V_{\delta_0,\pi/2+\varepsilon_0}$ and $A_\lambda^{-1} = (A - \lambda I)^{-1}$ (see Theorem 3.9). It follows from (3.9) and Theorem 3.9 that

$$\mathbb{U}(t) = \mathbb{V}(t)_\Omega + \mathbb{W}(t)\,,$$

where

$$\mathbb{V}(t) = \frac{i}{2\pi}\int\limits_\gamma \mathrm{e}^{-\lambda t} A_\lambda^{-1}\,\mathrm{d}\lambda\,, \qquad \mathbb{W}(t) = \frac{i}{2\pi}\int\limits_\gamma \mathrm{e}^{-\lambda t} G_\lambda\,\mathrm{d}\lambda\,.$$

Under the above assumptions the asymptotic expansions

$$\mathbf{tr}\,\mathbb{V}(t)_\Omega \sim \sum_{j\in\mathbb{Z}_+} q_j t^{(j-n)/d} + \sum_{l\in\mathbb{N}} r_l t^l \ln t$$

and

$$\mathbf{tr}\,\mathbb{W}(t) = g_{1-n} t^{(1-n)/d} + \ldots + g_\nu t^{\nu/d} + \mathbf{O}\big(t^{(\nu_2-1/4)/d}\big)$$

hold as $t \to 0+$, where ν and ν_2 are the same as in Theorem 3.9.

The formulae for $\mathbf{tr}\exp(-tB)$ are used to compute the index of elliptic boundary value problems (see, for example, Atiyah M.F., Bott R., Patodi V.K. (1973), Gilkey P.B. (1974), Grubb G. (1986)). The investigation of the function $\mathbf{tr}\exp(-tB)$ is the basis for the parabolic equation method in the spectral theory of (pseudo)differential operators (see Rozenblum G.V., Solomyak M.Z., Shubin M.A. (1989), Minakshisundaram S., Pleijel A. (1949), Taylor M.E. (1981)). In the case where B is a realization of the Laplace-Beltrami operator on a Riemannian manifold, the coefficients of the asymptotic expansion of the function $\mathbf{tr}\exp(-tB)$ play an important role in spectral geometry (see Rozenblum G.V., Solomyak M.Z., Shubin M.A. (1989), Bérard P.H. (1986), Gilkey P.B. (1974) and the literature cited there).

Now let (3.7) be a parameter-elliptic operator for $\theta = \pi$, and let a realization $B = (A + G)_T$ be an invertible operator. Consider the operator B^z defined by

$$B^z = \frac{i}{2\pi}\int\limits_\gamma \lambda^z\,\mathcal{R}_\lambda\,\mathrm{d}\lambda\,,$$

where $\mathbf{Re}\,z < 0$ and

$$\gamma = \partial V_{\delta_0,\varepsilon_0}\,, \quad V_{\delta_0,\varepsilon_0} = \big\{\zeta \in \mathbb{C} : |\zeta| \le \delta_0 \text{ or } |\arg\zeta - \pi| \le \varepsilon_0\big\}$$

with sufficiently small $\delta_0, \varepsilon_0 > 0$. Also, the integrand λ^z, which is analytic in $\mathbb{C} \setminus \bar{\mathbb{R}}_-$, is chosen positive for $z \in \mathbb{R}_-$ and $\lambda \in \mathbb{R}_+$. The power B^z, $\mathbf{Re}\,z < 0$, of a realization B is a bounded operator in $L_2(E)$ which is also nuclear for

$\mathbf{Re}\, z < -n/d$. For $\mathbf{Re}\, z \geq 0$ the power B^z is defined by $B^z = B^k\, B^{z-k}$, where $k \in \mathbb{Z}_+$ is sufficiently large and the definition is independent of the choice of k. The function $\mathbf{tr} B^z$, which is defined for $\mathbf{Re}\, z < -n/d$, is called the zeta-function of the elliptic realization B. It admits a meromorphic continuation to the set $\{z \in \mathbb{C} : \mathbf{Re}\, z < (\nu_2 - 1/4)/d\}$ with simple poles contained in the set $\{z = (j-n)/d \,|\, j = 1, 2, \ldots, n+\nu\,, j \neq n\}$, where ν and ν_2 are the same as in Theorem 3.9.

3.6. Spectral Asymptotics (Grubb G. (1986)). For a given operator Y acting in Hilbert (or finite-dimensional) space, we denote by $N(t; Y)$ the number of eigenvalues (taking into account their multiplicity) of the operator $(Y^*Y)^{1/2}$ lying in $]0, t]$. If Y is a self-adjoint operator, we denote by $N^{\pm}(t; Y)$ the number of positive (in the case of $+$) and respectively negative (in the case of $-$) eigenvalues of Y in $[-t, t]$.

Assuming the bundle E to be Hermitian throughout in this section, we introduce the notation

$$\mathcal{C}(A, \Omega) = (2\pi)^{-n} \int\limits_{T^*\Omega} N\big(1; \sigma_A^*(x,\xi)\sigma_A(x,\xi)\big)\, \mathrm{d}\xi\, \mathrm{d}x\,.$$

Suppose that the elliptic realization $B = (A+G)_T$ of order $d \in \mathbb{Z}_+$ is normal. Then

$$N\big(t; B\big) = \mathcal{C}\big(A, \Omega\big)\, t^{n/d} + \mathbf{O}\big(t^{(n-\theta)/d}\big)$$

as $t \to +\infty$, where θ is an arbitrary number less than $1/2$ (or 1 when A is a scalar differential operator).

Let A be a self-adjoint and invertible (i.e. elliptic) pseudodifferential operator of even order d acting in the sections of a bundle E_0 ($E_0|_M = E$), and let its normal realization $B = (A+G)_T$ also be a self-adjoint operator. Then

$$N^{\pm}\big(t; B\big) = \mathcal{C}^{\pm}\big(A, \Omega\big)\, t^{n/d} + \mathbf{O}\big(t^{(n-\theta)/d}\big)$$

as $t \to +\infty$, where θ is as above, and

$$\mathcal{C}^{\pm}\big(A, \Omega\big) = (2\pi)^{-n} \int\limits_{T^*\Omega} N^{\pm}\big(1; \sigma_A(x,\xi)\big)\, \mathrm{d}\xi\, \mathrm{d}x\,.$$

In the case of odd-order operators it turns out that the asymptotic formula for a normal self-adjoint realization $B = (A+G)_T$ is

$$N^{\pm}\big(t; B\big)\, t^{-n/d} \to \mathcal{C}^{\pm}\big(A, \Omega\big) \quad \text{as } t \to +\infty\,.$$

The above results are used in the investigation of the problem

$$\begin{aligned} \lambda\big(A_{1,\Omega} + G_1\big)\, u &= \big(A_{0,\Omega} + G_0\big)\, u \\ Tu &= 0\,, \end{aligned}$$

where $A_{0,\Omega}+G_0$ and $A_{1,\Omega}+G_1$ are operators of orders r and $r+d$ respectively ($r \in \mathbb{Z}_+$, $d \in \mathbb{N}$) (see Levendorskij S.Z. (1984a), Levendorskij S.Z. (1984b), Grubb (1986)).

§4. Boundary Problems for Elliptic Pseudodifferential Operators without the Transmission Property

4.1. The Theory of Vishik and Eskin (Dolgonos E.I. (1968), Dynin A.S. (1969a), (1969b), Eskin G.I. (1973), Rabinovich V.S. (1972), Shargorodsky E.M. (1989a), (1989b), (1994), (1995), Vishik M.I., Eskin G.I. (1964), (1965), (1966), (1967a), (1967b), (1967c)).

First we note that the title of this paragraph should not be taken to mean that pseudodifferential operators with the transmission property will be totally ignored. It means only that these operators step into the background. In particular, the theory of Vishik and Eskin, to be discussed in the present section, covers boundary value problems for pseudodifferential operators either with or without the transmission property.

Here we change slightly the notation adopted in the two previous paragraphs. Let $r \in \mathbb{N}$ and $\mu \in \mathbb{C}$. We denote by as $\mathbf{O}^{\mathbf{r}}_{\mu}$ the algebra of positively-homogeneous functions of order μ whose their restrictions to the unit sphere $\mathbf{S^{n-1}} \subset \mathbb{R}^n$ belong to $C^r(\mathbf{S^{n-1}})$.

A symbol $A(\xi) \in \mathbf{O}^{\mathbf{r}}_{\mu} \otimes \mathbb{C}^{N\times N}$ is said to be elliptic if $\det A(\xi) \neq 0$ for $\xi \neq 0$ (in the scalar case $A(\xi) \neq 0$ for $\xi \neq 0$).

For an arbitrary elliptic symbol $A(\xi) \in \mathbf{O}^{\mathbf{r}}_{\mu} \otimes \mathbb{C}^{N\times N}$ we introduce the following quantities:

$\lambda_1, \dots, \lambda_N$ are the eigenvalues of the matrix $A^{-1}(0,\dots,0,+1) \times A(0,\dots,0,-1)$ (some of them may be equal),

$\delta_k = (2\pi i)^{-1} \ln \lambda_k$ (the branch of the logarithm being chosen arbitrarily),

$$\kappa(\omega) = -\frac{1}{2\pi}\, \Delta \arg \det \left[(1+|t|)^{-\mu} A(\omega,t)\right] \Bigg|_{t=-\infty}^{+\infty} - \\ - \sum_{1\le k\le N} \mathbf{Re}\, \delta_k\,, \quad \omega \in \mathbf{S^{n-2}} \subset \mathbb{R}^{n-1}\,. \tag{4.1}$$

It can be seen easily that $\kappa(\omega) \in \mathbb{Z}$, $\forall \omega \in \mathbf{S^{n-2}}$, and that the function κ is continuous on the unit sphere $\mathbf{S^{n-2}}$. If $n \geq 3$, the sphere $\mathbf{S^{n-2}}$ is connected. Hence for $n \geq 3$ we have $\kappa(\omega) = \kappa = \text{const}$. If $n = 2$, the sphere $\mathbf{S^{n-2}} = \mathbf{S}^0 = \{\pm 1\}$ is disconnected, and it may happen that $\kappa(-1) \neq \kappa(1)$. We consider this case below, but here we demand that

$$\kappa(-1) = \kappa(1) = \kappa = \text{const}\,.$$

We denote by $A = A(D)$ the pseudodifferential operator with symbol $A(\xi)$, and by $A'(\xi)$ the symbol $A\big(\langle\xi'\rangle \frac{\xi'}{|\xi'|}, \xi_n\big)$.

Let $A(\xi) \in \mathbf{O}_\mu^{[n/2]+3} \otimes \mathbb{C}^{N\times N}$ be an elliptic symbol, and let $1 < p < \infty$, $1 \le q \le \infty$, $s \in \mathbb{R}$. Consider the boundary value problem

$$\pi_+ A' u_+ + \sum_{1\le l\le m_-} \pi_+ K_l'\big(w_l(x') \times \delta(x_n)\big) = f(x)\,, \tag{4.2}$$

$$\gamma_0 T_j' u_+ + \sum_{1\le l\le m_-} L_{jl}' w_l(x') = g_j(x')\,,\ \ 1 \le j \le m_+\,, \tag{4.3}$$

where $T_j(\xi)$ and $K_l(\xi)$ are $N-$dimensional vectors and $L_{jl}(\xi')$ are scalar functions satisfying the following conditions:

$$T_j(\xi) = |\xi'|^{\beta_j-\beta_j'}\,|\xi|^{\beta_j'}\,T_{0j}(\xi)\,,\quad \beta_j\,,\ \beta_j' \in \mathbb{C}\,,$$
$$\mathbf{Re}\,\beta_j' < s - \frac{1}{p}\,,\quad 1 \le j \le m_+\,,$$
$$K_l(\xi) = |\xi'|^{\rho_l-\rho_l'}\,|\xi|^{\rho_l'}\,K_{0l}(\xi)\,,\quad \rho_l\,,\ \rho_l' \in \mathbb{C}\,,$$
$$\mathbf{Re}\,\rho_l' < \mathbf{Re}\,\mu - s - 1 + \frac{1}{p}\,,\quad 1 \le l \le m_-\,,$$
$$L_{jl}(\xi') = |\xi'|^{\delta_{jl}}\,L_{0jl}(\xi')\,,$$
$$\delta_{jl} = \beta_j + \rho_l - \mu + 1\,,\quad 1 \le j \le m_+\,,\ \ 1 \le l \le m_-\,,$$

$T_{0j}(\xi)$, $K_{0l}(\xi)$, $L_{0jl}(\xi')$ are (vector-)functions, homogeneous of order 0, such that the components of $T_{0j}'(\xi)$ and $K_{0l}'(\xi)$ satisfy inequality of the form

$$\sum_{|\theta|\le[n/2]+1}\ \sup_{\xi\in\mathbb{R}^n}\ \big|\ \xi^\theta D_\xi^\theta\, a(\xi)\ \big| < +\infty\,,\ \text{ where }\ \theta_1,\dots,\theta_n \le 1\,, \tag{4.4}$$

and $L_{0jl}'(\xi')$ satisfies the same inequality but with n and ξ replaced by $(n-1)$ and ξ' respectively.

Let

$$f \in H_p^{s-\mathbf{Re}\mu}\left(\overline{\mathbb{R}^n_+}, \mathbb{C}^N\right)\ \left(\text{or } B_{p,q}^{s-\mathbf{Re}\mu}\left(\overline{\mathbb{R}^n_+},\ \mathbb{C}^N\right)\right)$$

and

$$g_j \in B_{p,p}^{s-\mathbf{Re}\beta_j-1/p}\left(\mathbb{R}^{n-1}\right)\ \left(\text{or } B_{p,q}^{s-\mathbf{Re}\beta_j-1/p}\left(\mathbb{R}^{n-1}\right)\right)$$

be the given functions, and let

$$u_+ \in \tilde{H}_p^s\left(\overline{\mathbb{R}^n_+}, \mathbb{C}^N\right)\ \left(\text{or } \tilde{B}_{p,q}^s\left(\overline{\mathbb{R}^n_+},\ \mathbb{C}^N\right)\right)$$

and

$$w_l \in B_{p,p}^{s_l}\left(\mathbb{R}^{n-1}\right)\ \left(\text{or } B_{p,q}^{s_l}\left(\mathbb{R}^{n-1}\right)\right)$$

be the unknown functions, where

$$s_l = s - \mathbf{Re}\,\mu + \mathbf{Re}\,\rho_l + 1 - 1/p\,.$$

The left-hand sides of (4.2) and (4.3) define the continuous operator

$$P \; : \; H_1(s,p) = \begin{array}{c} \tilde{H}_p^s\left(\overline{\mathbb{R}^n_+},\, \mathbb{C}^N\right) \\ \bigoplus \\ \oplus_{1\le l\le m_-} B^{s_l}_{p,p}\left(\mathbb{R}^{n-1}\right) \end{array} \longrightarrow$$

$$H_2(s,p) = \begin{array}{c} H_p^{s-\mathbf{Re}\mu}\left(\overline{\mathbb{R}^n_+},\, \mathbb{C}^N\right) \\ \bigoplus \\ \oplus_{1\le j\le m_+} B^{s-\mathbf{Re}\beta_j-1/p}_{p,p}\left(\mathbb{R}^{n-1}\right) \end{array},$$

$$\left(\text{or } P : B_1(s,p,q) = \begin{array}{c} \tilde{B}_{p,q}^s\left(\overline{\mathbb{R}^n_+},\, \mathbb{C}^N\right) \\ \bigoplus \\ \oplus_{1\le l\le m_-} B^{s_l}_{p,q}\left(\mathbb{R}^{n-1}\right) \end{array} \longrightarrow \right.$$

$$\left. B_2(s,p,q) = \begin{array}{c} B_{p,q}^{s-\mathbf{Re}\mu}\left(\overline{\mathbb{R}^n_+},\, \mathbb{C}^N\right) \\ \bigoplus \\ \oplus_{1\le j\le m_+} B^{s-\mathbf{Re}\beta_j-1/p}_{p,q}\left(\mathbb{R}^{n-1}\right) \end{array} \right).$$

Now we fix an arbitrary $\omega \in \mathbf{S^{n-2}}$ and denote by $A_\omega(\xi)$ the symbol $A(\omega|\xi'|\,,\xi_n)$, and similarly for $T_{\omega j}(\xi)$, $K_{\omega l}(\xi)$, $L_{\omega jl}(\xi')$. The operator corresponding to these symbols we denote by P_ω:

$$P_\omega : H_1(s,p) \;\to\; H_2(s,p) \quad \Big(\text{or } B_1(s,p,q) \to B_2(s,p,q)\Big)\,.$$

The following boundary value problem on the half-axis is associated with the system (4.2)–(4.3):

$$\pi_+ A(\omega, D_n) u_+(x_n) + \sum_{1\le l\le m_-} w_l \pi_+ K_l(\omega, D_n)\delta(x_n) \;=\; f(x_n)\,, \tag{4.5}$$

$$\gamma_0 T_j(\omega, D_n) u_+(x_n) + \sum_{1\le l\le m_-} L_{jl}(\omega) w_l = g_j\,, \;\; 1\le j\le m_+\,, \tag{4.6}$$

where

$$f \in H_p^{s-\mathbf{Re}\mu}\left(\bar{\mathbb{R}}_+,\, \mathbb{C}^N\right) \;\; \left(\text{or } B_{p,q}^{s-\mathbf{Re}\mu}\left(\bar{\mathbb{R}}_+,\, \mathbb{C}^N\right)\right),$$

$$u_+ \in \tilde{H}_p^s\left(\bar{\mathbb{R}}_+,\, \mathbb{C}^N\right) \;\; \left(\text{or } \tilde{B}_{p,q}^s\left(\bar{\mathbb{R}}_+,\, \mathbb{C}^N\right)\right),$$

g_j and w_l are complex numbers, and $A(\omega, D_n)$, $T_j(\omega, D_n)$, $K_l(\omega, D_n)$ are pseudodifferential operators in x_n depending on the parameter $\omega \in \mathbf{S^{n-2}}$ with the respective symbols $A(\omega, \xi_n)$, $T_j(\omega, \xi_n)$, $K_l(\omega, \xi_n)$.

We shall also need the following notation:

$$\mathbb{Z}(A) = \left\{\mathbf{Re}\,\mu/2 \,-\, \mathbf{Re}\,\delta_k \,+\, m \;\middle|\; m\in\mathbb{Z}\,,\; k=1,\dots,N\right\},$$
$$s_+ = \min\left\{\mathbf{Re}\,\mu \,-\, 1 \,-\, \mathbf{Re}\,\rho_l'\,,\; t \;\middle|\; t\in\mathbb{Z}(A)\,,\; t\ge s\,-\,1/p\,,\; l=1,\dots,m_-\right\},$$
$$s_- = \max\left\{\mathbf{Re}\,\beta_j'\,,\; t \;\middle|\; t\in\mathbb{Z}(A)\,,\; t\le s\,-\,1/p\,,\; j=1,\dots,m_+\right\}.$$

It is easily seen that, if

$$s - \frac{\mathbf{Re}\,\mu}{2} + \mathbf{Re}\,\delta_k - 1/p \notin \mathbb{Z},\ k = 1, \ldots, N, \tag{4.7}$$

then $s_- < s - 1/p < s_+$.

Theorem 4.1. *The following five statements are equivalent:*

1) *an operator* $P\colon\ H_1(s,p) \to H_2(s,p)$ *is of Noether type,*
2) *an operator* $P\colon\ H_1(s,p) \to H_2(s,p)$ *is invertible,*
3) *operators* $P_\omega\colon\ H_1(s,p) \to H_2(s,p)$ *are of Noether type,* $\forall\,\omega \in \mathbf{S^{n-2}}$,
4) *operators* $P_\omega\colon\ H_1(s,p) \to H_2(s,p)$ *are invertible,* $\forall \omega \in \mathbf{S^{n-2}}$,
5) *the boundary problem* (4.5), (4.6) *is uniquely solvable for arbitrary right-hand sides,* $\forall\,\omega \in \mathbf{S^{n-2}}$.

In any of the statements 1)–4) *we may replace* $H_i(s,p)$ *by* $B_i(s,p,q)$, $H_i(s_1,p_1)$ *or* $B_i(s_1,p_1,q_1)$ $(i = 1, 2)$, *where* $s_1 \in \mathbb{R}$, $p_1 \in (1,\infty)$, $q_1 \in [1,\infty]$, *and*

$$s_- < s_1, -\frac{1}{p_1} < s_+\,. \tag{4.8}$$

The analogous property also holds for the statement 5).

A necessary condition for 1)–5) *to hold is that* (4.7) *is satisfied together with the equation*

$$m_- - m_+ = \kappa + N + \sum_{1\le k\le N} \left[s - \frac{\mathbf{Re}\,\mu}{2} + \mathbf{Re}\,\delta_k - \frac{1}{p}\right]. \tag{4.9}$$

We shall say some words about the proof. The proof of the equivalence of the statements 1) and 3), 2) and 4) is carried out with the help of the localization in $\xi'/|\xi'| \in \mathbf{S^{n-2}}$. The investigation of P_ω is based on the factorization of matrix-functions. In this way we have to examine the factor-multipliers, which have to be Fourier L_p-multipliers since, in general, they have singularities at $\xi' = 0$. The statements 3) and 4) of Theorem 4.1 (and thus statements 1) and 2), being respectively equivalent to them) turn out to be equivalent to the invertibility of some matrix, depending on $\omega \in \mathbf{S^{n-2}}$. In general, this matrix is discontinuous in ω and even its size may change while $\omega \in \mathbf{S^{n-2}}$ varies. The condition (4.9) ensures that it is a square matrix. Also it is proved with the help of the factorization of matrix-functions, that statement 5) is equivalent to the invertibility of this matrix. Note that, if the condition (4.7) is not fulfilled, then the problem (4.5), (4.6) is not of Noether type (see, for example, Duduchava R.V. (1979)).

The statement 5) and also the similar one which states that the leading boundary symbol of the Green operator is an isomorphism (see Section 2.7), are generalizations of the Shapiro-Lopatinskij condition in the theory of boundary value problems for elliptic differential operators (see, for example, Section 2.8, or the article of M.S. Agranovich in the present volume).

The above theorem provides the opportunity of reducing the study of boundary value problems to the simpler case of H_2^t-spaces. Thus we have to change p and s into $p_1 = 2$ and $s_1 = s - 1/p + 1/2$ respectively.

Note now one important fact. Considering the unknown function u as belonging to $\tilde{H}_p^s$ $(\tilde{B}_{p,q}^s)$, we require at the same time that the boundary conditions

$$\gamma_0 u = 0, \ldots, \gamma_{[s-1/p]^-} u = 0$$

are satisfied for $s > 1/p$. The number of such boundary conditions increases as s increases. We have to bear in mind that, unlike the case of pseudodifferential operators with the transmission property, in the case under consideration the solution to the boundary value problem (4.2), (4.3) may not be C^∞-smooth up to the boundary for C^∞-smooth data. The behavior of the solution is characterized by an asymptotic expansion of the form

$$u_+(x', x_n) \sim \sum_{j=0}^{\infty} \sum_{k=0}^{m_j} \zeta_{jk}(x') x_n^{\mu_j} \ln^k x_n\,, \quad \text{as } x_n \to 0+\,.$$

The method of localization in $\xi'/|\xi'|$ allows us to investigate boundary value problems for elliptic pseudodifferential operators with symbols which have the special feature of being discontinuous in ξ' (see Shargorodsky E.M. (1995)). The case of discontinuity in ξ_n is considered in (Schulze B.-W. (1983)).

We denote by $a(\omega, r, \xi_n)$ the function $A(\xi', \xi_n)$ written in the coordinates (ω, r, ξ_n), where $\omega = \xi'/|\xi'|$, $r = |\xi'|$. We say that a positively homogeneous function $A(\xi)$ of order μ belongs to $\mathcal{D}_\mu$ if

$$a(\omega, r, \xi_n) \in C^\infty\left(\mathbf{S^{n-2}} \times \bar{\mathbb{R}}_+ \times \mathbb{R} \setminus (\mathbf{S^{n-2}} \times \{0\} \times \{0\})\right)$$

and the transmission property is fulfilled:

$$A(0, \ldots, 0, -1) = \mathrm{e}^{\mu\pi i} A(0, \ldots, 0, +1)\,,$$

$$\frac{\partial^k a(\omega, 0, -1)}{\partial r^k} = (-1)^k \mathrm{e}^{\mu\pi i} \frac{\partial^k a(\omega, 0, +1)}{\partial r^k}\,, \quad \forall k \in \mathbb{N}\,. \tag{4.10}$$

The results to be mentioned also hold for elliptic pseudodifferential operators with symbols from $\mathcal{D}_\mu$. However, we can study boundary value problems for such symbols, where the desired function belongs to

$$H_p^s\left(\overline{\mathbb{R}_+^n}, \mathbb{C}^N\right) \text{ or } B_{p,q}^s\left(\overline{\mathbb{R}_+^n}, \mathbb{C}^N\right), \quad \text{for } s > 1/p - 1\,.$$

In this case the analogue of Theorem 4.1 holds, in which the conditions (4.7) and (4.8) are superfluous, and (4.9) becomes $m_- - m_+ = \kappa$, where we have to put $\delta_k = \mu/2$ while computing κ from (4.1) (see (4.10)), but we do not stop here.

It is also possible to consider boundary value problems of the form (4.2), (4.3) for elliptic pseudodifferential operators with anisotropic-homogeneous symbols:

$$A\left(t^{d_1}\xi_1, \dots, t^{d_n}\xi_n\right) = t^{\mu} A(\xi),$$
$$\forall t \in \mathbb{R}_+, \quad \forall \xi \in \mathbb{R}^n \setminus \{0\}\,; \quad (d_1, \dots, d_n \geq 1).$$

Standard elliptic and $2b$-parabolic (pseudo)differential operators are examples of such operators. For example, it follows from general results that the Cauchy problem $u\,|_{t=o}= \phi$ for the *heat equation*

$$\frac{\partial u}{\partial t} - (\Delta - 1)u = f\,, \quad t > 0$$

is uniquely solvable in the corresponding functional spaces, while for the *backward heat equation*

$$\frac{\partial u}{\partial t} + (\Delta - 1)u = f\,, \quad t > 0$$

the initial condition is superfluous, and this equation is uniquely solvable in the corresponding anisotropic Besov and Bessel-potential spaces.

The Noether property of boundary value problems for elliptic pseudodifferential operators, acting in sections of smooth vector bundles over a smooth compact manifold with a boundary, is investigated in a rather standard way with the help of a partition of unity, "freezing" of coefficients and straightening the boundary. We do not intend to go into these details nor to concentrate on the question of the index. We note only that it is sufficient to compute the index in the spaces H_2^t.

We now say something about applications. A number of mixed problems of mathematical physics, and also other boundary value problems with boundary data supported on disjoint surfaces, are reduced to elliptic pseudodifferential equations on manifolds with boundary. As a rule, the pseudodifferential operators obtained in this way are without the transmission property. That is why the theory of Vishik and Eskin is often used to investigate problems in the theory of elasticity, hydrodynamics, electrodynamics and others (see, for example, Gol'dshtein R.V., Entov V.M. (1989), Natroshvili D.G., Chkadua O.O., Shargorodsky E.M. (1990), Costabel M., Stephan E.P. (1987), Duduchava R., Natroshvili D.G., Shargorodsky E.M. (1990), Stephan E.P. (1984), (1987), Wendland W.L., Zhu J. (1991), and also Rempel S., Schulze B.-W. (1984a), Rempel S., Schulze B.-W. (1984b)).

The case $n = 2$ is an exceptional one from the point of view of the theory of boundary value problems for elliptic pseudodifferential operators. Indeed, it has been pointed out that, if $n \geq 3$, the index $\kappa(\omega)$, $\omega \in \mathbf{S^{n-2}}$, of an elliptic symbol is a constant. If $n = 2$, it may happen that $\kappa(-1) \neq \kappa(1)$, and we note that this is not a pathological situation. For example, for such a classical operator as

$$\overline{\partial} = \frac{1}{2}\left(\frac{\partial}{\partial x_1} + i\frac{\partial}{\partial x_2}\right),$$

we have $\kappa(-1) = 0$ and $\kappa(1) = -1$. If $\kappa(-1) \neq \kappa(1)$, none of the boundary value problems (4.2), (4.3) may be uniquely solvable (be of Noether type), because the condition (4.9) is surely violated (see Theorem 4.1.). This circumstance explains the fact that for *Bitsadze equation*

$$\frac{\partial^2 u}{\partial \bar{z}^2} = f(z), \quad z \in \Omega \subset \mathbb{C},$$

not only the Dirichlet an Neumann problems, but also the problem $Tu |_{\partial\Omega} = \phi$, where T is an arbitrary $\mathbb{C}$-linear differential operator (more generally, an arbitrary boundary problem from the Boutet de Monvel algebra), do not possess the Noether property.

In the case $\kappa(-1) \neq \kappa(1)$ we have to introduce into the boundary conditions either the complex conjugation operator or the analytic projectors $\Pi_{\pm} = \frac{1}{2}(I \pm S)$ (or both of them), where S is a singular integral operator with a Cauchy kernel. The boundary value problems obtained for elliptic pseudodifferential operators are generalizations of the classical Hilbert and linear conjugation problems from the theory of analytic functions. The analogue of the theory of Vishik and Eskin is constructed for such problems. These general results may be applied, for example, to the investigation of generalized analytic vectors on Riemennian surfaces, the Bitsadze equation, etc. In order to illustrate these remarks, we consider the equation

$$\frac{\partial^{l+m} u}{\partial \bar{z}^l \partial z^m} = f(z), \quad z = x_1 + i x_2 \in \Omega \subset \mathbb{C}, \tag{4.11}$$

where

$$\frac{\partial}{\partial z} = \frac{1}{2}\left(\frac{\partial}{\partial x_1} - i\frac{\partial}{\partial x_2}\right), \quad \frac{\partial}{\partial \bar{z}} = \frac{1}{2}\left(\frac{\partial}{\partial x_1} + i\frac{\partial}{\partial x_2}\right)$$

and Ω is a bounded finitely-connected domain with boundary $Y = \partial\Omega$ (it is supposed that the components of Y are simple closed curves).

For definiteness we consider $m \leq l$. In the case $m = l$ (4.11) is the polyharmonic equation $\Delta^m u = f(z)$, which possesses boundary value problems of Noether type from the Boutet de Monvel algebra and, in particular, boundary value problems of the form

$$\sum_{k=0}^{n_r} a_{rk}(z) \frac{\partial^{n_r} u}{\partial \nu^k \partial s^{n_r - k}} \Big|_{\partial\Omega} = \phi_r, \quad r = 1, \dots, m, \tag{4.12}$$

where $\partial/\partial\nu$ is the derivative along the inner normal, and $\partial/\partial s$ is the derivative along the positive tangent direction (the orientation of the components of the curve Y is chosen in such a way that, during motion in the positive direction, the domain Ω remains on the left-hand side). If $l \neq m$, then for the operator on the left-hand side of (4.11) we have $\kappa(-1) \neq \kappa(1)$, and so it does not possess a boundary value problem of Noether type from the Boutet de Monvel algebra. We therefore attach to (4.12) the boundary conditions

$$\mathbf{Re}\left(\sum_{k=0}^{n_r} a_{rk}(z) \frac{\partial^{n_r} u}{\partial \nu^k \partial s^{n_r - k}} \Big|_{\partial\Omega}\right) = \phi_r, \quad r = m+1, \dots, l, \tag{4.13}$$

or

$$\Pi_+ \left(\sum_{k=0}^{n_r} a_{rk}(z) \frac{\partial^{n_r} u}{\partial \nu^k \partial s^{n_r - k}} \Big|_{\partial\Omega} \right) = \phi_r \,, \quad r = m+1, \dots, l \,, \tag{4.14}$$

where **Re** is an ordinary real part, and

$$\Pi_\pm = \frac{1}{2}(I \pm S_Y) \,,$$

$$S_Y \, \psi(z) = \frac{1}{\pi i} \int_Y \frac{\psi(\zeta)}{\zeta - z} \, \mathrm{d}\zeta \,, \quad z \in Y \,.$$

(In the case $m > l$, r takes the values $1, \dots, l$ in (4.12) and the values $l+1, \dots, m$ in (4.13) and (4.14). Further, we have to take Π_- instead of Π_+ in (4.14).)

Let $1 < p < \infty$ and $1 \le q \le \infty$. We shall look for the solutions to the boundary problems (4.11)–(4.13) and (4.11), (4.12), (4.14) in the spaces $H_p^t(\overline{\Omega})$ $\left(\text{or } B_{p,q}^t(\overline{\Omega})\right)$, assuming that

$$f \in H_p^{t-l-m}(\overline{\Omega}) \quad \left(\text{or } B_{p,q}^{t-l-m}(\overline{\Omega})\right),$$
$$\phi_r \in B_{p,p}^{t-n_r-1/p}(Y) \quad \left(\text{or } B_{p,q}^{t-n_r-1/p}(Y)\right), \quad \text{for } r = 1, \dots, m,$$
$$\phi_r \in B_{p,p}^{t-n_r-1/p}(Y, \mathbb{R}) \quad \left(\text{or } B_{p,q}^{t-n_r-1/p}(Y, \mathbb{R})\right), \quad \text{for } r = m+1, \dots, l,$$

in the case of the boundary condition (4.13), and

$$\phi_r \in \Pi_+ \, B_{p,p}^{t-n_r-1/p}(Y) \quad \left(\text{or } \Pi_+ \, B_{p,q}^{t-n_r-1/p}(Y)\right), \quad \text{for } r = m+1, \dots, l,$$

in the case of the boundary condition (4.14); everywhere $t > \max\{n_r | r = 1, \dots, l\} + 1/p$.

Note that the spaces $\Pi_+ B_{p,q}^\sigma(Y)$ are the analogues of the boundary values of the Hardy-Smirnov classes $E_p(\Omega)$ (for example, see Privalov I.I. (1950)). Indeed, if the curve Y belongs to the C^1-class, then the boundary values of the functional class $E_p(\Omega)$ form (see Dyn'kin E.M. (1987), Dyn'kin E.M. (1989), Havin V.P. (1965)) the space $\Pi_+ L_p(Y)$.

We shall not formulate the restrictions on the smoothness of the coefficients a_{rk} and the curve Y. These restrictions have to provide the possibility of straightening the boundary and "freezing" the coefficients. So we shall consider a_{rk} and Y to be sufficiently smooth (in particular we may assume that they are C^∞-smooth).

Theorem 4.2. *The boundary problems* (4.11)–(4.13) *and* (4.11), (4.12), (4.14) *possess the Noether property if the following conditions are valid at every point* $\zeta \in \partial\Omega$*:*

$$\det \left\| \sum_{k=j-1}^{n_r} a_{rk}(\zeta) \frac{k!}{(k-j+1)!} \, i^k \right\|_{r,j=1}^{l} \neq 0 \,, \tag{4.15}$$

$$\det\left\| \sum_{k=j-1}^{n_r} a_{rk}(\zeta)\,\frac{k!}{(k-j+1)!}\,(-i)^k \right\|_{r,j=1}^{m} \neq 0\,. \tag{4.16}$$

This theorem is also valid for the boundary problems (4.11)–(4.13) and (4.11), (4.12), (4.14) containing lower order terms, which give rise under certain restrictions to compact operators.

4.2. Algebras of Boundary Value Problems (Rempel S., Schulze B.-W. (1982 a), (1982b), (1983, 1984), (1984a)). We shall consider operators

$$P = \begin{pmatrix} A_\Omega + W + G & K \\ T & L \end{pmatrix} : \begin{matrix} L_2(E) \\ \bigoplus \\ L_2(F) \end{matrix} \longrightarrow \begin{matrix} L_2(E') \\ \bigoplus \\ L_2(F') \end{matrix} \tag{4.17}$$

of zero order. Operators of nonzero order are reduced to them with the help of order reduction operators. We shall give the corresponding local expressions in order to explain the action of the particular components in (4.17). It is sufficient to explain the scalar case.

The operator A is a polyhomogeneous pseudodifferential operator of zero order, in general without the transmission property, and L also is a polyhomogeneous pseudodifferential operator of zero order (acting on the boundary $\Gamma = \partial M$). To define G, K and T we need the spaces $V^{\pm}$ which are the images of $L_2(\mathbb{R}_{\pm}) \subset L_2(\mathbb{R})$ under the Fourier transform F. It is easily seen that $V^+ \oplus V^- = L_2(\mathbb{R})$ and $V^{\pm} = \Pi^{\pm} L_2(\mathbb{R})$ (see (2.6) and Lemma 2.6).

The **trace operator** T is an operator of the form

$$Tu(x') = (2\pi)^{-n} \int\limits_{\mathbb{R}^n} \mathrm{e}^{ix'\cdot\xi'}\, t(x',\xi',\xi_n)\, \widehat{e_+ u}(\xi)\, \mathrm{d}\xi'\, \mathrm{d}\xi_n\,,$$

where $u \in C^\infty(\overline{\mathbb{R}^n_+})$ and the function $t \in C^\infty\big(\mathbb{R}^{n-1}_{x'} \times (\mathbb{R}^{n-1}_{\xi'} \setminus \{0\}), V^-_{\xi_n}\big)$ is positively-homogeneous of order $-1/2$ in (ξ',ξ_n).

The **potential operator** K is defined by the formula

$$Kv(x) = (2\pi)^{-n} \int\limits_{\mathbb{R}^n} \mathrm{e}^{ix\cdot\xi}\, k(x',\xi)\widehat{v}(\xi')\, \mathrm{d}\xi\,,$$

where $v \in C_0^\infty(\mathbb{R}^{n-1})$ and the function $k \in C^\infty\big(\mathbb{R}^{n-1}_{x'} \times (\mathbb{R}^{n-1}_{\xi'} \setminus \{0\}), V^+_{\xi_n}\big)$ is positively-homogeneous of order $-1/2$ in (ξ',ξ_n).

The **singular Green operator** G is an operator of the form

$$Gu(x) = (2\pi)^{-n-1} \int\limits_{\mathbb{R}^{n+1}} \mathrm{e}^{ix\cdot\xi}\, g(x',\xi',\xi_n,\eta_n)\, \widehat{e_+ u}(\xi',\eta_n)\, \mathrm{d}\eta_n\, \mathrm{d}\xi\,,$$

where $u \in C^\infty(\overline{\mathbb{R}^n_+})$ and the function $g \in C^\infty\big(\mathbb{R}^{n-1}_{x'} \times (\mathbb{R}^{n-1}_{\xi'} \setminus \{0\}), V^+_{\xi_n} \otimes_H V^-_{\eta_n}\big)$ is positively-homogeneous of order -1 in (ξ',ξ_n,η_n); here $\otimes_H$ denotes the tensor product of Hilbert spaces.

The new ingredient in P is the so-called **Mellin operator** W. In order to define this operator, we denote by $C_{1/2}$ the space of all functions $\varphi \in C^\infty(\{z \in \mathbb{C} : \mathbf{Re}\, z = 1/2\})$ which can be continued analytically into some strip

$$\Sigma_\delta \equiv \{z \in \mathbb{C} : 1/2 - \delta < \mathbf{Re}\, z < 1/2 + \delta\}, \quad \delta = \delta(\varphi) > 0,$$

and are such that

$$\sup_{z \in \Sigma_\delta} | z^l \varphi(z) | < \infty, \quad \forall l \in \mathbb{Z}_+ .$$

It is well known that the *Mellin transform*

$$m\, u(z) \equiv \int_0^{+\infty} t^{z-1}\, u(t)\, \mathrm{d}t, \quad z \in \mathbb{C}, \quad u \in C_0^\infty(\mathbb{R}_+)$$

can be extended to $L_2(\mathbb{R}_+)$ and

$$m : L_2(\mathbb{R}_+) \longrightarrow L_2(\{z \in \mathbb{C} : \mathbf{Re}\, z = 1/2\}) .$$

is the isometric isomorphism. We shall denote the inverse Mellin transform by m^{-1}.

Let $\omega \in C_0^\infty(\bar{\mathbb{R}}_+)$ with $\omega(\eta) = 1$ in some neighbourhood of the point $\eta = 0$, and let $h \in C^\infty(\mathbb{R}^{n-1}, C_{1/2})$. Then the **Mellin operator** with the symbol $h(x', z)$ is given by the formula

$$W u(x', x_n) = (2\pi)^{1-n} \int_{\mathbb{R}^{n-1}} \mathrm{e}^{ix' \cdot \xi'}\, \omega(|\xi'| x_n)\, m^{-1}_{z \to x_n} \big(h(x', z) \times$$

$$\times\, m_{y_n \to z} F_{y' \to \xi'}\, u(y)\big)\, \mathrm{d}\xi', \quad u \in C_0^\infty(\overline{\mathbb{R}^n_+}) .$$

Now we have to add the compact operators to the operators of the type (4.17) and close the resulting set in the operator norm. The theory analogous to that of Boutet de Monvel is developed in (Rempel S., Schulze B.-W. (1982b)) for the totality of operators obtained by this way. In particular, the composition formulae are stated. Accordingly, the closed set of operators forms an algebra. Also the parametrix construction for an elliptic operator is investigated, the index theorem is proved, the corresponding operator complexes are studied, etc.

The results of (Rempel S., Schulze B.-W. (1982b)) are applied in (Rempel S., Schulze B.-W. (1984a), (1984b)) to the conjugation (contact) problems and the mixed problems.

The calculus of operators of the type (4.17) depending on a parameter (including operators of nonzero order) is constructed in (Rempel S., Schulze B.-W. (1983, 1984)). It is used in the work with resolvents and in the definition of complex powers of operators.

A Brief Bibliographic Survey

Pseudodifferential operators with the transmission property were introduced in (Vishik M.I., Eskin G.I. (1965), Boutet de Monvel L. (1966)). The theory of boundary value problems for elliptic pseudodifferential operators with the transmission property, including the Index Theorem, was constructed in (Boutet de Monvel L. (1971)). The detailed description of the theory was presented in (Rempel S., Schulze B.-W. (1982a), also see Grubb G. (1986)). In our §2, devoted to this theory, we essentially use (Grubb G. (1989), Grubb G. (1990), Grubb G., Hörmander L. (1990)), as well as the above-mentioned work.

The material in §3 is taken from (Grubb G. (1986), also see Grubb G. (1989)). The sole exception is the first asymptotic formula of Theorem 3.9, which results directly from (Agranovich M.S. (1987), also see Agranovich M.S. (1990)). The theory of boundary value problems with a parameter for elliptic differential operators with applications to parabolic problems was elaborated in (Agmon S. (1962), Agranovich M.S., Vishik M.I. (1964), Agmon S. (1965)).

Boundary value problems for elliptic differential equations with pseudodifferential boundary conditions were evidently considered for the first time in (Dynin A.S. (1961)).

Later, boundary value problems for a considerably wider class of pseudodifferential equations were investigated in (Agranovich M.S. (1965)). The general theory of boundary value problems for elliptic pseudodifferential operators was created in the series of papers by Vishik and Eskin (see Vishik M.I., Eskin G.I. (1964), (1965), (1966), (1967a), (1967b), (1967c)). The monograph (Eskin G.I. (1973)) is also devoted to this theory. The theory of Vishik and Eskin is an L_2-theory, the boundary value problems being considered in the Sobolev–Slobodetckij spaces H_2^s. The first results on L_p-theory were announced in (Dynin A.S. (1969a), see also Rabinovich V.S. (1972)). A detailed exposition of the L_p-analogue of the Vishik and Eskin theory is contained in the articles (Shargorodsky E.M. (1989a), (1989b), (1994), (1995)), which we followed in Section 4.1. Here we also note the work on multidimensional singular integral operators on manifolds with boundary: (Simonenko I.B. (1965)) for the L_2-theory, (Duduchava R. (1984), Shamir E. (1967)) for the L_p-theory, (Duduchava R.V., Schneider R.(1987)) for the case of L_p-spaces with power weights. Unfortunately, as far as we are aware, a satisfactory theory of boundary value problems for elliptic pseudodifferential operators without the transmission property in Hölder spaces, is absent.

The references to the articles of S.Rempel and B.W.Schulze concerning the algebras of general boundary value problems for elliptic pseudodifferential operators without the transmission property in H_2^s-spaces, which in some respects combine the methods of M.I.Vishik, G.I.Eskin and L.Boutet de Monvel, have been included in Section 4.2.

Now we would like to introduce briefly some topics beyond the scope of the present survey. Boundary value problems on noncompact manifolds are studied in (Kryakvin V.D. (1986), Rabinovich V.S. (1972), Cordes H.O., Erkip A.K. (1980), Erkip A.K. (1987) and others). Applications of the C^*-algebra theory to pseudodifferential operators on manifolds with boundary are discussed in (Cordes H.O. (1979)). More recently, active research on boundary value problems with singularities is in progress (discontinuous symbols, manifolds with singularities, etc.). These ideas can be found in the monographs (Plamenevskij B.A. (1986), Rempel S., Schulze B.-W. (1989)) and the literature cited there (also see the article of B.A.Plamenevskij in the present volume and (Derviz A.O.(1990))). We also note the articles (Schneider R. (1989a), Schneider R. (1989b)) about the order reduction operators on Lipschitz manifolds (also see Duduchava R.V., Speck F.-O.(1990), (1993)). Boundary value problems of Sobolev type with boundary conditions on manifolds of different dimensions are indicated in (Rempel S., Schulze B.-W. (1982a), see also the bibliographic information there). Boundary value problems for formally hypoelliptic pseudodifferential operators in the Hörmander sense, which are elliptic at the boundary and possess the transmission property there, are considered in (Levendorskij S.Z. (1990)).

References[4]

Agmon S. (1962): On the eigenfunctions and on the eigenvalues of general elliptic boundary value problems. Commun. Pure Appl. Math. **15**, 119–147. Zbl. 109,327

Agmon S. (1965): Lectures on Elliptic Boundary Value Problems. Van Nostrand Math. Studies. D.Van Nostrand Publ. Co., Princeton. Zbl. 142,374

Agranovich M.S. (1965): Elliptic singular integro-differential operators. Usp. Mat. Nauk **20**, No. 5, 3–120. [English transl.: Russian Math. Surv. **20**, No 5 (1965) 1–121] Zbl. 149,361

Agranovich M.S. (1987): Some asymptotic formulas for elliptic pseudodifferential operators. Funkts. Anal. Prilozh. **21**, No. 1, 63–65. [English transl.: Funct. Anal. Appl. **21** (1987) 53–56] Zbl. 631.35073

Agranovich M.S. (1990): Elliptic operators on compact manifolds. Itogi Nauki Tekhn., Akad.Nauk SSSR, VINITI, Moscow. Ser. Sovremennye Problemy Matematiki. Fundamentalnye Napravleniya [Current Problems in Mathematics. Fundamental Directions] **63**, 5–129. [English transl.: Encyclopaedia of Mathematical Sciences **63**, Springer-Verlag, Berlin Heidelberg New York 1994, 1–130] Zbl. 739.58061

Agranovich M.S., Vishik M.I. (1964): Elliptic problems with a parameter and parabolic problems of general type. Usp. Mat. Nauk **19**, No. 3, 53–161. [English transl.: Russ. Math. Surv. **19** (1964), No. 3, 53 – 157] Zbl. 137,296

[4] For the convenience of the reader, references to reviews in Zentralblatt für Mathematik (Zbl.), compiled using the MATH database, and Jahrbuch über die Fortschritte der Mathematik (Jbuch) have, as far as possible, been included in this bibliography.

Atiyah M.F., Bott R. (1964): The index problem for manifolds with boundary. Coll.Differ.Anal., Tata Inst., Bombay, Oxford Univers. Press, Oxford, 175–186. Zbl. 136,346

Atiyah M.F., Bott R., Patodi V.K. (1973): On the heat equation and the index theorem. Invent. Math. **19**, 279–330. Zbl. 257.58008

Bérard P.H. (1986): Spectral geometry: direct and inverse problems. Lect. Notes Math. **1207**, 1–272. Zbl. 608.58001

Besov O.V.,Il'in V.P.,Nikolskij S.M. (1975): Integral Representations of Functions and Imbedding Theorems. Nauka, Moscow. [English transl.: Scripta Series in Math. Winston & Sons, Washington D.C., V.I 1978, V.II 1979] Zbl. 352.46023

Bliev N.K. (1985): Generalized Analytic Functions in Fractional Spaces. Nauka, Alma-Ata (Russian). Zbl. 591.30046

Boutet de Monvel L. (1966): Comportement d'un opérateur pseudo-différentiel sur une variétè á bord, I-II. J. Anal. Math. **17**, 241–304. Zbl. 161,79

Boutet de Monvel L. (1971): Boundary problems for pseudo-differential operators. Acta Math. **126**, 11–51. Zbl. 206,394

Cordes H.O. (1979): Elliptic pseudo-differential operators - an abstract theory. Lect. Notes Math. **756**. Zbl. 417.35004

Cordes H.O., Erkip A.K. (1980): The N-th order elliptic boundary problem for non-compact boundaries. Rocky Mt. J. Math. **10**, 7–24. Zbl. 433.35069

Costabel M., Stephan E.P. (1987): An improved boundary element Galerkin method for three-dimensional crack problems. Integral Equations Oper. Theory **10**, 467–504. Zbl. 632.73091

Derviz A.O. (1990): An algebra, generated by general pseudodifferential boundary problems in a cone. Probl. Mat. Anal. [Problems in Mathematical Analysis], LGU Publ., Leningrad, **11**, 133–161. [English transl.: J. Sov. Math. **64** (1990), 1313–1330] Zbl. 789.47035

Dolgonos E.I. (1968): Normal solvability of convolution equations in the spaces of Hölder functions. Litov. Mat. Sb. **8**, 747–751 (Russian). Zbl. 177,147

Duduchava R.V. (1979): Integral Equations in Convolution with Discontinuous Pre - symbols, Singilar Integral Equations with Fixed Singularities and their Applications to some Problems of Mechanics. Metsniereba, Tbilisi. [English transl.: Teubner, Leipzig 1979] Zbl. 439.45002

Duduchava R. (1984): On multidimensional singular integral operators. I,II. J. Oper. Theory **11**, 41–76, 199–214. Zbl. 537.45015, Zbl. 566.45019

Duduchava R.V., Schneider R. (1987): The algebra of non-classical singular integral operators on half-space. Integral Equations Oper. Theory. **10**, 531–553. Zbl. 629.45014

Duduchava R.V., Speck F.-O. (1990): Bessel potential operators for the quarter-plane. Univ. Stuttgart Math. Inst. A.Bericht Nr. 25. See Appl.Anal. **45** (1992), 49–68. Zbl. 783.47047

Duduchava R.V., Speck F.-O. (1993): Pseudodifferential operators on compact manifolds with Lipschitz boundary. Math. Nachr. **160**, 149–191. Zbl. 796.58051

Duduchava R., Natroshvili D.G., Shargorodsky E.M. (1990): Boundary value problems of the mathematical theory of cracks. Tbiliss. Gos. Univ. Tr. Inst. Prikl. Mat. [Proc. I.N.Vekua Inst. Appl. Math.]. **39**, 68–84.

Duduchava R.V., Saginashvili A.I., Shargorodsky E.M. (1995): About two-dimensional singular integral operators with a shift. Acad. Sci. Georgia. Proc. A.Razmadze Math. Inst., **103**, 3–13 (Russian).

Dynin A.S. (1961): Multidimensional elliptic boundary value problems with a single unknown function. Dokl. Akad. Nauk SSSR **141**, 285–287. [English transl.: Sov. Math., Dokl. **2** (1961), 1431–1433] Zbl. 125,358

Dynin A.S. (1969a): On the theory of pseudodifferential operators on a manifold with boundary. Dokl. Akad. Nauk SSSR **186**, 251–253. [English transl.: Sov. Math., Dokl. **10** (1969), 575–578] Zbl. 209,452

Dynin A.S. (1969b): On the index of families of pseudodifferential operators on manifolds with a boundary. Dokl. Akad. Nauk SSSR **186**, 506–508. [English transl.: Sov. Math., Dokl. **10** (1969), 614–617] Zbl. 209,453

Dynin A.S. (1972): Elliptic boundary problems for pseudodifferential complexes. Funkts. Anal. Prilozh. **6**, 75–76. [English transl.: Funct. Anal. Appl. **6** (1972) 67–68] Zbl. 266.35036

Dyn'kin E.M. (1987): Methods of the theory of singular integrals. I: Hilbert transform and Calderón-Zygmund theory. Itogi Nauki Tekhn., Akad.Nauk SSSR, VINITI, Moscow. Ser. Sovremennye Problemy Matematiki. Fundamentalnye Napravleniya [Current Problems in Mathematics. Fundamental Directions] **15**, 197–292. [English transl.: Encyclopaedia of Mathematical Sciences **15**, Springer-Verlag, Berlin Heidelberg New York 1991, 167–260] Zbl. 661.42009

Dyn'kin E.M. (1989): Methods of the theory of singular integrals II: Littlewood-Paley theory and its applications. Itogi Nauki Tekhn., Akad.Nauk SSSR, VINITI, Moscow. Ser. Sovremennye Problemy Matematiki. Fundamentalnye Napravleniya [Current Problems in Mathematics. Fundamental Directions] **42**, 105–198. [English transl.: Encyclopaedia of Mathematical Sciences **42**, Springer-Verlag, Berlin Heidelberg New York 1992, 97–194] Zbl. 717.42025

Dzhangibekov G. (1989): About the Noether property and the index for some two-dimensional singular integral operators. Dokl. Akad. Nauk SSSR **308**, 1037–1041. [English transl.: Sov. Math. Dokl. **40** (1990), 394–399] Zbl. 704.45003

Dzhuraev A.D. (1987): The Method of Singular Integral Equations. Nauka, Moscow (Russian). Zbl. 633.35002

Egorov Yu.V., Shubin M.A. (1988): Linear partial differential equations. Elements of the modern theory. Itogi Nauki Tekhn., Akad.Nauk SSSR, VINITI, Moscow. Ser. Sovremennye Problemy Matematiki. Fundamentalnye Napravleniya [Current Problems in Mathematics. Fundamental Directions] **31**, 5–125. [English transl.: Encyclopaedia of Mathematical Sciences **31**, Springer-Verlag, Berlin Heidelberg New York 1994, 1–120] Zbl. 686.35005

Eidelman S.D. (1990): Parabolic equations. Itogi Nauki Tekhn., Akad.Nauk SSSR, VINITI, Moscow. Ser. Sovremennye Problemy Matematiki. Fundamentalnye Napravleniya [Current Problems in Mathematics. Fundamental Directions] **63**, 201–313. [English transl.: Encyclopaedia of Mathematical Sciences **63**, Springer-Verlag, Berlin Heidelberg New York 1994, 203–316] Zbl. 719.35036

Erkip A.K. (1987): Normal solvability of boundary value problems in half-space. Lect. Notes Math. **1256**, 123-134. Zbl. 619.35041

Eskin G.I. (1973): Boundary Value Problems for Elliptic Pseudodifferential Equations. Nauka, Moscow [English transl.: Transl. Math. Monogr. **52** (1981), Am. Math. Soc. Providence] Zbl. 262.35001

Fedosov B.V. (1991): Index Theorems. Itogi Nauki Tekhn., Akad.Nauk SSSR, VINITI, Moscow. Ser. Sovremennye Problemy Matematiki. Fundamentalnye Napravleniya [Current Problems in Mathematics. Fundamental Directions] **65**, 165–268, 275 . [English transl.: Encyclopaedia of Mathematical Sciences **65**, Springer-Verlag, Berlin Heidelberg New York, in preparation]

Frank L.S. (1979): Coercive singular perturbations. I: A priori estimates. Ann. Math. Pure Appl., IV Ser. **119**, 41-113. Zbl. 468.35011

Frank L.S., Wendt W.D. (1982): Coercive singular perturbations. II: Reduction to regular perturbations and applications. Commun. Partial Differ. Equations **7**, 469-535. Zbl. 501.35007

Frank L.S., Wendt W.D. (1984): Coercive singular perturbations. III: Wiener - Hopf operators. J. Anal. Math. **43**, 88-135. Zbl. 572.35006

Franke J. (1985): Besov-Triebel-Lizorkin spaces and boundary value problems. Semin.Anal.1984/85, Inst.Math., Berlin, 89-104. Zbl. 595.35121

Franke J. (1986): Elliptische Randwertprobleme in Besov-Triebel-Lizorkin Räumen. Dissertation. Friedrich Schiller Univ., Jena.

Gilkey P.B. (1974): The index theorem and the heat equation. Princeton Univ. Math. Lect. Ser. **4**, Publish or Perish. Boston. Zbl. 287.58006

Gol'dshtein R.V., Entov V.M. (1989): Qualitative Methods in Continuum Mechanics. Nauka, Moscow (Russian). Zbl. 673.73001

Grubb G. (1986): Functional calculus of pseudo-differential boundary problems. Progress in Math. **65**, Birkhäuser, Boston. Zbl. 622.35001

Grubb G. (1989): Parabolic pseudo-differential boundary problems and applications. Copenh. Univ. Math. Dept. Report. Series No.4, 79 pp.; Lect. Notes Math. **1495** (1991), 46–117. Zbl. 763.35116

Grubb G. (1990): Pseudo-differential boundary problems in L_p-spaces. Commun. Partial Differ. Equations **15**, 289–340. Zbl. 723.35091

Grubb G., Hörmander L. (1990): The transmission property. Math. Scand. **67**, 273–289. Zbl. 766.35088

Grubb G., Solonnikov V.A. (1987): Reduction of basic initial-boundary problems for Navier-Stokes equations to initial-boundary value problems for parabolic systems of pseudo-differential equations. Zap. Nauchn. Semin. Leningrad. Otd. Mat. Inst. Steklova. **163**, 37–48, 187. [English transl.: J.Sov.Math. **49** (1990), 1140–1147] Zbl. 655.35065

Grubb G., Solonnikov V.A. (1988): A pseudo-differential treatment of general inhomogeneous initial-boundary value problems for the Navier-Stokes equation. Proc. Journ. Eq. Dér. Part., St. Jean de Monts 1988. Ecole Polytechn. Palaiseau, Exp.No.3, 8 pp. Zbl. 675.76028

Havin V.P. (=Khavin V.P.) (1965): Boundary properties of the Cauchy type integrals and of conjugate harmonic functions in regions with reflectable boundary. Mat. Sb. **68**, 499–517 (Russian). Zbl. 141,266

Hörmander L. (1966): Pseudo-differential operators and hypoelliptic equations. Proc.Symp.Pure Math. **10**, 138-183. Zbl. 167,96

Hörmander L. (1983, 1985): The analysis of linear partial differential operators. I-IV, Springer-Verlag, Berlin Heidelberg New York. Zbl. 521.35001/2; Zbl. 601.35001; Zbl. 612.35001/2

Ivrii V.Ya. (1980): The second term of the spectral asymptotics for a Laplace - Beltrami operator on manifolds with boundary. Funkts. Anal. Prilozh. **14**, No.2, 25–34. [English transl.: Funct. Anal. Appl. **14** (1980) 98–105] Zbl. 565.35002

Ivrii V.Ya. (1984): Precise spectral asymptotics for elliptic operators acting on fiberings over manifolds with boundary. Lect. Notes Math. **1100**. Zbl. 565.35002

Komyak I.I. (1980): Conditions of Noetherity and a formula for the index of a class of singular integral equations over a circular domain. Differ. Uravn. **16**, 328-343. [English transl.: Differ. Equations **16** (1980), 215–226] Zbl. 436.45006

Kryakvin V.D. (1986): Pseudodifferential operators in weighted Hölder spaces. Noetherity of a general boundary value problem in a domain with non-compact bounday. Novocherkassk Agr. Eng. Inst., Novocherkassk, Preprint regist. in VI - NITI 17.12.86., No 8655B. [R. Zh. Mat. (1987), 4B 989 DEP] (Russian).

Kudryavtsev L.D., Nikolskij S.M. (1988): Spaces of differentiable functions of several variables and imbedding theorems. Itogi Nauki Tekhn., Akad.Nauk SSSR, VINITI, Moscow. Ser. Sovremennye Problemy Matematiki. Fundamentalnye Napravleniya [Current Problems in Mathematics. Fundamental Directions] **26**, 5–157. [English transl.: Encyclopaedia of Mathematical Sciences **26** , Springer-Verlag, Berlin Heidelberg New York 1990, 1–140] Zbl. 656.46023

Lasiecka I., Triggiani R. (1983): Stabilization and structural assignment of Dirichlet boundary feedback parabolic equations. SIAM J. Control Optimization **21**, 766-803. Zbl. 518.93046

Levendorskij S.Z. (1984a): Asymptotic behavior of the spectrum of problems of the form $Au = tBu$ for the operators that are elliptic in the Douglis - Nirenberg sense. Funkts. Anal. Prilozh. **18**, 84–85. [English transl.: Funct. Anal. Appl. **18** (1987) 253–255] Zbl. 556.35105

Levendorskij S.Z. (1984b): Asymptotics of the spectrum of linear operator pencils. Mat. Sb., Nov.Ser., **124**, 251–271 [English transl.: Math. USSR, Sb. **52** (1985), 245–266] Zbl. 553.35067

Levendorskij S.Z. (1990): The algebra of boundary value problems for a class of pseudodifferential operators of variable order. Dokl. Akad. Nauk SSSR **313**, 795–798. [English transl.: Sov. Math. Dokl. **42** (1991), 77–81] Zbl. 767.35118

Mikhlin S.G. (1962): Multidimensional Singular Integrals and Integral Equations. Gosudarstv. Izdat. Fiz. - Mat. Lit., Moscow. [English transl.: Pergamon Press, New York, 1965] Zbl. 105,303

Milnor J.W. (1965): Topology from the Differentiable Viewpoint. Univ. Press of Virginia, Charlottesville. Zbl. 136,204

Minakshisundaram S., Pleijel A. (1949): Some properties of eigenfunctions of the Laplace operator on Riemannian manifolds. Can. J. Math. **1**, 242–256. Zbl. 41,427

Munkres R. (1966): Elementary differential topology. Ann.Math.Stud.**54**. Zbl. 161,202

Nambu T. (1979): Feedback stabilization and distributed parameter systems of parabolic type. J. Differ. Equations **33**, 167 -188. Zbl. 404.35058

Natroshvili D.G., Chkadua O.O., Shargorodsky E.M. (1990): Mixed problems for homogeneous anisotropic elastic media. Tbiliss. Gos. Univ. Tr. Inst. Prikl. Mat. [Proc. I.N.Vekua Inst. Appl. Math.] **39**, 133–181 (Russian).

Nikolskij S.M. (1977): Approximation of Functions of Several Variables and Imbedding Theorems. 2nd ed., Nauka, Moscow. [English transl. of the 1st ed.: Springer-Verlag, Berlin Heidelberg New York, 1975] Zbl. 496.46020

Palais R.S. (1965): Seminar on the Atiyah-Singer Index Theorem. Princeton Univ. Press, Princeton, New Jersey. Zbl. 137,170

Pedersen M. (1988): Pseudo-differential perturbations and stabilization of distributed parameter systems: Dirichlet feedback control problems. Roskilde Univ. Tekst.**161**; SIAM J. Control Optimization **29** (1991), 222–252. Zbl. 742.93067

Peetre J. (1959): Theorèmes de régularité pour quelques classes d'opérateurs différentiels. Thesis, Univ. Lund; Meddel. Lunds Univ. Mat. Semin. **16**, 1–121. Zbl. 139,284

Pillat U., Schulze B.-W. (1980): Elliptische Randwertprobleme für Komplexe von Pseudodifferentialoperatoren. Math. Nachr. **94**, 173-210. Zbl. 444.58017

Plamenevskij B.A. (1986): Algebras of Pseudodifferential Operators. Nauka, Moscow. [English transl.: Mathematics and its applications (Soviet series) **43** (1989), Kluwer Acad. Publ. Dordrecht] Zbl. 615.47038

Privalov I.I. (1950): Boundary Properties of Analytic Functions. Gosudarstv. Izdat. Techn. - Teor. Lit., Moscow - Leningrad (Russian). Zbl. 45,347

Prössdorf S. (1988): Linear integral equations. Itogi Nauki Tekhn., Akad.Nauk SSSR, VINITI, Moscow. Ser. Sovremennye Problemy Matematiki. Fundamentalnye Napravleniya [Current Problems in Mathematics. Fundamental Directions] **27**, 5–130, 239. [English transl.: Encyclopaedia of Mathematical Sciences **27**, Springer-Verlag, Berlin Heidelberg New York 1991, 1–126] Zbl. 677.45001

Pseudodifferential Operators (1967): Mir, Moscow (Russian).

Rabinovich V.S. (1972): Pseudodifferential operators on one class of non-compact manifolds. Mat. Sb., Nov. Ser. **89**, 46–60. [English transl.: Math. USSR, Sb. **18** (1972), 45–59] Zbl. 243.58005

Rempel S., Schulze B.-W. (1982a): Index Theory of Elliptic Boundary Problems. Akademie Verlag, Berlin. Zbl. 504.35002

Rempel S., Schulze B.-W. (1982b): Parametrices and boundary symbolic calculus for elliptic boundary problems without transmission property. Math. Nachr. **105**, 45-149. Zbl. 544.35095

Rempel S., Schulze B.-W. (1983, 1984): Complex powers for pseudo-differential boundary problems I; II. Math. Nachr. **111**, 41-109; **116**, 269-314. Zbl. 528.35090; Zbl. 585.58041

Rempel S., Schulze B.-W. (1984a): A theory of pseudo-differential boundary value problems with discontinuous conditions I; II. Ann. Global Anal. Geom. **2**, 163-251; 289-384. Zbl. 566.35099; Zbl. 567.35084

Rempel S., Schulze B.-W. (1984b): Mixed boundary value problems for Lamé's system in three dimensions. Math. Nachr. **119**, 265-290. Zbl. 572.35031

Rempel S., Schulze B.-W. (1989): Asymptotics for elliptic mixed boundary problems. Math. Research. Akademie Verlag, Berlin. **50**. Zbl. 689.35104

Rozenblum G.V., Shubin M.A., Solomyak M.Z. (1989): Spectral theory of differential operators . Itogi Nauki Tekhn., Akad.Nauk SSSR, VINITI, Moscow. Ser. Sovremennye Problemy Matematiki. Fundamentalnye Napravleniya [Current Problems in Mathematics. Fundamental Directions] **64**, 1–248. [English transl.: Encyclopaedia of Mathematical Sciences **64**, Springer-Verlag, Berlin Heidelberg New York 1994, 272 pp.] Zbl. 715.35057

Schneider R.(1989a): Bessel potential operators for canonical Lipschitz domains. Techn. Hochschule Darmstadt. Prepr. Nr. **1213**; Math. Nachr. **150** (1991), 277–299. Zbl. 737.46026

Schneider R.(1989b): Reduction of order for pseudodifferential operators on Lipschitz domains. Techn. Hochschule Darmstadt. Prepr. Nr. **1253**; Commun. Partial Differ. Equations **16** (1991), 1263–1286. Zbl. 747.35056

Schulze B.-W. (1979): Adjungierte elliptische Randwert - Probleme und Andwendungen auf über- und unterbestimmte Systeme. Math. Nachr. **89**, 225-245. Zbl. 455.35085

Schulze B.-W. (1983): Pseudo-differential boundary problems with discontinuous symbols. Math. Nachr. **110**, 263-277. Zbl. 535.35085

Shamir E. (1967): Elliptic systems of singular integral operators. I: The half-space case. Trans. Am. Math. Soc. **127**, 107-124. Zbl. 157,193

Shargorodsky E.M. (1989a): Boundary problems for elliptic pseudodifferential operators on manifolds. Acad. Sci. Georgia. Proc. A.Razmadze Math. Inst. **105** (1995), 108–132. Preprint regist. in BIVU GSSR 28.06.89., No 547-G 89 [R. Zh. Mat. (1989), 12B 406 DEP] (Russian).

Shargorodsky E.M. (1989b): Boundary problems for elliptic pseudodifferential operators : the case of two-dimensional manifolds. Acad. Sci. Georgia. Proc. A.Razmadze Math. Inst. **103** (1995) , 29–69. Preprint regist. in BIVU GSSR 29.06.89., No 548-G 89 [R. Zh. Mat. (1989), 12B 405 DEP] (Russian).

Shargorodsky E. (1994): An L_p–analogue of the Vishik-Eskin theory. Mem. Differ. Equations Math. Phys. **2**, 41–146.

Shargorodsky E.M. (1995): Boundary problems for elliptic pseudodifferential operators: the half-space case. Acad. Sci. Georgia. Proc. A.Razmadze Math. Inst. **99**, 44–80 (Russian).

Shubin M.A. (1978): Pseudodifferential Operators and Spectral Theory. Nauka, Moscow. [English transl.: Springer Series in Soviet Math., Springer-Verlag, Berlin Heidelberg New York, 1987] Zbl. 451.47064

Simonenko I.B. (1965): A new general method for investigating linear operator equations of singular integral equation type. I,II. Izv. Akad. Nauk SSSR, Ser. Mat. **29**, 567-586, 757-782 (Russian). Zbl. 146,131

Stephan E.P. (1984): Boundary integral equations for mixed boundary value problems, screen and transmission problems in $\mathbb{R}^3$. Habilitationsschrift. Techn. Hochschule Darmstadt. Prepr. Nr. 848.

Stephan E.P. (1987): Boundary integral equations for screen problems in $\mathbb{R}^3$. Integral Equations Oper. Theory. **10**, 236-257. Zbl. 653.35016

Taylor M.E. (1981): Pseudodifferential Operators. Princeton Univ. Press, Princeton, New Jersey. Zbl. 453.47026

Trèves F. (1980): Introduction to Pseudodifferential and Fourier Integral Operators. I,II. Plenum Press, New York. Zbl. 453.47027

Triebel H. (1978): Interpolation Theory, Function Spaces, Differential Operators. Deutscher Verlag der Wissenschaften, Berlin. Zbl. 387.46033

Triebel H. (1983): Theory of Function Spaces. Birkhäuser Verlag, Basel Boston Stuttgart. Zbl. 546.46027

Triggiani R. (1979): On Nambu's boundary stabilizability problem for diffusion processes. J. Differ. Equations **33**, 189–200. Zbl. 418.35059

Vekua I.N. (1988): Generalized Analytic Functions. 2nd ed. Nauka, Moscow. [English transl. of the 1st ed.: Pergamon Press, London] Zbl. 698.47036 (1st Russ.ed. (1959) Zbl. 92,297)

Vishik M.I., Eskin G.I. (1964): Boundary problems for general singular equations in a bounded domain. Dokl. Akad. Nauk SSSR **155**, 24–27. [English transl.: Sov. Math., Dokl. **5**, 325–329] Zbl. 104,324

Vishik M.I., Eskin G.I. (1965): Convolution equations in a bounded domain. Usp. Mat. Nauk **20**, No.3, 89–152. [English transl.: Russ. Math. Surv. **20**, (1965) 85–151] Zbl. 152,342

Vishik M.I., Eskin G.I. (1966): Convolution equations in a bounded domain in spaces with weighted norms. Mat. Sb., Nov. Ser. **69**, 65–110 [English transl.: Am. Math. Soc. Transl. II. Ser. **67** (1968), 33–82] Zbl. 152,343

Vishik M.I., Eskin G.I. (1967a): Elliptic equations in convolution in a bounded domain and their applications. Usp. Mat. Nauk **22**, No.1, 15–76. [English transl.: Russ. Math. Surv. **22** (1967), 13–75] Zbl. 167,448

Vishik M.I., Eskin G.I. (1967b): Convolution equations of variable order. Tr. Mosk. Mat. O.-va **16**, 25–50 [English transl.: Trans. Mosc. Math. Soc. **16** (1967), 27–52] Zbl. 189,421

Vishik M.I., Eskin G.I. (1967c): Normally solvable problems for elliptic systems of convolution equations. Mat. Sb., Nov. Ser. **74**, 326–356 [English transl.: Math. USSR, Sb. **3** (1967), 303–330] Zbl. 162,200

Wendland W.L., Zhu J. (1991): The boundary element method for three-dimensional Stokes flows exterior to an open surface. Math. Comput. Modelling, **15**, No.6, 19–41. Zbl. 742.76055

III. Elliptic Boundary Value Problems in Domains with Piecewise Smooth Boundary

B.A. Plamenevskij

Translated from the Russian
by B.A. Plamenevskij

Contents

Introduction

This paper is a sketch of the theory of general elliptic boundary value problems in domains with edges of various dimensions on the boundary. In particular, the class of admissible domains contains polygons, cones, lenses and polyhedrons. Discontinuities in the coefficients of the operators along edges are allowed. We discuss solvability of the problems and obtain asymptotic formulas for solutions near singularities of the boundary and of the coefficients.

The solutions lose their smoothness at the edges of the boundary. This fact gives rise to questions about the behaviour of solutions near edges and about the choice of special function spaces (with weighted norms) where the operator possesses "good" properties (i.e. is Fredholm). In essence the matter reduces to the study of "model" problems with frozen coefficients in a ν-dimensional cone K^ν, or in a wedge $K^\nu \times R^{n-\nu}$, etc. The asymptotic formulas for solutions contain eigenvalues and associated vectors of some operator pencils $\lambda \mapsto \Upsilon(\lambda)$ (i.e. polynomials with operator coefficients). These spectral characteristics are determined by the coefficients and the boundary in a neighborhood of singular points. In addition, some "global" properties of the asymptotics are clarified, such as its dependence upon data of the problem as a whole and the connection with the index of the problem.

The theory of general elliptic problems in domains with piecewise smooth boundary essentially arose from V.A.Kondratjev's fundamental paper. This theory has been developed by a number of mathematicians. Many papers are devoted to various specific problems, applications in mechanics and electrodynamics of continua and numerical methods. All these specific problems are beyond the framework of the present paper. The list of references is far from comprehensive. In (Kondrat'ev, Oleinik (1983)), (Nazarov, Plamenevskij (1994)), (Dauge (1988)), (Grisvard (1985)), (Kufner, Sänding (1987)) the reader can find additional references.

§1. Boundary Value Problems in a Cone

1.1. Dirichlet Problem for the Laplace Operator in an Angle. Let $K = \{x = (x_1, x_2) \in R^2 : r > 0, \omega \in (0, 2\alpha]\}$; where (r, ω) are polar coordinates. Let us consider

$$(\Delta u)(x) = f(x), x \in K \;; \quad u = g \text{ on } \partial K \setminus O \,. \tag{1.1}$$

With the change of variables $x \longmapsto (t, \omega), t = \ln r$, we obtain the problem

$$(\partial_t^2 + \partial_\omega^2) v(t, \omega) = F(t, \omega), (t, \omega) \in \Pi; v(t, \alpha_\pm) = G_\pm(t)\,, t \in R \tag{1.2}$$

in the strip $\Pi = R \times (0, \alpha)$, where $v(t, \omega) = u(x), F(t, \omega) = e^{2t} f(x), x = e^t(\cos\omega, \sin\omega), \alpha_+ = \alpha, \alpha_- = 0$ and $G_\pm(t) = g(x)$ for $x = e^t(\cos\alpha_\pm, \sin\alpha_\pm)$.

Let $F \in C_c^\infty(\bar{\Pi})$ and $G_\pm \in C_c^\infty(R)$. We apply the Fourier transform

$$\hat{w}(\lambda) = \frac{1}{\sqrt{2\pi}} \int_R e^{-i\lambda t} w(t)\, dt$$

to (1.2) and obtain the boundary value problem for the family of ordinary differential equations

$$(\partial_\omega^2 - \lambda^2) \hat{v}(\lambda, \omega) = \hat{F}(\lambda, \omega), \omega \in (0, \alpha); \hat{v}(0, \alpha_\pm) = G_\pm(\lambda). \tag{1.3}$$

If we find solutions of (1.3) for every $\lambda \in R$, then the inverse Fourier transform gives a solution of (1.2). Using the Green function of (1.3),

$$\Gamma(\lambda; \phi, \omega) = (2i\lambda \sin(i\lambda\alpha))^{-1} \{\cos(i\lambda(\phi + \omega - \alpha)) - \cos(i\lambda(|\phi - \omega| - \alpha))\}, \tag{1.4}$$

we get

$$\hat{v}(\lambda, \omega) = \int_0^\alpha \Gamma(\lambda; \omega, \phi) \hat{F}(\lambda, \phi)\, d\phi - \sum_\pm \pm \partial_\omega \Gamma(\lambda, \omega, \alpha_\pm) \hat{G}_\pm(\lambda). \tag{1.5}$$

Using (1.4) and (1.5), we can check that for $l \geq 0$

$$\sum_{j=0}^{l+2} (1 + |\lambda|)^{2(l+2-j)} ||\partial_\omega^j \hat{v}(\lambda, \cdot); L_2(0, \alpha)||^2$$

$$\leq c(\sum_{j=0}^{l}(1+|\lambda|)^{2(l-j)}||\partial_\omega^j\hat{F}(\lambda,\cdot);L_2(0,\alpha)||^2+(1+|\lambda|)^{2l+3}(|\hat{G}_+(\lambda)|^2+|\hat{G}_-(\lambda)|^2)) \tag{1.6}$$

where c depends on neither v nor $\lambda \in R$. Now Plancherel's theorem gives

$$||v;H^{l+2}(\Pi)|| \leq c\left(||F;H^l(\Pi)|| + \sum_{\pm}||G_\pm;H^{l+3/2}(R)||\right) \tag{1.7}$$

H^s being the Sobolev space. Because of the density of $C_0^\infty(\bar{\Pi})$ and $C_0^\infty(R)$ in $H^l(\Pi)$ and $H^{l+3/2}(R)$ respectively, the solvability of (1.2) for $F \in C_0^\infty(\bar{\Pi})$ and $G_\pm \in H^{l+3/2}(R)$, together with (1.7), implies

Proposition 1.1. *For every $F \in H^l(\Pi)$ and $G_\pm \in H^{l+3/2}(R)$ there exists a unique solution $v \in H^{l+2}(\Pi)$ of (1.2), and the estimate (1.7) holds.*

Now we note that the function $\lambda \longmapsto \Gamma(\lambda,\cdot)$ (see (1.4)) is meromorphic on the complex plane having its only singular points at $\lambda_j = -j\pi\alpha^{-1}i$ ($j = \pm1,\pm2,\ldots$). Therefore, for any $\lambda \neq \lambda_j$, there exists a unique solution of (1.4) defined by (1.5). Moreover, the estimate (1.6) holds on every line $R+i\beta = \{\lambda \in C : \lambda = \sigma + i\beta, \sigma \in R\}$ for real $\beta \neq j\pi\alpha^{-1}$.

The family (1.3) of problems with parameter $\lambda \in R+i\beta$ is connected with (1.2) by the "complex" Fourier transform. This transform and its inverse are defined by

$$\hat{v}(\lambda) = (2\pi)^{-1/2}\int_R e^{-i\lambda t}v(t)\,dt\ ,\ \lambda \in R+i\beta, \tag{1.8}$$

$$v(t) = (2\pi)^{-1/2}\int_{R+i\beta} e^{i\lambda t}\hat{v}(\lambda)\,d\lambda. \tag{1.9}$$

Parseval's equality has the form

$$\int_R \exp(2\beta t)|v(t)|^2\,dt = \int_{R+i\beta}|\hat{v}(\lambda)|^2\,d\lambda.$$

For $l = 0,1,\ldots$ and $\beta \in R$ we introduce the space $W_\beta^l(\Pi)$ with norm

$$||v;W_\beta^l(\Pi)|| = \left(\sum_{j+k\leq l}\int_\Pi \exp(2\beta t)|\partial_t^j\partial_\omega^k v(t,\omega)|^2\,dt\,d\omega\right)^{1/2} \tag{1.10}$$

which is equivalent to $||\exp(\beta t)v;H^l(\Pi)||$.

We denote the corresponding space of traces on $\partial\Pi$ by $W_\beta^{l-1/2}(\partial\Pi)$ (for $l = 1,2,\ldots$).

Our discussion above implies

Proposition 1.2. *Let $\beta \neq j\pi\alpha^{-1}$ ($j = \pm1,\pm2,\ldots$) and let $F \in W_\beta^l(\Pi)$ and $G_\pm \in W_\beta^{l+2}(R)$. Then there exists a unique solution $v \in W_\beta^{l+2}(\Pi)$ of the problem (1.2) subject to the estimate*

$$||v; W_\beta^{l+2}(\Pi)|| = c_\beta \left(||F; W_\beta^l(\Pi)|| + \sum_\pm ||G_\pm; W_\beta^{l+3/2}(R)|| \right). \qquad (1.11)$$

We now return to the original coordinates. The norm (1.10) must be replaced by the (equivalent) norm in $V_\gamma^l(K)$ defined by

$$||u; V_\gamma^l(K)|| = \left(\sum_{|\alpha| \le l} \int_K r^{2(\gamma - l + |\alpha|)} |\partial_x^\alpha u(x)|^2 \, dx \right)^{1/2}$$

where $\gamma = \beta + l - 1$. Denote by $V_\gamma^{l-1/2}(\partial K)$ the space of traces on ∂K of functions in $V_\gamma^l(K)$. We rephrase Proposition 1.2 for the problem (1.1).

Proposition 1.3. *Let $\gamma \neq j\pi\alpha^{-1} + l - 1, j = \pm 1, \pm 2, \dots$ and let $f \in V_\gamma^l(K)$ and $g \in V_\gamma^{l+3/2}(\partial K)$. Then there exists a unique solution $u \in V_\gamma^{l+2}(K)$ of the problem (1.1) subject to the inequality*

$$||u; V_\gamma^{l+2}(K)|| \le c(||f; V_\gamma^l(K)|| + ||g; V_\gamma^{l+3/2}(\partial K)||).$$

We conclude this section with a few remarks. The weighted norms $||\cdot; V_\gamma^l||$ and the complex Fourier transform are of use not only because they enable us to extend the scale of admissible spaces. Some problems necessitate the use of such norms. For example, the Green function

$$\Gamma(\lambda; \phi, \omega) = (2i\lambda \sin(i\lambda\alpha))^{-1} \{\cos(i\lambda(\omega + \phi - \alpha)) + \cos(i\lambda(|\omega - \phi| - \alpha))\}$$

for the Neumann problem in the strip Π has poles at the points $\lambda_k = -ik\pi\alpha^{-1}$ where $k = 0, \pm 1, \dots$. Proposition 1.1 has no analog for the Neumann problem because of the pole $\lambda_0 = 0$. However, Propositions 1.2 and 1.3 remain valid for $\beta \neq k\pi\alpha^{-1}$. Finally, we can use the representation of solutions obtained from (1.5) and (1.9) in order to find the asymptotics of the solution $v(t, \omega)$ as $t \to -\infty$ or $t \to +\infty$. To this end we replace the line of integration $R + i\beta$ by a new line with another β and evaluate the corresponding residues at the poles of the Green function.

The above scheme is applicable to general elliptic boundary value problems in a cone. Instead of the argument related to the Green function for the problem (1.3) we shall use the theory of elliptic boundary value problems with a complex parameter λ.

1.2. General Elliptic Problems in a Cone. Solvability. Let K be an open cone in R^n with boundary ∂K and vertex O. We suppose that K cuts out on the unit sphere S^{n-1} with center at O an open set Ω with smooth $(n-2)$-dimensional boundary $\partial\Omega$.

Introduce the space $V_\beta^l(K)$ $(l = 0, 1, \dots; \beta \in R)$ of functions in K obtained by completing the set $C_0^\infty(\overline{K} \setminus O)$ in the norm

$$||u; V_\beta^l(K)|| = \left(\int_K \sum_{|\alpha|=0}^{l} r^{2(\beta-l+|\alpha|)} |D_x u(x)|^2 \, dx \right)^{1/2} \tag{1.12}$$

with $r = |x|$. The norm (1.12) is equivalent to the norm

$$\left(\int_0^\infty \sum_{k=0}^{l} ||(rD_r)^k u(r, \cdot); H^{l-k}(\Omega)||^2 r^{2(\beta-l)+n-1} \, dr \right)^{1/2} \tag{1.13}$$

where $u(r, \omega) \equiv u(x)$ and $\omega \in S^{n-1}$. We define the Mellin transform

$$\tilde{v}(\lambda) = \frac{1}{\sqrt{2\pi}} \int_0^\infty r^{-i\lambda-1} v(r) \, dr \tag{1.14}$$

for $v \in C_0^\infty(R_+)$. For the transform (1.14) the inverse formula

$$v(r) = \frac{1}{\sqrt{2\pi}} \int_{R+i\beta} r^{i\lambda} \tilde{v}(\lambda) \, d\lambda \tag{1.15}$$

holds together with Parseval's equality

$$\int_0^\infty |v(r)|^2 r^{2\beta-1} \, dr = \int_{R+i\beta} |\tilde{v}(\lambda)|^2 \, d\lambda \,. \tag{1.16}$$

These formulae are obtained from the corresponding property of the one-dimensional Fourier transform by the change of variable $t = \ln r$. By (1.16) the norms (1.12) and (1.13) are equivalent to the norm

$$\left(\int_{R+i(\beta-l+n/2)} \sum_{k=0}^{l} ||\tilde{u}(\lambda, \cdot); H^{l-k}(\Omega)||^2 |\lambda|^{2k} \, d\lambda \right)^{1/2} . \tag{1.17}$$

We also introduce the space $V_\beta^{l-1/2}(\partial K)$ $(l = 1, 2, \ldots)$ of traces on ∂K of functions in $V_\beta^l(K)$ with norm $||u; V_\beta^{l-1/2}(\partial K)|| = \inf\{||v; V_\beta^l(K)|| : v = u \text{ on } K\}$. It can be shown that this norm is equivalent to the norm

$$\left(\int_{R+i(\beta-l+n/2)} (|\lambda|^{2l-1} ||\tilde{u}(\lambda, \cdot); L_2(\partial\Omega)||^2 + ||\tilde{u}(\lambda, \cdot); H^{l-1/2}(\partial\Omega)||^2 \, d\lambda \right)^{1/2} . \tag{1.18}$$

We call a scalar differential operator $\mathcal{P}(x, D_x)$ in K a model operator if it admits the representation

$$\mathcal{P}(x, D_x) = r^{-l} \sum_{k=0}^{l} p_k(\omega, D_\omega)(rD_r)^k \equiv r^{-l} P(\omega, D_\omega, rD_r) \tag{1.19}$$

where $p_k(\omega, D_\omega)$ is a differential operator in Ω of order no higher than $l - k$ with coefficients smooth in $\overline{\Omega}$.

The operator $\mathcal{P} : V^s_\beta(K) \to V^{s-l}_\beta(K)$ is continuous for $s \geq l = ord\,\mathcal{P}$.

Let $\mathcal{L}(x, D_x)$ and $\mathcal{B}(x, D_x)$ be matrix differential operators in K of sizes $k \times k$ and $m \times k$, respectively, with elements $\mathcal{L}_{hj}$ and $\mathcal{B}_{qj}$ which are model operators. The orders of the operators $\mathcal{L}_{hj}$ and $\mathcal{B}_{qj}$ are equal to $s_h + t_j$ and $\sigma_q + t_j$, where $\{s_h\}, \{t_j\}$ and $\{\sigma_q\}$ are collections of integers and $s_1 + t_1 + \ldots + s_k + t_k = 2m$ with $t_j \geq 0$ and $max\,\{s_1, \ldots, s_k\} = 0$ (in the scalar case $k = 1, t_1 = 2m, s_1 = 0, \sigma_q = m_q - 2m$).

We consider the elliptic boundary value problem

$$\mathcal{L}(x, D_x)u(x) = f(x)\ ,\ x \in K, \mathcal{B}(x, D_x)u(x) = g(x)\ ,\ x \in \partial K \setminus O \tag{1.20}$$

where $u = (u_1, \ldots, u_k), f = (f_1, \ldots, f_k), g = (g_1, \ldots, g_m)$. We suppose that the ellipticity condition (1.20) is fulfilled everywhere on $\overline{K} \setminus O$.

Introduce the spaces of vector-functions

$$\mathcal{D}^l_\beta V(K) = \prod_{j=1}^{k} V^{l+t_j}_\beta(K),\ \ \mathcal{R}^l_\beta V(K) = \prod_{j=1}^{k} V^{l-s_j}_\beta(K) \times \prod_{q=1}^{m} V^{l-\sigma_q-1/2}_\beta(\partial K) \tag{1.21}$$

where $l \geq \max\{1 + \max \sigma_q, 0\}$. The operator $A = \{\mathcal{L}, \mathcal{B}\}$ of the boundary value problem (1.19) implements a continuous mapping

$$A : \mathcal{D}^l_\beta V(K) \to \mathcal{R}^l_\beta V(K). \tag{1.22}$$

Applying (1.19) to the entries of the matrices $\mathcal{L}(x, D_x)$ and $\mathcal{B}(x, D_x)$ we rewrite the problem (1.20) in the form

$$\sum_{j=1}^{k} r^{(-s_h+t_j)} L_{hj}(\omega, D_\omega, rD_r)u_j(r, \omega) = f_h(r, \omega)\ ,\ h = 1, \ldots, k, \tag{1.23}$$

$$\sum_{j=1}^{k} r^{-(\sigma_q+t_j)} B_{qj}(\omega, D_\omega, rD_r)u_j(r, \omega) = g_q(r, \omega)\ ,\ q = 1, \ldots, m. \tag{1.24}$$

We commute the factors r^{-t_j} with the operators $L_{hj}(\omega, D_\omega, rD_r)$ and $B_{qj}(\omega, D_\omega, rD_r)$. Since $rD_r(r^{-t_j}u) = r^{-t_j}(rD_r + it_j)u$, we obtain, for example, $r^{-t_j}L_{hj}(\omega, D_\omega, rD_r) = L_{hj}(\omega, D_\omega, rD_r - it_j)r^{-t_j}$. Further we multiply the h-th equation (1.23) by r^{s_h} and the q-th equation (1.24) by r^{σ_q}. We now have in place of (1.23) and (1.24) the equations

$$\sum_{j=1}^{k} L_{hj}(\omega, D_\omega, rD_r - it_j)r^{-t_j}u_j(r, \omega) = r^{s_h} f_h(r, \omega), \tag{1.25}$$

$$\sum_{j=1}^{k} B_{qj}(\omega, D_\omega, rD_r - it_j)r^{-t_j}u_j(r, \omega) = r^{\sigma_q} g_q(r, \omega). \tag{1.26}$$

Denote by Υ the operator pencil $C \ni \lambda \mapsto \Upsilon(\lambda) = \{L(\lambda), B(\lambda)\}$ where the matrices $L(\lambda)$ and $B(\lambda)$ are defined by

$$L(\lambda) = (L_{hj}(\omega, D_\omega, \lambda - it_j))_{h,j=1}^{k}\ , \quad B(\lambda) = (B_{qj}(\omega, D_\omega, \lambda - it_j))_{q,j=1}^{m,k}. \quad (1.27)$$

Because the problem (1.20) is elliptic, the pencil Υ turns out to be elliptic in Ω (i.e. Υ is the operator of an elliptic boundary value problem with a parameter). For ease of notation we shall denote the entries of $L(\lambda)$ and $B(\lambda)$ by $L_{hj}(\lambda)$ and $B_{qj}(\lambda)$.

From the ellipticity of Υ it follows that the mapping $\Upsilon(\lambda) : \mathcal{D}^l H(\Omega) \to \mathcal{R}^l H(\Omega)$ is an isomorphism for all λ with the exception of certain isolated points (the spaces are defined by (1.21) with H^s in place of V^s_β). These isolated points, with possibly finitely many exceptions, are situated inside a double angle $\{\lambda \in C : |Im\,\lambda| > c|Re\,\lambda|\}$. If λ does not belong to this angle and $|\lambda| > R$, where R is large enough, then for all $v = (v_1, \dots, v_k) \in \mathcal{D}^l H(\Omega)$ we have the inequality (see (Agranovich, Vishik (1964)))

$$\sum_{j=1}^{k}\sum_{\nu=0}^{l+t_j} |\lambda|^{2\nu} ||v_j; H^{l+t_j-\nu}(\Omega)||^2 \le c\{\sum_{j=1}^{k}\sum_{\nu=0}^{l-s_j} |\lambda|^{2\nu} ||(L(\lambda)v)_j; H^{l-s_j-\nu}(\Omega)||+$$

$$+\sum_{q=1}^{m} ||(B(\lambda)v)_q; H^{l-\sigma_q-1/2}(\partial\Omega)||^2 + |\lambda|^{2(l-\sigma_q)-1} ||(B(\lambda)v)_q; L_2(\partial\Omega)||^2\}.$$

(1.28)

Theorem 1.4. *The operator (1.22) is an isomorphism if and only if the line $R + i(\beta - l + n - 2)$ is free from the spectrum of the pencil Υ. If this line contains a point of the spectrum, then the set $Im\,A$ is nonclosed.*

In order to verify this theorem we apply the Mellin transform (1.14) to the equations (1.25) and (1.26), make use of (1.28), and take into account (1.12), (1.13), (1.17) and (1.18).

We now present analogous theorems for L_p and the Hölder classes. Let $V^l_{p.\beta}(K)$ be the completion of $C_c^\infty(K \setminus O)$ in the norm

$$||u; V^l_{p,\beta}(K)|| = \left(\int_K \sum_{|\alpha|=0}^{l} r^{p(\beta-l+|\alpha|)} |D^\alpha_x u(x)|^p \, dx \right)^{1/p}$$

where $\beta \in R, l = 0, 1, \dots,$ and $p > 1$. Replacing V^s_β in (1.21) by $V^s_{p,\beta}$ we define the spaces $\mathcal{D}^l_{p,\beta} V(K)$ and $\mathcal{R}^l_{p,\beta} V(K)$.

Theorem 1.5. *The operator $A = \{\mathcal{L}, \mathcal{B}\} : \mathcal{D}^l_{p,\beta} V(K) \to \mathcal{R}^l_{p,\beta} V(K)$ is an isomorphism if and only if the line $R + i(\beta - l + np^{-1})$ is free from the spectrum of the pencil Υ. If the line contains a point of the spectrum then the subspace $Im\,A$ is nonclosed.*

We introduce the norm

$$||v; \Lambda^{l,\alpha}(K)|| = \sup_{x\in K} \sum_{|\alpha|\le l} |x|^{-l-\alpha+|\gamma|} |D_x^\gamma v(x)| +$$

$$+ \sup_{x,y\in K} \sum_{|\gamma|\le l} |x-y|^{-\alpha} |D_x^\gamma v(x) - D_y^\gamma v(y)|$$

for functions in the cone K. Let $\Lambda_\beta^{l,\alpha}(K)$ denote the completion of the set of compactly supported functions in $\overline{K} \setminus O$ in the norm $||u; \Lambda_\beta^{l,\alpha}(K)|| = ||r^\beta u; \Lambda^{l,\alpha}(K)||$. We define similary the function space $\Lambda_\beta^{l,\alpha}(\partial K)$ on ∂K.

Theorem 1.6 *The operator $A : \mathcal{D}_\beta^{l,\alpha}\Lambda(K) \to \mathcal{R}_\beta^{l,\alpha}\Lambda(K)$ is continuous (cf. (1.21)). This operator is an isomorphism if and only if the line $R+i(\beta-l-\alpha)$ is free from the spectrum of Υ. If the line contains a point of the spectrum then the subspace $Im\, A$ is nonclosed.*

Corresponding theorems are also valid for elliptic boundary value problems in the cylinder $\Pi = \{(\omega, t) : \omega \in \Omega, t \in R\}$ with coefficients independent of t. We apply the Fourier transform $F_{t\to\lambda}$ instead of the Mellin transform. The formulations are obvious.

1.3. Eigenvectors and Associated Vectors of the Operator Pencil. We recall some notions related to the spectrum of the operator pencil $\lambda \to \Upsilon(\lambda)$ which will be of use in the description of the asymptotics of solutions of the problem (1.20) as $r \to 0$ or $r \to +\infty$.

Let $\lambda \to \Upsilon(\lambda)$ be a holomorphic operator-valued function in a neighborhood of λ_0 whose values are continuous linear operators $\Upsilon(\lambda) : E_1 \to E_2, E_j$ being a Banach space. Let also a vector-function $\lambda \mapsto \phi(\lambda) \in E_1$ be holomorphic at the point λ_0. If $\phi(\lambda_0) \neq 0$ and the function $\lambda \mapsto \Upsilon(\lambda)\phi(\lambda)$ vanishes at λ_0 then ϕ is called a root function of the operator Υ at the point λ_0. If Υ has at least one root function at λ_0, then λ_0 is called an eigenvalue of Υ. By the multiplicity of the root function ϕ we mean the multiplicity of the zero of the function $\lambda \mapsto \Upsilon(\lambda)\phi(\lambda)$, and the vector $\phi_0 = \phi(\lambda_0)$ is called an eigenvector corresponding to λ_0. Let ϕ be a root function at the point λ_0 of multiplicity κ and $\phi(\lambda) = \sum(\lambda-\lambda_0)^j\phi_j,\ j \ge 0$. Then $\phi_1, \dots, \phi_{r-1}$ are called associated vectors of the eigenvector ϕ_0, and the ordered set $\phi_0, \phi_1, \dots, \phi_{r-1}$ is called a Jordan chain corresponding to the eigenvalue λ_0. The rank of the eigenvector ϕ_0 (rank ϕ_0) is the maximal multiplicity of all root functions such that $\phi(\lambda_0) = \phi_0$.

If $\lambda \mapsto \Upsilon(\lambda)$ is an elliptic operator pencil then all eigenvalues turn out to be isolated (see Subsect. 1.2). In addition $\dim\ker\Upsilon(\lambda_0) < \infty$ for each eigenvalue λ_0 and the ranks of all eigenvectors are finite.

Let $J = \dim\ker\Upsilon(\lambda_0)$ and let $\phi^{(0,1)}, \dots, \phi^{(0,J)}$ be a system of eigenvectors such that rank $\phi^{(0,1)}$ is maximal among the ranks of all eigenvectors corresponding to λ_0, and rank $\phi^{(0,j)}$, where $j = 2, \dots, J$, is maximal among the ranks of all eigenvectors in any direct complement in $\ker \Upsilon(\lambda_0)$ of the linear span $\mathcal{L}(\phi^{(0,1)}, \dots, \phi^{(0,j-1)})$. The numbers $\kappa_j = \operatorname{rank} \phi^{(0,j)}$ are called the partial multiplicities of the eigenvalue λ_0, and the total multiplicity of λ_0 is, by definition, the sum $\kappa_1 + \ldots + \kappa_J$. If for each $j = 1, \dots, J$ the vectors $\phi^{(0,j)}, \dots, \phi^{(\kappa_j-1,1)}$ form a Jordan chain, then the collection of vectors $\{\phi^{(0,j)}, \dots, \phi^{(\kappa_j-1,j)}; j = 1, \dots, J\}$ is called a canonical system of Jordan chains corresponding to the eigenvalue λ_0.

In what follows we restrict ourselves to consideration of the pencil $\Upsilon(\lambda) = \{L(\lambda), B(\lambda)\}$ with L and B defined by (1.27). The eigenvectors and associated vectors $\phi^{(0,j)}, \dots, \phi^{(\kappa_j-1,j)}$ corresponding to λ_0 satisfy the equalities

$$\sum_{q=0}^{\nu} (1/q!) \partial_\lambda^q L(\lambda_0) \phi^{(\nu-q,j)} = 0 \quad \text{on} \quad \Omega$$

$$\sum_{q=0}^{\nu} (1/q!) \partial_\lambda^q B(\lambda_0) \phi^{(\nu-q,j)} = 0 \quad \text{on} \quad \partial\Omega$$

where $\nu = 0, \dots, \kappa_j - 1$ and $j = 1, \dots, J$.

For an analog of the following assertion (the elliptic problems for a scalar equation of order $2m$) we refer the reader to (1.37),(1.38) in Sect.1.5.

Proposition 1.7. *Suppose that for the operator $\{\mathcal{L}(x, D_x), \mathcal{B}(x, D_x)\}$ of (1.20) there is the Green formula*

$$(\mathcal{L}u, v)_K + (\mathcal{B}u, \mathcal{T}v)_{\partial K} = (u, \mathcal{L}^* v)_K + (\mathcal{S}u, \mathcal{Q}v)_{\partial K} \tag{1.28}$$

for all $u = (u_1, \dots, u_k), v = (v_1, \dots, v_k)$ with components in $C_c^\infty(\overline{K} \setminus O)$ where $\mathcal{L}^* \equiv \mathcal{L}(x, D_x)^*$ is the operator formally adjoint to $\mathcal{L}(x, D_x)$, and $\mathcal{T}, \mathcal{S}, \mathcal{Q}$ are matrix differential operators with model entries (see (1.19)), $\operatorname{ord} \mathcal{B}_{h,j} + \operatorname{ord} \mathcal{T}_{h,j} \le s_j + t_j - 1$, and $\operatorname{ord} \mathcal{S}_{hj} + \operatorname{ord} \mathcal{Q}_{hi} \le s_i + t_j - 1$. Then for the pencil $\Upsilon(\lambda) = \{L(\lambda), B(\lambda)\}$ there is the Green formula

$$(L(\lambda)\phi, \psi)_\Omega + (B(\lambda)\phi, T(\overline{\lambda})\psi)_{\partial\Omega} = (\psi, L^*(\overline{\lambda})\psi)_\Omega + (S(\lambda)\phi, Q(\overline{\lambda})\psi)_{\partial\Omega}\,. \tag{1.29}$$

Here $B(\lambda) = (B_{hj}(\lambda - it_j)), S(\lambda) = (S_{hj}(\lambda - it_j)), T(\overline{\lambda}) = (T_{hj}(\overline{\lambda} - is_j)), Q(\overline{\lambda}) = (Q_{hj}(\overline{\lambda} - is_j))$ are differential operators depending polynomially on λ, and $L^*(\overline{\lambda}) = L(\lambda)^*$ is the operator formally adjoint to $L(\lambda)$. The operators in (1.28) and (1.29) are linked by relations

$$\mathcal{L}_{qj}(x.D_x)^* = r^{-(s_q+t_j)} L_{qj}^*(\omega, D_\omega, rD_r + i(n - s_q - t_j)), \quad q, j = 1, \dots, k$$
$$\mathcal{T}_{hj}(x, D_x) = r^{-(s_j - \sigma_h - 1)} T_{hj}(\omega, D_\omega, rD_r - in)$$
$$\mathcal{S}_{hj}(x, D_x) = r^{-(t_j + \delta_h)} S_{hj}(\omega, D_\omega, rD_r)$$
$$\mathcal{Q}_{hj}(x, D_x) = r^{-(s_j - \delta_h - 1)} Q_{hj}(\omega, D_\omega, rD_r - in)$$

where $\delta_h = \operatorname{ord} S_{hj} - t_j, h = 1, \dots, m$ and $q, j = 1, \dots, k$. Conversely, (1.29) implies (1.28).

We call $\Upsilon^*(\lambda) \equiv \{L^*(\lambda), Q(\lambda)\}$ the adjoint pencil to $\Upsilon(\lambda)$. Note that the connection between $\{\mathcal{L}(x, D_x), \mathcal{B}(x, D_x)\}$ and Υ is somewhat different from that between the operators $\{\mathcal{L}(x, D_x)^*, \mathcal{Q}(x, D_x)\}$ adjoint to $\{\mathcal{L}(x, D_x), \mathcal{B}(x, D_x)\}$ with respect to the Green formula (1.28) and the pencil Υ^*.

If λ_0 is an eigenvalue of the pencil $\lambda \mapsto \Upsilon(\lambda)$ then $\overline{\lambda}_0$ turns out to be an eigenvalue of the pencil $\lambda \mapsto \Upsilon^*(\lambda)$. The sets of partial multiplicities of λ_0 and $\overline{\lambda}_0$ coincide.

We make use of the Jordan chains of Υ and Υ^* subject to the biorthogonality and normalization condition given by the following assertion.

Proposition 1.8. *Suppose that to the eigenvalue λ_0 of the elliptic pencil $\lambda \mapsto \Upsilon(\lambda) = \{L(\lambda), B(\lambda)\}$ there corresponds the canonical system of Jordan chains $\{\phi^{(0,j)}, \dots, \phi^{(\kappa_j-1,j)}; j = 1, \dots, J\}$. Then there exists a canonical system of Jordan chains $\{\psi^{(o,j)}, \dots, \psi^{(\kappa_j-1,j)}; j = 1, \dots, J\}$ of the pencil $\lambda \mapsto \Upsilon^*(\lambda) = \{L^*(\lambda), Q(\lambda)\}$ corresponding to λ_0 such that*

$$\sum_{p=0}^{\mu}\sum_{q=0}^{k}(1/(\mu+k+1-p-q)!)\{(\partial_\lambda^{\mu+k+1-p-q}L(\lambda_0)\phi^{(q,\sigma)}, \psi^{(p,\zeta)})_\Omega +$$

$$+(\partial_\lambda^{\mu+k+1-p-q}B(\lambda_0-)\phi^{(q,\sigma)}, \sum_{r=0}^{p}(1/r!)\partial_\lambda^r T(\overline{\lambda}_0)\psi^{(p-r,s)})_{\partial\Omega}\} = \delta_{\sigma,\zeta}\delta_{\kappa_\sigma-k-1,\mu} \tag{1.30}$$

We conclude this section with the description of the principal part of the function $\lambda \mapsto \Upsilon(\lambda)^{-1}$ in a neighborhood of a pole.

Proposition 1.9. *Let λ_0 be an eigenvalue of the elliptic pencil $\lambda \mapsto \Upsilon(\lambda)$. Then in a neighborhood of λ_0*

$$\Upsilon(\lambda)^{-1} = \sum_{j=1}^{J}\sum_{k=1}^{\kappa_j}(\lambda-\lambda_0)^{-k}\sum_{q=0}^{\kappa_j-k}(\,\cdot\,, \psi^{(q,j)})\phi^{(\kappa_j-k-q,j)} + \Gamma(\lambda) \tag{1.31}$$

where Γ is the holomorphic part of $\Upsilon(\lambda)^{-1}$, and $\{\phi^{(0,j)}, \dots, \phi^{(\kappa_j-1,j)}\}$, $\{\psi^{(0,j)}, \dots, \psi^{(\kappa_j-1,j)}\}$ are canonical systems of Jordan chains of the pencils Υ and Υ^* respectively, subject to (1.30).

In (Maz'ya, Plamenevskij (1975b)) an analog of Proposition 1.8 was proven without requirements on the existence of the Green formula (1.28).

1.4. Asymptotics of Solutions of Elliptic Boundary Value Problems in a Cone. We denote the vector $(r^{t_1+\lambda}v_1, \dots, r^{t_k+\lambda}v_k)$ by $r^{\vec{t}+\lambda}v$. The function

$$u(r,\omega) = r^{\vec{t}+i\lambda_0}\sum_{q=0}^{s}\frac{1}{q!}(i\ln r)^q\phi^{(s-q)}(\omega) \tag{1.32}$$

satisfies the homogeneous problem (1.20) if and only if λ_0 is an eigenvalue of the pencil $\lambda \mapsto \Upsilon(\lambda) = \{L(\lambda), B(\lambda)\}$ while $\phi^{(0)}, \dots, \phi^{(s)}$ is a Jordan chain corresponding to λ_0. Each solution of the form (1.32) of the homogeneous problem (1.20) is called a power solution (of order s) corresponding to λ_0.

Proposition 1.10. *1) Let $\{\phi^{(0,j)}, \dots, \phi^{(\kappa_j-1,j)}, j = 1, \dots, J\}$ be a canonical system of Jordan chains for some eigenvalue λ_0 of Υ. Then the functions*

$$u^{(k,j)}(r,\omega) = r^{\vec{t}+i\lambda_0} \sum_{s=0}^{k} \frac{1}{s!}(i \ln r)^s \phi^{(k-s,j)}(\omega)$$

where $k = 0, \dots, \kappa_j - 1$ and $j = 1, \dots, J$, form a basis in the space of power solutions of the homogeneous problem (1.20) corresponding to λ_0.

2) Let $\{\phi^{(0,j)}, \dots, \phi^{(\kappa_j-1,j)}, j = 1, \dots, J\}$ be a canonical system of chains for the eigenvalue $\overline{\lambda}_0$ of the pencil $\lambda \mapsto \Upsilon^*(\lambda)$. Then the functions

$$v^{(p,j)}(r,\omega) = r^{\vec{s}-n+i\overline{\lambda}_0} \sum_{q=0}^{p} \frac{1}{q!}(i \ln r)^q \psi^{(p-q,j)}(\omega)$$

form a basis in the space of power solutions of the homogeneous problem

$$\begin{aligned} \mathcal{L}(x.D_x)^* v(x) &= 0, \quad x \in K \\ \mathcal{Q}(x, D_x) v(x) &= 0, \quad x \in \partial K \setminus O \end{aligned} \tag{1.33}$$

adjoint to the problem (1.20) relative to the Green formula (1.29).

The following assertion gives the asymptotics of solutions of the problem (1.20).

Theorem 1.11. *Let $\{f, g\} \in \mathcal{R}^l_\beta V(K) \cap \mathcal{R}^l_\gamma V(K)$ and let the lines $R + i(\beta - l + n/2)$ and $R + i(\gamma - l + n/2)$ contain no points of the spectrum of Υ. Suppose that the strip between these lines contains the eigenvalues $\lambda_1, \dots, \lambda_N$ of the pencil Υ and $u \in \mathcal{D}^l_\beta V(K)$ is a solution of the problem (1.20). Then*

$$u(r,\omega) = \sum_{\nu=1}^{N} \sum_{j=1}^{J_\nu} \sum_{k=0}^{\kappa_{j,\nu}-1} c_\nu^{(k,j)}(f,g) u_\nu^{(k,j)}(r,\omega) + v(r,\omega), \tag{1.34}$$

where

$$u_\nu^{(k,j)}(r,\omega) = r^{\vec{t}+i\lambda_\nu} \sum_{s=0}^{k} \frac{1}{s!}(i \ln r)^s \phi_\nu^{(k-s,j)}(\omega), \tag{1.35}$$

$\{\phi_\nu^{(0,j)}, \dots, \phi_\nu^{(\kappa_{j,\nu}-1,j}; j = 1, \dots, J_\nu\}$ being a canonical system of chains of the pencil Υ corresponding to the eigenvalue λ_ν (thus, the functions (1.35) form a basis in the space of power solutions of the homogeneous problem (1.20); the mappings $\{f, g\} \mapsto c_\nu^{(k,j)}(f, g)$ are continuous linear functionals on $\mathcal{R}^l_\beta V(K) \cap \mathcal{R}^l_\gamma V(K)$; finally, v belongs to $\mathcal{D}^l_\gamma V(K)$ and satisfies the problem (1.20)).

For definitness, consider $\beta > \gamma$. Then (1.34) provides the asymptotics for the solution $u \in \mathcal{D}^l_\beta V(K)$ as $r \to 0$ and the asymptotics for the solution $v \in \mathcal{D}^l_\gamma V(K)$ as $r \to \infty$.

The formula (1.34) and the expressions for $c_\nu^{(k,j)}(f,g)$ given by Theorem 1.12 below can be derived from the representation

$$r^{-\overrightarrow{t}} u = \int_{R+i(\beta-l+n/2)} r^{i\lambda} \Upsilon(\lambda)^{-1} (r^{\overrightarrow{s}} f)\tilde{}(\lambda, \cdot\,)\, d\lambda$$

of the solution $u \in \mathcal{D}^l_\beta V(K)$ of the problem (1.20) (see (1.25), (1.26)). To this end we replace the line of integration by the line $R+i(\gamma-l+n/2)$ and make use of (1.31) to evaluate the residues. Another way of proving Theorems 1.11 and 1.12 was described in (Maz'ya, Plamenevskij (1975b)), (Maz'ya, Plamenevskij (1977a)).

Theorem 1.12. *Suppose that the conditions of Theorem 1.11 are satisfied and the Green formula (1.28) is valid for the problem (1.20). Then the functionals* $\mathcal{R}^l_\beta V(R) \cap \mathcal{R}^l_\gamma V(K) \ni \{f,g\} \mapsto c_\nu^{(k,j)}(f,g)$ *in (1.34) are defined by*

$$c_\nu^{(k,j)}(f,g) = (f, iv_\nu^{(\kappa_{j,\nu}-k-1,j)})_K + (g, i\mathcal{T} v_\nu^{(\kappa_{j,\nu}-k-1,j)})_{\partial K} \tag{1.36}$$

where $v_\nu^{(p,j)}$ *is the solution of the homogeneous problem (1.33) given by*

$$v_\nu^{(p,j)}(r,\omega) = r^{i\overline{\lambda}-n+\overrightarrow{s}} \sum_{q=0}^{p} \frac{1}{q!}(i \ln r)^q \psi_\nu^{(p-q,j)}(\omega)$$

and $\{\psi_\nu^{(p,j)}; j = 1, \dots J_\nu, p = 0, \dots, \kappa_{j,\nu} - 1\}$ *is a canonical system of chains of the pencil* $\lambda \mapsto \Upsilon^*(\lambda)$ *corresponding to the eigenvalue* $\overline{\lambda_\nu}$. *The Jordan chains* $\phi_\nu^{(0,j)}, \dots, \phi_\nu^{(\kappa_{j,\nu}-1,j)}$ *in (1.35) and* $\psi_\nu^{(o,j)}, \dots, \psi_\nu^{(\kappa_{j,\nu}-1,j)}$ *are subject to the orthogonality and normalization conditions (1.30).*

Theorems 1.11 and 1.12 with obvious modification remain valid for the spaces $V^s_{p,\beta}$ and $\Lambda^{s,\alpha}_\beta$.

1.5. Fundamental Solutions of Elliptic Boundary Value Problems in a Cone. In this section we restrict ourselves to the discussion of boundary value problems for a scalar equation of order $2m$ with normal boundary conditions (we consider $k = 1, t_1 = 2m, s_1 = 0$, and $\sigma_q = m_q - 2m$). In this case we write the operator $\mathcal{L}(x, D_x)$ in the cone K and the boundary operators $\mathcal{B}_q(x, D_x)$ in the form $\mathcal{L}(x, D_x) = r^{-2m} L(\omega, D_\omega, rD_r)$ and $\mathcal{B}_q = r^{-m_q} B_q(\omega, D_\omega, rD_r)$ and introduce the pencil

$$\Upsilon(\lambda) = \{L(\omega, D_\omega, \lambda), B_1(\omega, D_\omega, \lambda), \dots, B_m(\omega, D_\omega, \lambda)\} \tag{1.37}$$

Note that the pencil (1.37) differs from the pencil (1.27) by the shift from $\lambda - 2mi$ to λ. We write the Green formula in the form

$$(\mathcal{L}u, v)_K + \sum_{j=1}^{m} (\mathcal{B}_j u, \mathcal{T}_j v)_{\partial K} = (u, \mathcal{L}^* v)_K + \sum_{j=1}^{m} (\mathcal{S}_j u, \mathcal{Q}_j v)_{\partial K}$$

and set

$$\begin{aligned} &L(\omega, D_\omega, rD_r + i(n-2m))^* = r^{2m} \mathcal{L}(x, D_x)^*, \\ &T_j(\omega, D_\omega, rD_r + i(n-2m)) = r^{2m-m_j-1} \mathcal{T}_j(x, D_x), \\ &S_j(\omega, D_\omega, rD_r) = r^{ord\, \mathcal{S}_j} \mathcal{S}_j(x, D_x), \\ &Q_j(\omega, D_\omega, rD_r + i(n-2m)) = r^{ord\, \mathcal{Q}_j} \mathcal{Q}_j(x, D_x). \end{aligned} \tag{1.38}$$

The adjoint pencil is defined by $\Upsilon^*(\lambda) = \{L^*(\lambda), Q_1(\lambda), \dots, Q_m(\lambda)\}$ where $L^*(\lambda) = L(\overline{\lambda})^*$, and the following Green formula is valid:

$$(L(\lambda)u, v)_\Omega + \sum_{j=1}^{m} (B_j(\lambda)u, T_j(\overline{\lambda})v)_{\partial\Omega} = (u, L^*(\overline{\lambda})v)_\Omega + \sum_{j=1}^{m} (S_j(\lambda)u, Q_j(\overline{\lambda})v)_{\partial\Omega}.$$

Theorem 1.13. *If the line $R + i(\beta + n/p - 2m)$ is free from the spectrum of the pencil (1.37) then the following assertions are valid:*

1. There exists a unique solution $x \mapsto \mathcal{G}(x, y)$ of the boundary value problem

$$\begin{aligned} &\mathcal{L}(x, D_x)\mathcal{G}(x, y) = \delta(x-y), \quad x, y \in K, \\ &\mathcal{B}_j(x, D_x)\mathcal{G}(x, y) = 0, \quad x \in \partial K \setminus O, \ y \in K, \ j = 1, \dots, m, \end{aligned} \tag{1.39}$$

such that the function $x \mapsto \eta(|x||y|^{-1})\mathcal{G}(x, y)$ belongs to $V^{2m+s}_{p,\beta+s}(K), s = 0, 1, \dots,$ for every fixed $y \in K$. Here $\eta \in C^\infty(R_+), \eta(t) = 1$ for $t > 2$ and $2t < 1$, $\eta(t) = 0$ for $t \in (3/4, 1)$.

2. The function $(x, y) \mapsto \mathcal{G}(x, y)$ is infinitely differentiable in $x, y \in K \setminus O, x \neq y$. If $|x|/2 < |y| < 2|x|$, then

$$|D_x^\alpha D_y^\gamma \mathcal{G}(x, y)| \le c_{\alpha\gamma}(|x-y|^{2m-n-|\alpha|-|\gamma|} + |y|^{2m-n-|\alpha|-|\gamma|})$$

for $2m - n \neq |\alpha| + |\gamma|$, and

$$|D_x^\alpha D_y^\gamma \mathcal{G}(x, y)| \le c_{\alpha\gamma}(|\log \frac{|x-y|}{|x|}| + 1) \tag{1.40}$$

for $2m - n = |\alpha| + |\gamma|$.

3. G is a unique solution of the problem

$$\begin{aligned} &\mathcal{L}^*(y, D_y)\overline{\mathcal{G}(x, y)} = \delta(x-y), \quad y \in K, \\ &\mathcal{Q}_j(y, D_y)\overline{\mathcal{G}(x, y)} = 0, \quad y \in \partial K \setminus O, \ j = 1, \dots, m, \end{aligned} \tag{1.41}$$

such that $y \mapsto \eta(|x||y|^{-1})\mathcal{G}(x, y)$ is in $V^{2m+s}_{p',-\beta+2m+s}(K), s = 0, 1, \dots, p' = p(p-1)^{-1}$.

The equalities (1.39) and (1.41) are equivalent to

$$\overline{v(y)} = (\mathcal{G}(\,\cdot\,,y), \mathcal{L}^* v)_K + \sum_{j=1}^{m} (\mathcal{S}_j \mathcal{G}(\,\cdot\,,y), \mathcal{Q}_j v)_{\partial K} \tag{1.42}$$

$$w(x) = (\mathcal{L}w, \overline{\mathcal{G}(x,\,\cdot\,)})_K + \sum_{j=1}^{m} (\mathcal{B}_j w, \mathcal{T}_j \overline{\mathcal{G}(x,\,\cdot\,)})_{\partial K} \tag{1.43}$$

where $v, w \in C_c^\infty(\overline{K} \setminus O)$. The representations (1.42) and (1.43) are extended by continuity to all $v \in V^{2m}_{p',-\beta+2m}(K)$ and $w \in V^{2m}_{p,\beta}(K)$.

From Theorem 1.13(3), it follows that solutions $w \in V^{2m}_{p,\beta}(K)$ and $v \in V^{2m}_{p',-\beta+2m}(K)$ of the problem $\{\mathcal{L}, \mathcal{B}_1, \ldots, \mathcal{B}_m\}w = \{f, f_1, \ldots, f_m\}$ and $\{\mathcal{L}^*, \mathcal{Q}_1, \ldots, \mathcal{Q}_m\}v = \{g, g_1, \ldots, g_m\}$ can be written as

$$w(x) = \int_K \mathcal{G}(x,y) f(y)\, dy + \sum_{j=1}^{m} \int_{\partial K} \overline{\mathcal{T}_j(y, D_y)\overline{\mathcal{G}(x,y)}} f_j(y)\, dy, \tag{1.44}$$

$$v(x) = \int_K \mathcal{G}(x,y) g(y)\, dy + \sum_{j=1}^{m} \int_{\partial K} \overline{\mathcal{S}_j(y, D_y)\mathcal{G}(y,x)} g_j(y)\, dy. \tag{1.45}$$

The theorems in Sect.1.4 enable us to find the asymptotics of the Green function $(x,y) \mapsto \mathcal{G}(x,y)$ near the vertex of the cone. Roughly speaking, if x approaches the vertex we use Theorem 1.11. If y tends to the vertex then because of Theorem 1.12 the asymptotics depend on the behavior of the solutions $v_\nu^{(p,j)}$ of the adjoint problem (see (1.36)).

Suppose that there are no points of the spectrum of the pencil (1.37) on the line $R + i(\beta + np^{-1} - 2m)$. Denote by $\lambda_1^+, \lambda_2^+, \ldots$ all the eigenvalues of $\varUpsilon$ lying above $R + i(\beta + np^{-1} - 2m)$ numbered in increasing order of their imaginary parts. The ordering of the eigenvalues with equal imaginary parts is arbitrary. Similarly, $\lambda_1^-, \ldots, \lambda_n^-$ are the eigenvalues of $\varUpsilon$ lying below $R + i(\beta + np^{-1} - 2m)$ numbered in decreasing order of their imaginary parts. We denote by κ_μ^+ the multiplicity of λ_μ^+ and by $\phi_\mu^{(0,\sigma;+)}, \ldots, \phi_\mu^{(\kappa_{\sigma,\mu}-1,\sigma;+)}$ the Jordan chains of the pencil $\varUpsilon$ corresponding to the eigenvalue $\lambda_\mu^+, \sigma = 1, \ldots, J_\mu^+; \kappa_{1,\mu}^+ + \ldots + \kappa_{J_\mu^+,\mu}^+ = \kappa_\mu^+$. Let $\psi_\mu^{(0,\sigma;+)}, \ldots, \psi_\mu^{(\kappa_{\sigma,\mu}^+-1,\sigma;+)}$ be the Jordan chains of the pencil $\varUpsilon^*(\lambda) = \{L^*(\lambda), Q_1(\lambda), \ldots, Q_m(\lambda)\}$ for $\overline{\lambda_\mu^+}$ satisfying the orthogonality and normalization conditions (1.30). We denote analogously the chains associated with the eigenvalues λ_μ^- and $\overline{\lambda_\mu^-}$.

Theorem 1.14. *Let G be the Green function in Theorem 1.13. Then the following holds:*

1) For $2|x| < |y|$

$$\begin{aligned}\mathcal{G}(x,y) = -i \sum_{\mu=1}^{N} \frac{|x|^{i\lambda_\mu^-}}{|y|^{i\lambda_\mu^- + n - 2m}} \sum_{\sigma=1}^{J_\mu^-} \sum_{l=0}^{\kappa_{\sigma,\mu}^- - 1} \frac{1}{(\kappa_{\sigma,\mu}^- - l - 1)!} \times \\ \times (i \ln(|x|/|y|))^{\kappa_{\sigma,\mu}^- - l - 1} \sum_{t=0}^{l} \phi_\mu^{(t,\sigma;-)}(x/|x|) \overline{\psi_\mu^{(l-t,\sigma;-)}(y/|y|)} + R(x,y)\end{aligned} \tag{1.46}$$

where N is a positive integer such that $\operatorname{Im}\lambda_N^- > \operatorname{Im}\lambda_{N+1}^-$ *and*

$$|D_x^\gamma D_y^\delta R(x,y)| \le \text{const}\,\frac{|x|^{-Im\,\lambda_{N+1}^- - |\gamma|}}{|y|^{-Im\,\lambda_{N+1}^- - 2m+n+|\delta|}}|\ln(|y|/|x|)|^{\kappa_-}$$

(γ, δ are arbitrary multi-indices and κ_- is the largest of the lengths of the Jordan chains corresponding to the eigenvalues of the pencil (1.37) situated on the line $R + i\operatorname{Im}\lambda_{N+1}^-$).

2) For $2|y| < |x|$

$$\mathcal{G}(x,y) = i\sum_{\mu=1}^{N}\frac{|x|^{i\lambda_\mu^+}}{|y|^{i\lambda_\mu^+ + n-2m}}\sum_{\sigma=1}^{J_\mu^+}\sum_{l=0}^{\kappa_{\sigma,\mu}^+ -1}\frac{1}{(\kappa_{\sigma,\mu}^+ - l - 1)!}\times$$
$$\times\,(i\ln(|x|/|y|))^{\kappa_{\sigma,\mu}^+ - l-1}\sum_{t=0}^{l}\phi_\mu^{(t,\sigma;+)}(x/|x|)\overline{\psi_\mu^{(l-t,\sigma;+)}(y/|y|)} + R_+(x,y) \tag{1.47}$$

where N is a positive integer such that $\operatorname{Im}\lambda_N^+ < \operatorname{Im}\lambda_{N+1}^+$ *and*

$$|D_x^\gamma D_y^\delta R_+(x,y)| \le \text{const}\,\frac{|x|^{-Im\,\lambda_{N+1}^+ - |\gamma|}}{|y|^{-Im\,\lambda_{N+1}^+ - 2m+n+|\delta|}}|\ln(|x|/|y|)|^{\kappa_+}$$

(γ, δ are arbitrary multi-indices and κ_+ is the largest of the lengths of the Jordan chains corresponding to the eigenvalues of the pencil (1.37) situated on the line $R + i\operatorname{Im}\lambda_{N+1}^+$).

The asymptotic representations analogous to (1.46) and (1.47) hold for the Poisson kernels $\mathcal{P}_j(x,y) = \overline{\mathcal{T}_j(y,D_y)\overline{\mathcal{G}(x,y)}}$ of the problem (1.20) (see (Maz'ya, Plamenevskij (1979))).

All the results of this section are valid with obvious changes for the spaces $\Lambda_\beta^{s,\alpha}$.

1.6. Example: the Oblique Derivative Problem for the Laplace Operator in an Angle. We consider the operator

$$Au = \{\Delta u, (a^\pm u_{\tau_\pm} + b^\pm u_{\nu_\pm})|_{\omega=\pm\alpha}\}$$

in an angle $K = \{(r,\omega) : 0 < r < \infty, |\omega| < \alpha\}$, where $a^\pm, b^\pm$ are real numbers, $\tau^\pm$ is the direction of the ray $\omega = \alpha^\pm$ and $\nu^\pm$ is the outward directed normal on the side $\omega = \pm\alpha$ of the angle K. Suppose that $b^\pm \ne 0$; then A turns out to be elliptic with normal boundary conditions. Since

$$Au = \{u_{rr} + r^{-1}u_r + r^{-2}u_{\omega\omega}, (a^\pm u_r \pm b^\pm r^{-1}u_\omega)|_{\omega=\pm\alpha}\},$$

the pencil (1.37) takes the form

$$\Upsilon(\lambda)v = \{L(\lambda)v, B^\pm(\lambda)v\} = \{v_{\omega\omega} - \lambda^2 v, (\pm b^\pm v_\omega + i\lambda a^\pm v)|_{\omega=\pm\alpha}\}.$$

It can be checked directly that the eigenvalues of the pencil Υ are equal to $0, (\kappa^+ + \kappa^- - k\pi)$ where $\kappa^\pm = \text{arctg}\,(a^\pm/b^\pm)$ and $k = 0, \pm 1, \ldots$ Except zero, all these eigenvalues are simple and have the eigenfunctions

$$\phi(\omega) = \cos\frac{1}{2}[(\kappa^+ + \kappa^- + k\pi)\frac{\omega}{2} + \kappa^+ - \kappa^- - k\pi].$$

The eigenvalue $\lambda = 0$ proves to be simple in the case $a^+b^- + a^-b^+ \neq 0$ and the corresponding eigenfunction is identically 1. If $a^+b^- + a^-b^+ = 0$ then the multiplicity of the eigenvalue $\lambda = 0$ is equal to 2; there is the eigenfunction 1 and the associated function $a^+\omega/ib^+$. The Green formula for $\Upsilon(\lambda)$ has the form

$$\int_{-\alpha}^{\alpha} (u_{\omega\omega} - \lambda^2 u)\overline{v}\, d\omega - \int_{-\alpha}^{\alpha} u\overline{(v_{\omega\omega} - \overline{\lambda}^2 v)}\, d\omega =$$
$$= \{(\pm b^\pm u_\omega + i\lambda a^\pm u)\frac{\overline{v}}{\pm b^\pm} - \frac{u}{b^\pm}\overline{(\pm b^\pm v_\omega - i\overline{\lambda} a^\pm v)}\}\big|_{-\alpha}^{\alpha}.$$

Therefore,

$$\Upsilon^*(\lambda)v = \Upsilon(\overline{\lambda})^* v = \{v_{\omega\omega} - \lambda^2 v, (\pm b^\pm v_\omega - i\lambda a^\pm v)\big|_{\omega=\pm\alpha}\}.$$

To the nonzero eigenvalues $\lambda = -(\kappa^+ + \kappa^- + k\pi)/2i\alpha$ of the pencil $\lambda \mapsto \Upsilon^*(\lambda)$ there correspond the eigenfunctions $\psi(\omega) = c_k\phi(\omega)(c_k = \text{const})$. Under the condition $a^+b^- + a^-b^+ \neq 0$ the simple eigenvalue $\lambda = 0$ has the eigenfunction $c_0 = \text{const}$. In the case $a^+b^- + a^-b^+ = 0$ the eigenfunction c_0 and associated function $c_1 - c_0 a^+\omega/ib^+$ correspond to the eigenvalue $\lambda = 0$.

For $a^+b^- + a^-b^+ \neq 0$ the condition (1.30) takes the form

$$\int_{-\alpha}^{\alpha} \partial_\lambda L\phi\overline{\psi}\, d\omega + (\partial_\lambda B^+\phi)(\alpha)\overline{(T^+\psi)(\alpha)} - (\partial_\lambda B^-\phi)(-\alpha)\overline{(T^-\psi)(-\alpha)} = 1 \tag{1.48}$$

where $T^\pm\psi = (\mp b^\pm)^{-1}\psi$. The equality (1.48) is equivalent to

$$-2\lambda\int_{-\alpha}^{\alpha} \phi\overline{\psi}\, d\omega - i\frac{a^+}{b^+}\phi(\alpha)\overline{\psi(\alpha)} - i\frac{a^-}{b^-}\phi(-\alpha)\overline{\psi(-\alpha)} = 1\,.$$

In the case $\lambda \neq 0$, substituting for ϕ and ψ their expressions we obtain

$$\overline{c_k} i\left\{\frac{\kappa^+ + \kappa^- + k\pi}{\alpha}\int_{-\alpha}^{\alpha} \cos^2\frac{1}{2}[(\kappa^+ + \kappa^- + k\pi)\frac{\omega}{\alpha} - \kappa^- + \kappa^+ - k\pi]\, d\omega\right\} = 1\,.$$

It follows that $c_k = -i(\kappa^+ + \kappa^- + k\pi)^{-1}, k = \pm 1, \pm 2, \ldots$. If $\lambda = 0$ then $c_0 = -i(tg\,\kappa^+ + tg\,\kappa^-)^{-1}$.

Now let $a^+b^- + a^-b^+ = 0$ and $\lambda = 0$. The conditions (1.30) become

$$\begin{aligned}&\int_{-\alpha}^{\alpha} (\partial_\lambda L\phi^{(1)})\overline{\psi^{(0)}}\, d\omega + \frac{1}{2}\int_{-\alpha}^{\alpha} (\partial_\lambda^2 L\phi^{(0)})\overline{\psi^{(0)}}\, d\omega + \\ &(\partial_\lambda B^+\phi^{(1)})(\alpha)\overline{(T^+\psi^{(0)})(\alpha)} - (\partial_\lambda B^-\phi^{(1)})(-\alpha)\overline{T^-\psi^{(0)}(-\alpha)} = 1,\end{aligned} \tag{1.49}$$

$$\int_{-\alpha}^{\alpha}(\partial_\lambda L\phi^{(1)})\overline{\psi^{(1)}}\,d\omega+\frac{1}{2}\int_{-\alpha}^{\alpha}(\partial_\lambda^2 L\phi^{(1)})\overline{\psi^{(0)}}\,d\omega+\frac{1}{2}\int_{-\alpha}^{\alpha}(\partial_\lambda^2 L\phi^{(0)})\overline{\psi^{(1)}}\,d\omega+$$
$$+(\partial_\lambda B^+\phi^{(1)})(\alpha)(T^+\psi^{(1)})(\alpha)-\partial_\lambda B^-\phi^{(1)}(-\alpha)T^-\psi^{(1)}(-\alpha)=0. \tag{1.50}$$

We set $\phi^{(0)}=1,\psi^{(0)}=c_0,\phi^{(1)}=a^+\omega/ib^+,\psi^{(1)}=-ic_0a^+\omega/ib^++c_1$. As a result we have $c_0=(b^+)^2/\{2\alpha[(a^+)^2+(b^+)^2]\}$ and $c_1=0$. Thus, in the case $a^+b^-+a^-b^+=0$, to the eigenvalue $\lambda=0$ there correspond the eigenfunction $\psi^{(0)}=(b^+)^2/\{2\alpha[(a^+)^2+(b^+)^2]\}$ and the associated function $\psi^{(1)}=a^+b^+\omega/\{2\alpha[(a^+)^2+(b^+)^2]\}$ satisfying the orthogonality and normalization conditions (1.30).

We consider the operator

$$A: V^{2+s}_{p,\beta+s}(K)\to V^s_{p,\beta+s}(K)\times V^{s+1-1/p}_{p,\beta+s}(\partial K)$$

where β and p satisfy one of the inequalities

$$\frac{(k-1)\pi+\kappa^+-\kappa^-}{2\alpha}<2-\frac{2}{p}-\beta<\frac{k\pi+\kappa^++\kappa^-}{2\alpha}$$

while $0\notin((k-1)\pi+\kappa^++\kappa^-)(2\alpha)^{-1},(k\pi+\kappa^++\kappa^-)(2\alpha)^{-1})$. If the last condition is not fulfilled, then $\beta+2p^{-1}-2$ belongs to one of the intervals $(0,(\pi-\kappa^+-\kappa^-)(2\alpha)^{-1})$ and $((-\kappa^+-\kappa^-)(2\alpha)^{-1},0)$ for $\kappa^++\kappa^->0$ and to either $(0,(-\kappa^+-\kappa^-)(2\alpha)^{-1})$ or $((-\kappa^+-\kappa^--\pi)(2\alpha)^{-1},0)$ for $\kappa^++\kappa^-<0$.

Let $\mathcal{G}$ be the Green function for A. We consider the case $k\pi+\kappa^+-\kappa^->0$. For $2|x|<y$ we have

$$\mathcal{G}(x,y)=\sum_{q=k}^{N}\left(\frac{|x|}{|y|}\right)^{\frac{\kappa^++\kappa^-+q\pi}{2\alpha}}\frac{1}{\kappa^++\kappa^-+q\pi}\cos\frac{1}{2}[\frac{\kappa^++\kappa^-+q\pi}{2\alpha}\omega_x-\kappa^-+$$
$$+\kappa^+-q\pi]\cos\frac{1}{2}[\frac{\kappa^++\kappa^-+q\pi}{2\alpha}\omega_y-\kappa^-+\kappa^+-q\pi]+R_N(x,y) \tag{1.51}$$

where ω_x,ω_y are angular coordinates of x,y and the remainder R_N has the same order as the first term omitted.

If $k\pi+\kappa^++\kappa^-\le 0$ then the term corresponding to the eigenvalue $\lambda=0$ must be added to the right-hand side of (1.51). For $\kappa^++\kappa^-\neq 0$ this term is $(\operatorname{tg}\kappa^++\operatorname{tg}\kappa^-)^{-1}$ and for $\kappa^++\kappa^-=0$ it is

$$\frac{\cos^2\kappa^+}{2\alpha}\ln\frac{|x|}{|y|}-\frac{\sin 2\kappa^+}{4\alpha}(\omega_x+\omega_y).$$

For $2|y|<|x|$ and $\kappa^++\kappa^-+(k-1)\pi<0$ the representation of $\mathcal{G}(x,y)$ can be obtained by summation in (1.51) from $q=-N$ to $q=k+1$. If $\kappa^++\kappa^-+(k-1)\pi<0$ then the term corresponding to the eigenvalue $\lambda=0$ must be added to the right-hand side.

§2. Boundary Value Problems in Domains with Conical Points on the Boundary

2.1. Domain, Function Spaces, Operators. Fredholm Property. Let G be an open subset of R^n with compact closure $\overline{G}$ and boundary ∂G. Suppose that there exists a finite set $\mathcal{S} = \{x^1, \ldots, x^T\}$ of the points $x^\tau \in \partial G$ such that $\partial G \setminus \mathcal{S}$ is a smooth $(n-1)$-dimensional submanifold of R^n. We assume that every point $x^\tau \in \mathcal{S}$ has a neighborhood U^τ in R^n diffeomorphic to the open unit ball $B^n(x^\tau)$ with center at x^τ, while the image of $U^\tau \cap \overline{G}$ is $B^n(x^\tau) \times \overline{K^\tau}$ where K^τ is an open n-dimensional cone with vertex x^τ. The cone K^τ cuts out on the sphere $\partial B^n(x^\tau)$ a set Ω^τ with smooth boundary $\partial\Omega^\tau$. The points x^τ of the set $\mathcal{S}$ are called conical. From the definitions of the norms and admissible operators given below it follows that without loss of generality we can require the coincidence of $U^\tau \cap \overline{G}$ with $B^n(x^\tau) \cap \overline{K^\tau}$. We associate to the point x^τ a real number β^τ and denote by β the vector $(\beta^1, \ldots, \beta^T)$. Let $\zeta_\tau \in C^\infty(R^n)$, with $\operatorname{supp} \zeta_\tau \subset U^\tau$ and $0 \le \zeta_\tau \le 1$, while $\zeta_\tau = 1$ near $x^\tau, \tau = 1, \ldots, T$. Set $\zeta_0 = 1 - \zeta_1 - \ldots - \zeta_T$ and define the norm

$$||u; V_\beta^l(G)|| = (||\zeta_0 u; H^l(G)||^2 + \sum_{\tau=1}^{T} ||\zeta_\tau u; V_{\beta^\tau}^l(K^\tau)||^2)^{1/2} \tag{2.1}$$

where H^l is the Sobolev space and the norm $||\cdot; V_{\beta^\tau}^l(K^\tau)||$ is given by (1.12), $l = 0, 1, \ldots$. For $l = 1, 2, \ldots$ we denote by $V_\beta^{l-1/2}(\partial G)$ the space of traces on $\partial G \setminus \mathcal{S}$ of functions in $V_\beta^l(G)$. The embedding operators $V_\beta^l(G) \to V_{\beta-1}^{l-1}(G)$ (for $l \ge 1$) and $V_\beta^{l-1/2}(\partial G) \to V_{\beta-1}^{l-3/2}(\partial G)$ (for $l \ge 2$) are continuous. The embedding operators $V_\beta^l(G) \to V_\beta^{l-1}(G)$ (for $l \ge 1$) and $V_\beta^{l-1/2}(\partial G) \to V_\beta^{l-3/2}(\partial G)$ (for $l \ge 2$) are compact.

Let $\mathcal{L}$ and $\mathcal{B}$ be matrix differential operators in G of dimensions $k \times k$ and $m \times k$ with elements $\mathcal{L}_{hj}$ and $\mathcal{B}_{qj}$, $\operatorname{ord} \mathcal{L}_{hj} = s_h + t_j, \operatorname{ord} \mathcal{B}_{qj} = \sigma_q + t_j$, where s_h, t_j, σ_q are the same as in section 1.2. We suppose that the coefficients of the operators $\mathcal{L}_{hj}$ and $\mathcal{B}_{qj}$ belong to the class $C^\infty(\overline{G} \setminus \mathcal{S})$. We shall describe $\mathcal{L}_{hj}$ and $\mathcal{B}_{qj}$ near a conical point. We call a scalar differential operator $\mathcal{P}$ of order m admissible in a neighborhood U^τ of the conical point x^τ if in this neighborhood

$$\mathcal{P}(x, D_x) = \sum_{|\alpha| \le m} p_\alpha(x) D_x^\alpha, \quad p_\alpha(x) = r^{|\alpha|-m} p_\alpha^0(r, \omega)$$

where (r, ω) are local spherical coordinates with origin at x^τ and the functions

$$[0, \delta] \times \overline{\Omega^\tau} \ni (r, \omega) \mapsto r^\mu D_r^\mu D_\omega^\gamma p_r^0(r, \omega)$$

are continuous for $\mu, |\gamma| = 0, 1, \ldots$. The principal part $\mathcal{P}^0$ of the operator $\mathcal{P}$ at the point x^τ is the operator in the cone K^τ obtained from $\mathcal{P}$ by replacing the coefficients $p_\alpha(x)$ by $r^{|\alpha|-m}p^0_\alpha(0,\omega)$. We assume that the operators $\mathcal{L}_{hj}$ and $\mathcal{B}_{qj}$ are admissible near every conical point.

We introduce the spaces $\mathcal{D}^l_\beta V(G)$ and $\mathcal{R}^l_\beta V(G)$ by (1.21) with the change of $V^s_\beta(K)$ for $V^s_\beta(G)$, etc. It is clear that the operator

$$\mathcal{A} = \{\mathcal{L}(x, D_x), \mathcal{B}(x, D_x)\} : \mathcal{D}^l_\beta V(G) \to \mathcal{R}^l_\beta V(G) \tag{2.2}$$

is continuous.

The operator $\{\mathcal{L}, \mathcal{B}\}$ is called elliptic in G if the following conditions are satisfied:

1) this operator is elliptic at every point of the set $\overline{G} \setminus \mathcal{S}$;

2) the operator $\{\mathcal{L}^0, \mathcal{B}^0\}_\tau$ in K^τ consisting of the principal parts $\mathcal{L}$ and $\mathcal{B}$ at the point x^τ is elliptic on $\overline{K}^\tau \setminus \{x^\tau\}, \tau = 1, \dots, T$.

It is easy to show that the elements of $\mathcal{L}^0$ and $\mathcal{B}^0$ are model, i.e. admit representations of the form (1.19). As in (1.27) we assign to $\{\mathcal{L}^0, \mathcal{B}^0\}_\tau$ the (elliptic) operator pencil $\lambda \mapsto \Upsilon^\tau(\lambda) = \{L^\tau(\lambda), B^\tau(\lambda)\}$ in the domain Ω^τ.

Theorem 2.1. *Let $\{\mathcal{L}(x, D_x), \mathcal{B}(x, D_x)\}$ be an elliptic operator in the domain G. Then the operator (2.2) is Fredholm if and only if the line $R + i(\beta^\tau - l + n/2)$ contains no eigenvalues of the pencil $\lambda \mapsto \Upsilon^\tau(\lambda), \tau = 1, \dots, T$.*

The analogous assertions are valid for the spaces $\mathcal{D}^l_{p,\beta}V(G)$ and $\Lambda^{l,\alpha}_\beta(G)$; their formulations are left to the reader (see Theorems 1.5 and 1.6).

The proof of these theorems follows the usual scheme in the theory of elliptic problems (frozen coefficients, regularizers). The conical points can be included in the scheme by means of Theorems 1.4-1.6.

2.2. Asymptotics of Solutions near the Conical Points. To obtain the asymptotic formulas we need some further conditions (in comparison with Sect.2.1) on the coefficients of the problem. We choose a conical point x^τ and for ease of notation assume that $x^\tau = 0$. A scalar differential operator

$$\mathcal{P}(x, D_x) = \sum_{|\alpha| \le m} p_\alpha(x) D^\alpha_x$$

is called δ-admissible in a neighborhood $\mathcal{U}^\tau$ of the point x^τ if the coefficients p_α can be represented as

$$p_\alpha(x) = r^{|\alpha|-m} p^0_\alpha(\omega) + r^{|\alpha|-m+\delta} p^1_\alpha(r, \omega)$$

in $\mathcal{U}^\tau$, where δ is a positive number, $p^0_\alpha \in C^\infty(\overline{\Omega})$, and the functions

$$[0, \epsilon] \times \overline{\Omega} \ni (r, \omega) \mapsto r^\mu D^\mu_r D^\nu_\omega p^\alpha_\alpha(r, \omega)$$

are continuous for $\mu, |\nu| = 0, 1, \dots$.

Theorem 2.2. *Suppose that the elements of the matrices $\mathcal{L}(x, D_x)$ and $\mathcal{B}(x, D_x)$ are δ-admissible in a neighborhood $\mathcal{U}^\tau$. Let β^τ and γ^τ be real numbers* *such that* $0 < \beta^\tau - \gamma^\tau < \delta$. *We denote by* η *a function in* $C_c^\infty(\mathcal{U}^\tau)$ *equal to one near the point* x^τ. *Let* u *be a solution of the problem*

$$\begin{aligned} \mathcal{L}(x, D_x)u(x) &= f(x), \quad u \in G, \\ \mathcal{B}(x, D_x)u(x) &= g(x), \quad x \in \partial G \setminus \mathcal{S} \end{aligned} \tag{2.3}$$

subject to $\eta u \in \mathcal{D}^l_{\gamma^\tau} V(K^\tau)$. *We assume that* $\eta\{f, g\} \in \mathcal{R}^l_{\gamma^\tau} V(K^\tau)$. *Finally, we suppose that the lines* $R + i(\beta^\tau - l + n/2)$ *and* $R + i(\gamma^\tau - l + n/2)$ *contain no eigenvalues of the pencil* Υ^τ, *while the strip* $\gamma^\tau - l + n/2 < Im\,\lambda < \beta^\tau - l + n/2$ *contains the eigenvalues* $\lambda_1, \dots, \lambda_N$ *of* Υ^τ.

Then in the neighborhood $\mathcal{U}^\tau$ *the formula*

$$u(r, \omega) = \sum_{\nu=1}^{N} \sum_{j=1}^{J_\nu} \sum_{k=0}^{\kappa_{j,\nu}-1} c^{(k,j)}_{\nu,\tau} u^{(k,j)}_{\nu,\tau}(r, \omega) + v(r, \omega) \tag{2.4}$$

holds; here $u^{(k,j)}_{\nu,\tau}$ *are defined by (1.35) with the Jordan chains of the pencil* Υ^τ *corresponding to the eigenvalue* λ^τ, $c^{(k,j)}_{\nu,\tau}$ *are constant coefficients, and* $\eta v \in \mathcal{D}^l_{\gamma^\tau} V(K^\tau)$.

Proof W e have

$$\begin{aligned} \mathcal{L}(x, D_x)\eta u &= \eta f + [\mathcal{L}, \eta]u \ \text{ in } \ G, \\ \mathcal{B}(x, D_x)\eta u &= \eta g + [\mathcal{B}, \eta]u \ \text{ on } \ \partial G \setminus S. \end{aligned}$$

We can regard these relations as equations in the cone K^τ. We rewrite them in the form

$$\mathcal{L}^0(x, D_x)\eta u = \eta f + [\mathcal{L}, \eta]u + (\mathcal{L}^0 - \mathcal{L})\eta u \equiv f' \ \text{ in } \ K^\tau, \tag{2.5}$$

$$\mathcal{B}^0(x, D_x)\eta u = \eta g + [\mathcal{B}, \eta]u + (\mathcal{B}^0 - \mathcal{B})\eta u \equiv g' \ \text{ on } \ \partial K^\tau \setminus O \tag{2.6}$$

where $\mathcal{L}^0, \mathcal{B}^0$ are the principal parts of $\mathcal{L}, \mathcal{B}$ at the point x^τ. The coefficients of the commutators $[\mathcal{L}, \eta]$ and $[\mathcal{B}, \eta]$ vanish near the vertex of the cone. Therefore, $\{[\mathcal{L}, \eta]u, [\mathcal{B}, \eta]u\} \in \mathcal{R}^l_{\gamma^\tau} V(K^\tau)$. Since the elements of the matrices $\mathcal{L}$ and $\mathcal{B}$ are δ-admissible, we have the inclusion $\{(\mathcal{L}^0 - \mathcal{L})\eta u, (\mathcal{B}^0 - \mathcal{B})\eta u\} \in \mathcal{R}^l_{\gamma^\tau} V(K^\tau)$. Finally, $\eta\{f, g\} \in \mathcal{R}^\gamma_{\gamma^\tau} V(K^\tau)$ by assumption. Thus, $\{f', g'\} \in \mathcal{R}^l_{\gamma^\tau} V(K^\tau)$. Now the formula (2.4) follows from Theorem 1.11.

Theorem 1.12 allows us to represent the coefficients $c^{(k,j)}_{\nu,\tau}$ in (2.4) as functionals on $\{f', g'\}$. We emphasize that f' and g' depend on the solution u. In what follows the formulas for $c^{(k,j)}_{\nu,\tau}$ will be given where $c^{(k,j)}_{\nu,\tau}$ are functionals on the vectors $\{f, g\}$ forming the right-hand side of the problem (2.3). These functionals depend on the data of the problem in the whole domain G.

2.3. Properties of the Kernel and Cokernel of the Problem. Index. Let $S' = \{x^1, \dots, x^{T_1}\}$ be a subset of the set S of conical points, $T_1 \le T$. Denote by β and γ the vectors $(\beta^1, \dots, \beta^{T_1}, \beta^{T_1+1}, \dots, \beta^T)$ and $(\gamma^1, \dots, \gamma^{T_1},$

$\beta^{T_1+1}, \dots, \beta^{T})$ where $0 < \beta^\tau - \gamma^\tau < \delta$ and $\tau = 1, \dots, T_1$. Suppose that the matrices $\mathcal{L}$ and $\mathcal{B}$ have δ-admissible elements in a neighborhood of S'.

Proposition 2.3. *Let the line $R + i(\gamma^\tau - l + n/2)$ contains no eigenvalues of the pencil $\Upsilon^\tau(\tau = 1, \dots, T_1)$, while the strip $\gamma^\tau - l + n/2 < Im\,\lambda < \beta^\tau - l + n/2$ contains the eigenvalues $\lambda_1^\tau, \dots, \lambda_{N^\tau}^\tau$ of Υ^τ. Let*

$$\kappa = \sum_{\tau=1}^{T_1} \sum_{\nu=1}^{N^\tau} \sum_{j=1}^{J_\nu^\tau} \kappa_{j\nu}^\tau \tag{2.7}$$

be the sum of total multiplicities of $\lambda_1^{\tau_1}, \dots, \lambda_{N^\tau}^\tau, \tau = 1, \dots, T_1$. Suppose that the line $R + i(\beta^\tau - l + n/2)$ is free from the spectrum of $\Upsilon^\tau, \tau = 1, \dots, T$. Then the homogeneous problem (2.3) can have no more than κ solutions in $\mathcal{D}_\beta^l V(G)$ linearly independent modulo the space $\mathcal{D}_\gamma^l V(G)$.

Let $u_{\gamma,\tau}^{(j,k)}$ be a power solution of the homogeneous problem with the operator $\{\mathcal{L}^0, \mathcal{B}^0\}_\tau$ in K^τ (see (2.4)). Denote by η_τ a function in $C^\infty(\overline{G})$ such that $\operatorname{supp} \eta_\tau \subset \mathcal{U}^\tau$ and $\eta_\tau = 1$ near x^τ. We extend each of the functions $\eta_\tau u_{\nu,\tau}^{(j,k)}$ by zero to the domain G. We order the set of the functions $\{\eta_\tau u_{\nu,\tau}^{(j,k)}; \tau = 1, \dots, T_1; \nu = 1, \dots, N^\tau; j = 1, \dots, J_\nu^\tau; k = 0, \dots, \kappa_{j\nu}^\tau - 1\}$ arbitrarily and denote elements of the set by $U_1, \dots, U_\kappa$. By Theorem 2.2 any solution $Z \in \mathcal{D}_\beta^l V(G)$ of the homogeneous problem (2.3) satisfies the congruence

$$Z \equiv \sum_{j=1}^{\kappa} c_j U_j \pmod{\mathcal{D}_\gamma^l V(G)}, \ c_j = \text{const.}, \tag{2.8}$$

which implies Proposition 2.3.

Introduce the vectors $U = (U_1, \dots, U_\kappa)$ and $Z = (Z_1, \dots, Z_d)$, where $0 \le d \le \kappa$. The components of Z form a maximal collection of solutions of the homogeneous problem (2.3) in $\mathcal{D}_\beta^l V(G)$ linearly independent modulo $\mathcal{D}_\gamma^l V(G)$ (a basis modulo $\mathcal{D}_\gamma^l V(G)$). According to (2.8), $Z \equiv CU (\operatorname{mod} \mathcal{D}_\gamma^l V(G))$ where C is a $d \times \kappa$ matrix with rank equal to d. It may be assumed that $C = (\mathcal{D}_1, \mathcal{D}_2)$ while $\mathcal{D}_1$ is a nondegenerate $d \times d$ matrix. Therefore,

$$\mathcal{D}_1^{-1} Z \equiv (\mathbf{1}, \mathcal{D}_1^{-1}\mathcal{D}_2) U \pmod{\mathcal{D}_\gamma^l V(G)}.$$

We suppose with no loss of generality that

$$Z_j \equiv (U_j + \sum_{k=d+1}^{\kappa} c_{jk} U_k) \pmod{\mathcal{D}_\gamma^l V(G)}, j = 1, \dots, d. \tag{2.9}$$

Any basis $\{Z_1, \dots, Z_d\}$ modulo $\mathcal{D}_\gamma^l V(G)$ in the space of solutions in $\mathcal{D}_\beta^l V(G)$ of the homogeneous problem (2.3) is called canonical if it admits a representation of the form (2.9).

We now suppose that the Green formula (1.28) (with K replaced by G) holds for the operator of the problem (2.3). (In (Maz'ya, Plamenevskij

(1975b)) and (Maz'ya, Plamenevskij (1977a)) the corresponding questions were discussed with no Green's formula.) We also suppose that the operator $\{\mathcal{L}(x, D_x)^*, \mathcal{Q}(x, D_x)\}$ of the adjoint problem possesses all the properties of the operator $\{\mathcal{L}, \mathcal{B}\}$. Denote by $V_1, \dots, V_\kappa$ the set $\{\eta_\tau v_{\nu,\tau}^{(j,k)}\}$ where $v_{\nu,\tau}^{(j,k)}$ are defined in the same way as in Theorem 1.12 for the model operator in K^τ generated by the adjoint problem. By definition, the sets $\{U_1, \dots, U_\kappa\}$ and $\{V_1, \dots, V_\kappa\}$ are ordered compatibly if $U_h = \eta_\tau u_{\nu,\tau}^{(k,j)}$ and $V_h = \eta_\tau v_{\nu,\tau}^{(\kappa_j, \zeta-k-1,j)}$, while the Jordan chains in $u_{\nu,\tau}^{(k,j)}$ and $v_{\nu,\tau}^{(p,l)}$ are subject to the condition (1.30).

Proposition 2.4. *Suppose the hypotheses of Proposition 2.3 are satisfied and the sets $\{U_1, \dots, U_\kappa\}$ and $\{V_1, \dots, V_\kappa\}$ are ordered compatibly. Further suppose that $\{Z_1, \dots, Z_d\}$ is a canonical basis modulo $\mathcal{D}_\gamma^l V(G)$ in the space of solutions in $\mathcal{D}_\beta^l V(G)$ of the homogeneous problem (2.3).*

Then there exist solutions $\phi_h, h = d+1, \dots, \kappa$, of the problem

$$\mathcal{L}(x, D_x)^* v = 0 \text{ in } G, \quad \mathcal{Q}(x, D_x) v = 0 \text{ on } \partial G \setminus \mathcal{S} \tag{2.10}$$

satisfying the congruences

$$\phi_h \equiv (V_h - \sum_{j=1}^{d} \overline{c_{jh}} V_j) \ (\mathrm{mod} \ \prod_{q=1}^{k} V_{-\beta+2l}^{l+s_q}(G))$$

where c_{jh} are the coefficients in (2.9) and $-\beta + 2l = (-\beta^1 + 2l, \dots, -\beta^T + 2l)$, l being a large number. The functions $\phi_{d+1}, \dots, \phi_\kappa$ form a basis modulo $\prod V_{-\beta+2l}^{l+s_q}(G)$ in the space of solutions in $\prod V_{-\gamma+2l}^{l+s_q}(G)$ of the problem (2.10).

We come now to describe the increment of the index of the boundary value problem (2.3) resulting from a variation of the weight exponent β. Recall that the index of a Fredholm operator $\mathcal{A}$ is defined by $\operatorname{Ind} \mathcal{A} = \dim \ker \mathcal{A} - \dim \operatorname{coker} \mathcal{A} = \dim \ker \mathcal{A} - \dim \ker \mathcal{A}^*$.

Suppose that the lines $R + i(\gamma^\tau - l + n/2)$ $(\tau = 1, \dots, T_1)$ and $R + i(\beta^\tau - l + n/2)$ $(\tau = 1, \dots, T)$ contain no eigenvalues of the pencil $\varUpsilon^\tau$. Let κ be the same number as in (2.7). We consider the operators $\mathcal{A}_\gamma \equiv \{\mathcal{L}, \mathcal{B}\} : \mathcal{D}_\beta^l V(G) \to \mathcal{R}_\gamma^l V(G)$ and $\mathcal{A}_\beta \equiv \{\mathcal{L}, \mathcal{B}\} : \mathcal{D}_\beta^l V(G) \to \mathcal{R}_\beta^l V(G)$. Thus, $\mathcal{A}_\gamma$ and $\mathcal{A}_\beta$ are the operators of the same boundary value problem (2.3) acting on the different spaces. By Theorem 2.1 each of them is Fredholm.

Theorem 2.5. $\operatorname{Ind} \mathcal{A}_\beta = \operatorname{Ind} \mathcal{A}_\gamma + \kappa$.

We outline the proof. By Proposition 2.3, $\dim \ker \mathcal{A}_\beta = \dim \ker \mathcal{A}_\gamma + d$ with some $d, 0 \le d \le \kappa$. From Proposition 2.4 it follows that $\dim \ker \mathcal{A}_\beta^* = \dim \ker \mathcal{A}_\gamma^* - (\kappa - d)$. Therefore $\operatorname{Ind} \mathcal{A}_\beta = \dim \ker \mathcal{A}_\beta - \dim \ker \mathcal{A}_\beta^* = \dim \ker \mathcal{A}_\gamma - \dim \ker \mathcal{A}_\gamma^* + \kappa = \operatorname{Ind} \mathcal{A}_\gamma + \kappa$.

2.4. Formulas for the Coefficients in (2.4). We recall that $\{U_1, \dots, U_\kappa\}$ stands for the set of functions $\{\eta_\tau u_{\nu,\tau}^{(k,j)}\}$. Therefore the representation (2.4) can be rewritten in the form

$$u \equiv (\sum_{j=1}^{\kappa} c_j U_j) \,(\mathrm{mod}\ \mathcal{D}^l_\gamma V(G)).$$

Thus we shall discuss the formulas for the coefficients c_j.

Theorem 2.6. *Suppose that the hypotheses of Proposition 2.3 are satisfied. Let $\{Z_1, \dots, Z_d\}$ be a canonical basis modulo $\mathcal{D}^l_\gamma V(G)$ in the space of solutions in $\mathcal{D}^l_\beta V(G)$ of the homogeneous problem (2.3) subject to the congruences (2.9). Further, we assume that the problem (2.3) with right-hand side $\{f, g\} \in \mathcal{R}^l_\gamma V(G)$ is solvable in the space $\mathcal{D}^l_\beta V(G)$.*

Then for any constants $c_1, \dots, c_d$ there exists a solution $u \in \mathcal{D}^l_\beta V(G)$ of the problem (2.3) satisfying

$$u \equiv (\sum_{j=1}^{d} c_j U_j + \sum_{k=d+1}^{\kappa} b_k U_k) \quad (\mathrm{mod}\ \mathcal{D}^l_\gamma V(G)).$$

The constants b_k are defined by

$$b_k = (f, i\phi_k)_G + (g, i\mathcal{T}\phi_k)_{\partial G} + \sum_{h=1}^{d} c_h c_{hk}, \quad d+1 \le k \le \kappa,$$

where c_{hj} are the coefficients in (2.9) and $\phi_{d+1}, \dots, \phi_\kappa$ the solutions of the homogeneous problem (2.10) indicated in Proposition (2.4).

If the coefficients and right-hand side of the problem (2.3) admit expansions in asymptotic series of the form $\sum r^{i\mu_q} f_q(\ln r, \omega), q \ge 0$, where $\operatorname{Im}\mu_q \ge \operatorname{Im}\mu_{q+1}, \operatorname{Im}\mu_q \to -\infty$ and $z \mapsto f_q(z, \omega)$ are polynomials whose coefficients are smooth functions in ω, then the solutions of the problem can be expanded in a series of the same type. Another way of making the asymptotics more precise, and obtaining a remainder with any desired decrease near the vertex of the cone, will be indicated in Theorem 2.9.

2.5. On the Asymptotics of Solutions near the Conical Points Again. The results of this section will be of use in Sect.2.5 where we describe the fundamental solutions. As in Sect.1.5, we consider a scalar equation of order $2m$ with normal boundary conditions. Let us write the problem in the form

$$\begin{aligned}&\mathcal{L}(x, D_x)u(x) = f(x),\ x \in G;\ \mathcal{B}_j(x, D_x)u(x) = f_j(x),\ x \in \partial G \setminus O,\\ &j = 1, \dots, m.\end{aligned} \tag{2.11}$$

For the sake of simplicity we suppose that there is only one conical point O on ∂G. Assume that the coefficients of $\mathcal{L}$ and $\mathcal{B}_j$ are δ-admissible near O. Also, we require the problem (2.11) to be uniquely solvable in $V^{2m}_{p,\beta}(G)$ for any $f \in V^0_{p,\beta}(\partial G)$ and $f_j \in V^{2m-m_j-1/p}_{p,\beta}(\partial G)$ with some p, β.

Proposition 2.7. *1) Let λ_μ be an eigenvalue of the pencil (1.37) (corresponding to the principal part of the operator (2.11) at the point O), and let*

$Im\,\lambda_\mu > \beta + np^{-1}$. Then there exist κ_μ solutions $\mathcal{U}_\mu^{(k,j)}$ $(k = 0,\dots,\kappa_{j,\mu}-1; j = 1,\dots,J_\mu; \kappa_\mu = \kappa_{1,\mu} + \dots + \kappa_{J_\mu,\mu})$ that are smooth on $\overline{G}\setminus O$ and admit the representations

$$\mathcal{U}_\mu^{(k,j)}(x) = u_\mu^{(k,j)}(x) + R_\mu^{(k,j)}(x) \tag{2.12}$$

near the point O; here

$$u_\mu^{(k,j)}(x) \equiv u_\mu^{(k,j)}(r,\omega) = r^{i\lambda_\mu}\sum_{s=0}^{k}\frac{1}{s!}(i\ln r)^s\phi_\nu^{(k-s,j)}(\omega). \tag{2.13}$$

(To obtain (2.13) we take into account the difference in the definitions of the pencil (1.27) and (1.37) and adapt (1.35) for the case under consideration.) The inequalities

$$|D_x^\gamma R_\mu^{(k,j)}(x)| \le c_\gamma |x|^{-Im\,\lambda_\mu - |\gamma| + \epsilon} \tag{2.14}$$

hold for any multi-index γ and small positive ϵ.

2) Let λ_μ be an eigenvalue of the pencil (1.37) such that $\operatorname{Im}\lambda_\mu < \beta + np^{-1} - 2m$. Then there exist κ_μ solutions $\mathcal{V}_\mu^{(k,j)}$ of the homogeneous problem with operator $\{\mathcal{L}(x,D_x)^*, \mathcal{Q}_1(x,D_x),\dots,\mathcal{Q}_m(x,D_x)\}$ (adjoint to (2.11) with respect to the Green formula) that are smooth on $\overline{G}\setminus O$ and admit the representations

$$\mathcal{V}_\mu^{(k,j)}(x) = v_\mu^{(k,j)}(x) + T_\mu^{(k,j)}(x) \tag{2.15}$$

where

$$v_\mu^{(k,j)}(x) \equiv v_\mu^{(k,j)}(r,\omega) = r^{i\overline{\lambda}_\mu - n + 2m}\sum_{s=0}^{k}\frac{1}{s!}(i\ln r)^s\psi_\mu^{(k-s,j)}(\omega) \tag{2.16}$$

and

$$|D_x^\gamma T_\mu^{(k,j)}(x)| \le c_\gamma |x|^{Im\,\lambda_\mu - n + 2m - |\gamma| + \epsilon}. \tag{2.17}$$

We are now in a position to apply Theorems 2.2 and 2.6 to obtain the following assertion.

Theorem 2.8. *1) Let λ_μ be an eigenvalue of the pencil (1.37) such that $\operatorname{Im}\lambda_\mu < \beta + np^{-1} - 2m$. Denote by $\lambda_\nu, \nu = 1,\dots,M$, the eigenvalues of this pencil satisfying $\operatorname{Im}\lambda_\mu \le \operatorname{Im}\lambda_\nu < \beta + np^{-1} - 2m$. Suppose that $F_\mu^{(k,\sigma)}$ are functions in $C_c^\infty(\overline{G}\setminus O)$ subject to*

$$(F_\mu^{(k,\sigma)}, \mathcal{V}_\nu^{(\kappa_{\zeta,\nu}-q-1,\zeta)})_G = \delta_{\mu,\nu}\delta_{\sigma,\zeta}\delta_{k,q}$$

where $\nu = 1,\dots,M; \sigma,\zeta = 1,\dots,J_\nu; k,q = 0,\dots,\kappa_{\zeta,\nu}-1$. Then near the point O the solution $P_\mu^{(k,\sigma)} \in V_{p,\beta}^{2m}(G)$ of the problem (2.1) for $f = F_\mu^{(k,\sigma)}, f_j = 0$, admits the representation $P_\mu^{(k,\sigma)}(x) = u_\mu^{(k,\sigma)}(x) + R_\mu^{(k,\sigma)}(x)$, where $u_\mu^{(k,\sigma)}$ is defined by (2.13) and $R_\mu^{(k,\sigma)}$ satisfies (2.14).

2) Let λ_μ *be an eigenvalue of the pencil (1.37),* $\operatorname{Im}\lambda_\mu > \beta + np^{-1} - 2m$ *and* $\lambda_1, \dots, \lambda_M$ *the eigenvalues satisfying* $\operatorname{Im}\lambda_\mu \geq \operatorname{Im} d_\nu > \beta + np^{-1} - 2m$. *Denote by* $\Phi_\mu^{(k,\sigma)}$ *functions in* $C_c^\infty(\overline{G} \setminus O)$ *subject to*

$$(\Phi_\mu^{(k,j)}, \mathcal{U}_\nu^{(\kappa_{\zeta,\nu}-q-1,\zeta)})_G = \delta_{\mu,\nu}\delta_{\sigma,\zeta}\delta_{k,q}.$$

Then in a neighborhood of the point O *the solution* $S_\mu^{(k,\sigma)} \in V^{2m}_{p',-\beta+2m}(G)$ *of the problem*

$$\mathcal{L}(x, D_x)^* S_\mu^{(k,\sigma)} = \Phi_\mu^{(k,\sigma)}, \quad \mathcal{Q}_j S_\mu^{(k,\sigma)} = 0, \ 1 \leq j \leq m$$

admits the representation $S_\mu^{(k,\sigma)} = v_\mu^{(k,\sigma)}(x) + T_\mu^{(\sigma,k)}(x)$ *where* $v_\mu^{(k,\sigma)}$ *is the same function as in (2.16) and* $T_\mu^{(\sigma,k)}$ *satisfies (2.17).*

Let us enumerate eigenvalues of the pencil (1.37) taking into account their multiplicities. The eigenvalues lying above the line $R+i(\beta+np^{-1}-2m)$ are allocated negative numbers in such a way that $\operatorname{Im}\lambda_\mu > \operatorname{Im}\lambda_\nu$ implies $|\mu| > |\nu|$. The ordering of the eigenvalues with equal imaginary parts is arbitrary. The eigenvalues lying below $R+i(\beta+np^{-1}-2m)$ are allocated nonnegative numbers in decreasing order of their imaginary parts. Thus, to every eigenvalue with number $j < 0$ there corresponds only one function of the form $\mathcal{U}_\mu^{(k,\sigma)}$ which will be denoted by $\mathcal{U}_j$. We adjust the notation $P_\mu^{(k,\sigma)}$ similarly. If $\mathcal{U}_\mu^{(k,\sigma)}$ and $P_\mu^{(k,\sigma)}$ correspond to a number q then $S_\mu^{(\kappa_{\sigma,\mu}-k-1,\sigma)}$ and $\mathcal{V}_\mu^{(\kappa_{\sigma,\mu}-k-1,\sigma)}$ are denoted by the same number q. The functions $F_\mu^{(k,\sigma)}$ and $\Phi_\mu^{(k,\sigma)}$ are denoted by the same numbers as $P_\mu^{(k,\sigma)}$ and $S_\mu^{(k,\sigma)}$.

The following theorem gives the asymptotic representation for a solution of the problem (2.11) in terms of P_q (which was mentioned at the end of Sect.2.3).

Theorem 2.9 *Let* $f \in V^0_{p,\beta'}(G)$, $f_j \in V^{2m-m_j-p^{-1}}_{p,\beta'}(\partial G)$ *and* $\beta' < \beta$. *Suppose that the line* $R + i(\beta' + np^{-1} - 2m)$ *is free from the spectrum of the pencil (1.37) while the strip* $\beta' + np^{-1} - 2m < Im\,\lambda < \beta + np^{-1} - 2m$ *contains the eigenvalues* $\lambda_0, \lambda_1, \dots, \lambda_N$. *Then a solution* $u \in V^{2m}_{p,\beta}(G)$ *of the problem (2.11) admits the representation*

$$u = \sum_{q=0}^{N} c_q P_q + R_N, \quad R_N \in V^{2m}_{p,\beta'}(G). \tag{2.18}$$

Proof W e choose a number β'' such that $\beta'' < \beta$ and the strip $\beta'' + np^{-1} - 2m \leq \operatorname{Im}\lambda < \beta+np^{-1}-2m$ contains no eigenvalues of the pencil (1.37). Then the solution u belongs to $V^{2m}_{p,\beta''}(G)$ as well. If, for some positive δ, there are no eigenvalues of the pencil on the line $R+i(\beta''-\delta/2+np^{-1}-2m)$ while the strip $\beta''-\delta/2+np^{-1}-2m < \operatorname{Im}\lambda < \beta''+np^{-1}-2m$ contains the eigenvalues $\lambda_0, \dots, \lambda_M$ then by Theorems 2.2 and 2.6

$$v = (u - \sum_{q=0}^{M} c_q P_q) \in V^{2m}_{p,\beta''-\delta/2}(G).$$

Since $\mathcal{L}v \in V^0_{p,\beta'}(G)$ and $\mathcal{B}_j v \in V^{2m-m_j-p^{-1}}_{p,\beta'}(\partial G)$, we can apply the same reasoning to the function v, while $\beta'' - \delta/2$ plays the role of β. We proceed to obtain (2.18) in finitely many steps.

Theorem 2.10. *The coefficients c_q in (2.18) are defined by*

$$c_q = -i \sum_{r=0}^{q} \overline{d}_{q,r}[(f, \mathcal{V}_r)_G + \sum_{j=1}^{m}(f_j, \mathcal{T}_j \mathcal{V}_r)_{\partial G}]$$

where J_j are operators in the Green formula (compare to the Green formula at the beginning of Sect.1.5), $d_{qq} = 1$, and the numbers $d_{q,r}$ are calculated by the recursion relation

$$d_{q,r} = -i \sum_{p=r}^{q-1} (F_p, \mathcal{V}_q)_G d_{p,r}, \quad r < q.$$

2.6. Asymptotics of the Fundamental Solutions of the Problem (2.11) near a Conical Point. We follow the hypotheses and notation of Sect.2.4.

Theorem 2.11. *1) There exists a unique solution $\mathcal{G}(x, y)$ of the boundary value problem*

$$\mathcal{L}(x, D_x)\mathcal{G}(x, y) = \delta(x, y), \; x, y \in G,$$
$$\mathcal{B}_j(x, D_x)\mathcal{G}(x, y) = 0, \; x \in \partial G \setminus O, \; y \in G$$

such that the function $x \mapsto \mathcal{G}(x, y)$ belongs to $V^{2m+s}_{p,\beta+s}(G)$ for any fixed $y \in G$; here η is a smooth function on $G \times G$ equal to zero in a neighborhood of the diagonal, and $s = 0, 1, \ldots$

2) The function $(x, y) \mapsto \mathcal{G}(x, y)$ is smooth for $x, y \in \overline{G} \setminus O, x \neq y$.

3) G is a unique solution of the problem

$$\mathcal{L}(y, D_y)^* \overline{\mathcal{G}(x, y)} = \delta(x - y), \quad y \in G,$$
$$\mathcal{Q}_j(y, D_y)\overline{\mathcal{G}(x, y)} = 0, \quad 1 \le j \le m, \; y \in \partial G \setminus O$$

such that the function $y \mapsto \eta(x, y)\mathcal{G}(x, y)$ belongs to $V^{2m+s}_{p',-\beta+2m+s}(G)$, and $s = 0, 1, \ldots$

The formulas similar to (1.42) and (1.45) are valid.

Theorem 2.12. *Let $\mathcal{G}$ be the Green function in Theorem 2.11. Then the following assertions are valid:*

1) For $2|x| < |y|$

$$\mathcal{G}(x,y) = -i\sum_{q=0}^{N}\sum_{r=0}^{q}\overline{d}_{q,r}\overline{\mathcal{V}_r(y)}P_q(x) + R_{N+1}(x,y) \tag{2.19}$$

where N *is an integer such that* $Im\,\lambda > Im\,\lambda_{N+1}$, *and*

$$|D_x^\gamma D_y^\delta R_N(x,y)| \le c_{\gamma\delta}\frac{|x|^{-Im\,\lambda_{N+1}-|\gamma|}}{|y|^{-Im\,\lambda_{N+1}+n-2m+|\delta|}}(\ln\frac{|y|}{|x|})^{\mu_{N+1}}\,, \tag{2.20}$$

where γ, δ *are any multi-indices and* μ_{N+1} *is the largest of the lengths of the Jordan chains corresponding to the eigenvalues lying on the line* $R+i\mathrm{Im}\,\lambda_{N+1}$.

2) For $2|y| < |x|$

$$\mathcal{G}(x,y) = i\sum_{q=1}^{N}\sum_{r=1}^{q} l_{q,r}\mathcal{U}_{-r}(x)S_{-q}(y) + R_{-N-1}(x) \tag{2.21}$$

where N *is an integer such that* $Im\,\lambda_{-N} < Im\,\lambda_{-N-1}, l_{qq} = 1$ *and*

$$l_{q,r} = -\sum_{p=r}^{q}(\Phi_{-p},\mathcal{U}_{-q})l_{p,r},\ r < q.$$

The remainder R_{-N-1} *satisfies (2.20) with* $N+1$ *replaced by* $-N-1$.

If we restrict attention to the first terms in (2.19) and (2.21) corresponding to the eigenvalues in a sufficient narrow strip, then we can obtain more explicit asymptotic formulas.

Corollary 2.13. *Let* $\lambda_1^-,\dots,\lambda_N^-$ *be the eigenvalues of the pencil (1.37) in the strip* $\beta'+np^{-1}-2m < Im\,\lambda < \beta''+np^{-1}-2m$ *numbered in decreasing order of their imaginary parts. We suppose that* $\beta' < \beta'' \le \beta$ *and* $\beta''-\beta' < \delta$. *(Recall that the coefficients of* $\mathcal{L}(x,D_x)$ *and* $\mathcal{B}(x,D_x)$ *are* δ*-admissible near the point* O.*) Assume that the line* $R+i(\beta'+np^{-1}-2m)$ *and the strip* $\beta''+np^{-1}-2m \le \mathrm{Im}\,\lambda < \beta+np^{-1}-2m$ *do not contain points of the spectrum of the pencil (1.37). Then for* $2|x| < |y|$

$$\mathcal{G}(x,y) = -i\sum_{\mu=1}^{M}\sum_{\sigma=1}^{J_\mu}\sum_{k=0}^{\kappa_{\sigma,\mu}^- -1}\overline{\mathcal{V}_\mu^{(\kappa_{\sigma,\mu}^- -k-1,\sigma;-)}(y)}u_\mu^{(k,\sigma;-)}(x) + R_-(x,y)$$

where $u_\mu^{(k,\sigma;-)}$ *are defined by equalities of the form (2.13) and* $\mathcal{V}_\mu^{(k,\sigma;-)}$ *are constructed by (2.15) and correspond to the eigenvalue* λ_μ^-. *The estimate*

$$|D_x^\alpha D_y^\gamma R_-(x,y)| \le c_{\alpha\gamma}\frac{|x|^{-Im\,\lambda_N^-+\epsilon-|\alpha|}}{|y|^{-Im\,\lambda_N^-+\epsilon+n-2m+|\gamma|}}$$

holds with a positive and sufficiently small ϵ, *and* α *and* γ *are arbitrary multi-indices.*

In the zone $2|y| < |x|$ *the asymptotic formula can be formulated similarly.*

The asymptotic formulas for the Green functions of elliptic systems of equations were given in (Maz'ya, Plamenevskij (1979)).

2.7. Boundary Value Problems in Function Spaces with Nonhomogeneous Norms. The "homogeneous" norms $|| \cdot ; V^l_{p,\beta}||$ and $|| \cdot ; \Lambda^{l,\alpha}_\beta||$ turn out to be inadequate for certain problems. In general, the spaces $V^l_{p,\beta}(G)$ do not contain the Sobolev spaces $W^l_p(G)$. Therefore, we introduce the spaces $W^l_{p,\beta}(G)$ with "nonhomogeneous" norms and prove the Fredholm property for elliptic problems in such a scale of spaces.

We suppose that the set $\mathcal{S}$ of conical points contains only one point O and near this point G coincides with a cone K. Denote by $W^l_{p,\beta}(G)$ the space endowed with the norm

$$||u; W^l_{p,\beta}(G)|| = (\sum_{|\alpha|=0}^{l} \int_G r^{p\beta}|D^\alpha u(x)|^2\, dx)^{1/p}, \quad r = |x|.$$

We clarify the connection between $W^l_{p,\beta}$ and $V^l_{p,\beta}$ in Theorems 2.14 and 2.15.

Theorem 2.14 *1) If either $\beta < -np^{-1}$ or $\beta > l - np^{-1}$ then the spaces $V^l_{p,\beta}(G)$ and $W^l_{p,\beta}(G)$ coincide.*

2) If $\nu - np^{-1} < \beta < \nu + 1 - np^{-1}$ for some $\nu = 0, 1, \dots, l-1$, then $W^l_{p,\beta}(G)$ is the direct sum $W^l_{p,\beta} \dot{+} \Pi_{l-\nu-1}$ where $\Pi_{l-\nu-1}$ is the space of polynomials (in x) of degree at most $l-\nu-1$. The functionals $u \mapsto D^l u(0), |\alpha| = 0, \dots, l-\nu-1$, defined originally for smooth functions, are continuous on $W^l_{p,\beta}(G)$. The set $C^\infty_c(\overline{G} \setminus O) \dot{+} \Pi_{l-\nu-1}$ is dense in $W^l_{p,\beta}(G)$. The norm in $W^l_{p,\beta}(G)$ is equivalent to

$$||u - \mathcal{P}_{l-\nu-1}(\,\cdot\,; u); V^l_{p,\beta}(G)|| + \sum_{|\alpha|=0}^{l-\nu-1} |D^\alpha u(0)|$$

where the polynomial $x \mapsto \mathcal{P}_{l-\nu-1}(x; u)$ is a projection of $u \in W^l_{p,\beta}(G)$ onto $\Pi_{l-\nu-1}$.

Introduce the space $W^{l-1/p}_{p,\beta}(\partial G)$ of traces on ∂G of functions in $W^l_{p,\beta}(G)$, $l = 1, 2, \dots$ By Theorem 2.14, if $\beta < -np^{-1}$ or $\beta > l - np^{-1}$ then the spaces $V^{l-1/p}_{p,\beta}(\partial G)$ and $W^{l-1/p}_{p,\beta}(\partial G)$ coincide. Suppose that for some ν the inequalities $\nu - n/p < \beta < \nu + 1 - np$ are satisfied. We denote by $Y_{l-\nu-1}$ the space of polynomials of degree at most $l - \nu - 1$ that are not identically zero on the cone $\partial K = \partial\Omega \times \overline{R}_+$ (the case $Y_{l-\nu-1} = \Pi_{l-\nu-1}$ is not excluded). We denote by π the projection of $\Pi_{l-\nu-1}$ onto $Y_{l-\nu-1}$. Note that, for any $u \in W^{l-1/p}_{p,\beta}(\partial G)$, the polynomial $\pi\mathcal{P}_{l-\nu-1}(\mathcal{U})$ does not depend on $\mathcal{U}$, where $\mathcal{U} \in W^l_{p,\beta}(G)$ is an arbitrary extension of u to G.

Theorem 2.15. *The space $W^{l-1/p}_{p,\beta}(\partial G)$ is the direct sum*

$$V^{l-1/p}_{p,\beta}(\partial G) \dot{+} Y_{l-\nu-1}.$$

The projection of u onto the subspace $Y_{l-\nu-1}$ is a polynomial $\pi\mathcal{P}_{l-\nu-1}(\mathcal{U})$. The norm in $W^{l-1/p}_{p,\beta}(\partial G)$ is equivalent to

$$||u - \pi\mathcal{P}_{l-\nu-1}(\mathcal{U}); V^{l-1/p}_{p,\beta}(\partial G)|| + \max\{|\pi\mathcal{P}_{l-\nu-1}(x;\mathcal{U})|; x \in \partial G\}.$$

Let $\mathcal{M}(x, D_x)$ be a differential operator on G with admissible coefficients (see Sect.2.1) and let $\mu = \operatorname{ord}\mathcal{M}$.

Proposition 2.16. *1) For $\beta < -n/p$ or $\beta > l - n/p$ the operator $\mathcal{M} : W^l_{p,\beta}(G) \to W^{l-\mu}_{p,\beta}(G)$ is continuous when $l \geq \mu$.*

2)If $\nu - n/p < \beta < \nu + 1 - n/p$ for some $\nu = 0, \ldots, l-1$ then the inclusion $\mathcal{M}(W^l_{p,\beta}(G)) \subset W^{l-\mu}_{p,\beta}(G)$ holds if and only if $\mathcal{M}\Pi_{l-\nu-1} \subset W^{l-\mu}_{p,\beta}(G)$. If the last condition is satisfied then the operator $\mathcal{M} : W^l_{p,\beta}(G) \to W^{l-\nu}_{p,\beta}(G)$ is continuous.

3) If $\beta < -n/p$ or $\beta > l - n/p$ then the operator $W^l_{p,\beta}(G) \ni u \to \mathcal{M}u|\partial G \in W^{l-\mu-1/p}_{p,\beta}(\partial G)$ is continuous. In the case $\nu - n/p < \beta < \nu - n/p + 1$ this operator is continuous if and only if $\mathcal{M}\Pi_{l-\nu-1}|\partial G \subset W^{l-\mu-1/p}_{p,\beta}(\partial G)$.

Let $\{\mathcal{L}(x, D_x), \mathcal{B}(x, D_x)\}$ be the operator of an elliptic boundary value problem (see Sect.2.1) and $\mathcal{D}^l_{p,\beta}W$ and $\mathcal{R}^l_{p,\beta}W$ the spaces defined by (2.1), where V^k_p and K are replaced by $W^s_{p,\beta}$ and G. We set $\Pi(l, \overrightarrow{t}, \nu) = \Pi_{l+t_1-\nu-1} \times \ldots \times \Pi_{l+t_k-\nu-1}$ (by definition $\Pi_q = \{0\}$ if $q < 0$). According to Proposition 2.16, the operator

$$A \equiv \{\mathcal{L}, \mathcal{B}\} : \mathcal{D}^l_{p,\beta}W(G) \to \mathcal{R}^l_{p,\beta}W(G) \tag{2.22}$$

is continuous if

$$\beta < -n/p \quad \text{or} \quad \beta > l + \max\{t_j\} - n/p. \tag{2.23}$$

If for some $\nu = 0, 1, \ldots, l + \max\{f_j\} - 1$ the inequalities

$$\nu - n/p < \beta < \nu - n/p + 1 \tag{2.24}$$

are satisfied, then the operator (2.22) is continuous if and only if $\{\mathcal{L}, \mathcal{B}\}(\Pi(l, \overrightarrow{t}, \nu)) \subset \mathcal{R}^l_{p,\beta}W(G)$.

In what follows we suppose that one of the conditions (2.23) and (2.24) is fulfilled. Let $\operatorname{im}\{\mathcal{N}, E\}$ and $\ker\{\mathcal{N}, E\}$ stand for the image and the kernel of the operator $\mathcal{N}$ defined on the space E.

Proposition 2.17. *1) The problem (2.3) is solvable in $\mathcal{D}^l_{p,\beta}W(G)$ if and only if there exists a polynomial $\mathcal{P} \in \Pi(l, \overrightarrow{t}, v)$ such that $A\mathcal{P} - \{f, g\} \in \operatorname{im}\{A, \mathcal{D}^l_{p,\beta}(G)\}$.*

2) Any element $u \in \ker\{A, \mathcal{D}^l_{p,\beta}W(G)\}$ has the form $u = z + v - \mathcal{P}$ where $z \in \ker\{\mathcal{L}, \mathcal{B}; \mathcal{D}^l_{p,\beta}V(G)\}$, v is a solution of the problem $Av = A\mathcal{P}$

belonging to the space $\mathcal{D}^l_{p,\beta}V(G)$, *and* $\mathcal{P}$ *is any element of* $\Pi(l, \overrightarrow{t}, v)$ *satisfying* $A\mathcal{P} \in \operatorname{im}\{A, \mathcal{D}^l_{p,\beta}V(G)\}$.

We conclude this section with the following theorem on the Fredholm property of the problem (2.3) in the scale $W^s_{p,\beta}$.

Theorem 2.18. *Let* $\Pi_0(l, \overrightarrow{t}, \nu)$ *be the subspace of* $\Pi(l, \overrightarrow{t}, \nu)$ *consisting of elements* $\mathcal{P}$ *such that* $A\mathcal{P} \in \mathcal{D}^l_{p,\beta}W(G)$ *where* $A = \{\mathcal{L}, \mathcal{B}\}$. *The operator*

$$A : (\mathcal{D}^l_{p,\beta}V(G) \dot{+} \Pi_0(l, \overrightarrow{t}, \nu)) \to \mathcal{R}^l_{p,\beta}W(G) \tag{2.25}$$

is continuous. If the line $R + i(\beta - l + n/p)$ *is free from the spectrum of the pencil (1.27) then the operator (2.25) is Fredholm.*

2.8. Self-adjoint Problems in Domains with Outlets to Infinity. Special attention is drawn to self-adjoint problems because the self-adjoint -ness allows us to get further information and consider new settings of the problems where the old ones turn out to be inadequate. Sometimes, physically meaningful problems arise in domains with cylindrical outlets to infinity in place of conical points. (It goes without saying that change of variables converts a conical point to a cylindrical outlet.) We consider the situation where the cylindrical outlets are waveguides in some sense. In other words, there exist functions of power growth ("waves") satisfying the homogeneous problem in the cylinder. This fact justifies the statement of the problem relating to "radiation conditions". We introduce incoming and outgoing waves, the "intrinsic" radiation conditions, and the scattering matrix (which is unitary).

Let Π^r be a cylinder in R^{n+1} with smooth n-dimensional boundary $\partial\Pi^r$ and cross-section Ω^r orthogonal to the axis, $r = 1, \dots, N$. Assume that a domain G in R^{n+1} coincides with the union of $\Pi^r_+ = \{(y^r, t^r) \in \Omega^r \times R = \Pi^r : t^r > 0\}$ outside a large ball. We suppose that the sets Π^r_+ have no intersections, and the boundary ∂G is smooth. Let $\mathcal{L}(x, D_x)$ be a formally self-adjoint $k \times k$ matrix of differential operators, $\operatorname{ord}\mathcal{L}_{ij} = s_i + t_j$, $t_j = \tau_j + \max\tau_j$, and $s_j = \tau_j - \max\tau_i$, where $\{\tau_1, \dots, \tau_k\}$ is a collection of nonnegative integers. We denote by $\mathcal{L}^r$ the operator $\mathcal{L}$ written in the coordinates (y^r, t^r) in Π^r. We assume that there exists an operator L^r whose coefficients do not depend on t^r, while the coefficients of $\mathcal{L}^r - L^r$ and all their derivatives are equal to $o(\exp(-\delta' t^r))$ as $t^r \to +\infty$; here $r = 1, \dots, N$ and δ' is a positive number. Further, we suppose that the Green formula

$$(\mathcal{L}u, v)_G + (\mathcal{B}u, \mathcal{Q}v)_{\partial G} = (u, \mathcal{L}v)_G + (\mathcal{Q}u, \mathcal{B}v)_{\partial G} \tag{2.26}$$

holds for $u, v \in C^\infty_c(\overline{G})$, the coefficients of the operators $\mathcal{B}$ and $\mathcal{Q}$ stabilize exponentially as $t \to +\infty$, and $ord\,\mathcal{B}_{hj} = \sigma_h + t_j$. Let B^r and Q^r stand for the principal parts of $\mathcal{B}$ and $\mathcal{Q}$ at infinity. The coefficients of B^r and Q^r may depend on y^r only. We assume that the problems $\{\mathcal{L}, \mathcal{B}\}$ and $\{L^r, B^r\}, r = 1, \dots, N$, in the domains G and Π^r are elliptic.

Let ρ_β be a smooth positive function on $\overline{G}$, equal to $\exp(\beta t^r)$ on Π^r_+. Denote by $W^l_p(G)$ the space of functions on G with norm $||\rho_\beta u; H^l(G)||_G$ and define the spaces $\mathcal{D}^l_\beta W(G)$ and $\mathcal{R}^l_\beta W(G)$ in the usual fashion (see (1.21)). For the operators $\{L^r(y, D_y, D_t), B^r(y, D_y, D_t)\}$ of the "limiting" problems in the cylinders Π^r we introduce the pencils $\lambda \mapsto \Upsilon^r(\lambda) = \{L^r(y, D_y, \lambda), B^r(y, D_y, \lambda)\}$ in $\Omega^r, r = 1, \ldots, N$. The assertions similar to the results of Sect.2.1-2.3 are valid for the operator

$$A(\beta) \equiv \{\mathcal{L}, \mathcal{B}\} : \mathcal{D}^l_\beta W(G) \to \mathcal{R}^l_\beta W(G). \tag{2.27}$$

For the sake of convenience we shall give the necessary summary in Theorem 2.19. Let δ be a small positive number such that the strip $\{\lambda \in C : |\operatorname{Im} \lambda| \le \delta\}$ contains the real eigenvalues of the elliptic pencils $\lambda \mapsto \Upsilon^r(\lambda), r = 1, \ldots, N$, only, and $\delta < \delta'$. (The δ' is connected with the description of the properties of the coefficients.) Denote by $\lambda_{\nu r}$ the real eigenvalues of $\Upsilon^r, \nu = 1, \ldots, I_r$. Let $\{\phi^{(k,j)}_{\nu,r}, k = 0, \ldots, \kappa_{\nu jr} - 1, j = 1, \ldots, J_{\nu r}\}$ be canonical systems of Jordan chains. We assume that $\chi \in C^\infty(R), \chi(t) = 0$ for $t < t_0$ and $\chi(t) = 1$ for $t > 2t_0$ where t_0 is a large positive number. We define the functions $U^{(k,j)}_{\nu,r}$ on $G \cap \Pi^r_+$ by

$$U^{(k,j)}_{\nu,r}(y,t) = \chi(t) e^{i\lambda_{\nu r} t} \sum_{q=0}^{k} \frac{1}{q!} (it)^q \phi^{(k-q,j)}_{\nu,r}(y) \tag{2.28}$$

and extend them by zero to G.

Theorem 2.19. *1) The operator (2.27) is Fredholm if and only if the line $R + i\beta$ is free from the spectra of $\Upsilon^r, r = 1, \ldots, N$.*

2) Let $u \in \mathcal{D}^l_{-\delta} W(G), \{f, g\} \in \mathcal{R}^l_\delta W(G)$ and

$$\mathcal{L}(x, D_x) u = f \quad \text{on} \quad G, \quad \mathcal{B}(x, D_x) u = g \quad \text{on} \quad \partial G. \tag{2.29}$$

Then

$$u - \sum_{r=1}^{N} \sum_{\nu=1}^{I_\nu} \sum_{j=0}^{J_{\nu r}} \sum_{k=0}^{\kappa_{\nu jr}-1} c^{(k,j)}_{\nu,r} U^{(k,j)}_{\nu,r} \in \mathcal{D}^l_\delta W(G)$$

where $c^{(k,j)}_{\nu,r} =$ const.

3) The equality $\operatorname{Ind} \mathcal{A}(-\delta) = \operatorname{Ind} \mathcal{A}(\delta) + \kappa$ holds, where $\kappa = \sum \kappa_{\nu jr}$ is the sum of the multiplicities of the real eigenvalues of $\Upsilon^r, r = 1, \ldots, N$.

4) Suppose that the line $R + i\beta$ contains no eigenvalues of the pencils Υ^r. Then the problem (2.29) with right-hand side $\{f, g\} \in \mathcal{R}^l_\beta W(G)$ has a solution $u \in \mathcal{D}^l_\beta W(G)$ if and only if the relation

$$(f, z)_G + (g, \mathcal{Q} z)_{\partial G} = 0$$

holds for any $z \in \ker \mathcal{A}(-\beta)$.

Let $z^1, \ldots, z^d$ be a basis in the subspace *ker* $\mathcal{A}(\delta)$ (the functions z^k satisfy the homogeneous problem (1.27) and decay exponentially at infinity). Since

$\ker \mathcal{A}(-\delta) \supset \ker \mathcal{A}(\delta)$, a basis in $\ker \mathcal{A}(-\delta)$ can be obtained by adding $\zeta^j \in \mathcal{D}^l_{-\delta}W(G) \setminus \mathcal{D}^l_\delta W(G), j = 1, \dots, T$, to the functions $z^1, \dots, z^d$. By Theorem 2.19, part 4,

$$\dim \operatorname{coker} A(-\delta) = \dim \ker A(\delta) = d,$$

$$\dim \operatorname{coker} A(\delta) = \dim \ker A(-\delta) = d + T.$$

Therefore $\operatorname{Ind} A(-\delta) = -\operatorname{Ind} A(\delta) = T$. In view of Theorem 2.19, part 3, we now have the following assertion.

Proposition 2.20. *The total multiplicity κ of the real eigenvalues of $\Upsilon^r, r = 1, \dots, N$, is equal to $2T$.*

In what follows we need canonical systems of Jordan chains subject to certain conditions. For any Υ^r the Green formula

$$(L(\lambda)u, v)_\Omega + (B(\lambda)u, Q(\overline{\lambda})v)_{\partial\Omega} = (u, L(\overline{\lambda})v)_\Omega + (Q(\lambda)u, B(\overline{\lambda})v)_{\partial\Omega}$$

holds. (The superscript r is omitted because up to the end of the section we consider an arbitrary elliptic pencil generated by a self-adjoint problem in the cylinder $\Omega \times R$.) Thus, for real λ the operator $\Upsilon(\lambda) = \{L(\lambda), B(\lambda)\}$ is self-adjoint with respect to the Green formula.

Proposition 2.21. *Let $\lambda_0 \in R$ be an eigenvalue of Υ. Then there exists a canonical system of Jordan chains $\{\phi^{(o,j)}, \dots, \phi^{(\kappa_j - 1, j)}; j = 1, \dots, J\}$ corresponding to λ_0, such that*

$$\begin{aligned} &\sum_{p=0}^{\mu} \sum_{q=0}^{k} \frac{1}{(k+\mu+1-p-q)!} \{((\partial_\lambda^{k+\mu+1-p-q} L)(\lambda_0)\phi^{(q,\tau)}, \phi^{(p,\zeta)})_\Omega + \\ &+ ((\partial_\lambda^{k+\mu+1-p-q} B)(\lambda_0)\phi^{(q,\tau)}, \sum_{r=0}^{p} \frac{1}{r!} (\partial_\lambda^r Q)(\lambda_0)\phi^{(p-r,\zeta)})_{\partial\Omega}\} = \\ &= \pm \delta_{\tau,\zeta} \delta_{\kappa_\tau - k - 1, \mu}\,; \end{aligned} \tag{2.30}$$

here $\mu = 0, \dots, \kappa_\tau - 1; k = 0, \dots, \kappa_\zeta - 1; \tau, \zeta = 1, \dots, J$. The sign in the right-hand side can not be chosen arbitrarily: it depends on the number of the chain.

We note that the formulas (2.30) are not connected with (1.30). Two different systems of chains appear in (1.30), while (2.31) contains one system only.

We introduce outgoing and incoming waves. Let Ξ be the space spanned by $U^{(k,j)}_{\gamma,r}$ in (2.28) and let $\mathcal{L}_w = (\Xi \dot{+} \mathcal{D}^l_\delta W(G))/\mathcal{D}^l_\delta W(G)$. Elements of $\mathcal{L}_w$ are called waves. Clearly, $\dim \mathcal{L}_w = \kappa$.

For $u, v \in \Xi \dot{+} \mathcal{D}^l_\delta W(G)$ the bilinear form

$$q(u,v) = (\mathcal{L}u, v)_G + (\mathcal{B}u, \mathcal{Q}v)_{\partial G} - (u, \mathcal{L}v)_G - (\mathcal{Q}u, \mathcal{B}v)_{\partial G}$$

takes finite values. If u or v belongs to $\mathcal{D}^l_\delta W(G)$ then $q(u,v) = 0$. So, the form proves to be defined on $\mathcal{L}_w \times \mathcal{L}_w$, and $q(U,V) = -\overline{q(V,U)}$ for all $U, V \in \mathcal{L}_w$.

Thus, $Re\, q(U,V) = 0$ for any wave U. We call a wave U outgoing (incoming) if $iq(U,V)$ is a positive (negative) number. In the classical situation where the Sommerfeld principle and Mandelstam principle (of energy radiation) are applicable, the above definition of outgoing and incoming waves is in agreement with the definitions accepted in these principles.

Theorem 2.22. *There exists a basis $\{U_1, \dots, U_{2T}\}$ in the space $\mathcal{L}_w$ subject to the orthogonality and normalization conditions*

$$q(U_j, U_k) = 0 \quad \textit{for } j \neq k,$$

$$q(U_j, U_j) = -i, \; q(U_{j+T}, U_{j+T}) = i \;\; \textit{for } j = 1, \dots, T.$$

(Thus, the waves $U_1, \dots, U_T$ are outgoing and the waves $U_{T+1}, \dots, U_{2T}$ are incoming.) Any basis $W_1, \dots, W_{2T}$ consisting of orthogonal waves (i.e. waves satisfying $q(W_j, W_k) = 0$ for $j \neq k$) contains T outgoing waves and T incoming ones.

We construct the basis $U_1, \dots, U_{2T}$ in the following way. Let us start with the functions (2.28). We assume that the functions are defined by means of Jordan chains subject to (2.30). If the supports of the functions have no intersections or the functions correspond to different eigenvalues of the pencil Υ^τ then these functions are q-orthogonal. Consider the functions corresponding to the same eigenvalue and denote them by $v^{(k,\tau)}$. From (2.30) it follows that

$$q(v^{(j,\zeta)}, v^{(k,\tau)}) = \pm i \delta_{\zeta,\tau} \delta_{\kappa_\zeta - k - 1, j}; \tag{2.31}$$

the sign in (2.31) is the same as in (2.30). So, when constructing the basis, in essence we deal with the only Jordan chain. Let a length κ_ζ of the chosen Jordan chain be equal to $2m$ where m is a positive integer. Set

$$u_\pm^{(l,\zeta)} = \frac{1}{\sqrt{2}} (v^{(l,\zeta)} \pm v^{(\kappa_\zeta - l - 1, \zeta)}), \quad l = 0, 1, \dots, m-1. \tag{2.32}$$

Then by virtue of (2.31) $q(u_+^{(l,\zeta)}, u_+^{(l,\zeta)}) = \pm i$ (the sign is the same as in (2.31) for $t = \zeta$) and $q(u_-^{(l,\zeta)}, u_-^{(l,\zeta)}) = \mp i$ (the sign is opposite to the sign in (2.31)). Therefore a Jordan chain of length $2m$ provides $2m$ orthogonal waves, of which m are outgoing and m are incoming. Now let $\kappa_\zeta = 2m + 1$. The functions $u_\pm^{(l,\zeta)}$ for $l = 0, \dots, m-1$ are defined by (2.32) as before and $u^{(m,\zeta)} = v^{(m,\zeta)}$. From (2.31) it follows that $q(u^{(m,\zeta)}, u^{(m,\zeta)}) = \pm i$ and $q(u^{(m,\zeta)}, u_\pm^{(l,\zeta)}) = 0$ for $l = 0, \dots, m-1$. The type of wave $u^{(m,\zeta)}$ depends on the sign in (2.31). To define the scattering matrix we require a special basis in $ker\, A(-\delta)$.

Proposition 2.23. *Let $\{U_1, \dots, U_{2T}\}$ be the same basis in the space $\mathcal{L}_w$ as in Theorem 2.22. Then there exist bases $\{\zeta_1, \dots, \zeta_T\}$ and $\{\eta_1, \dots, \eta_T\}$ modulo $\mathcal{D}_\delta^l W(G)$ in $ker\, A(-\delta)$ such that*

$$\zeta_j = U_j + \sum_{k=1}^{T} t_{jk} U_{k+T}, \tag{2.33}$$

$$\eta^{\cdot}_j = U_{j+T} + \sum_{k=1}^{T} s_{jk} U_k \tag{2.34}$$

where $j = 1, \ldots, T$ *and* $x^{\cdot} \in \mathcal{L}_w$ *is the coset representative of* $x \in \Xi \dot{+} \mathcal{D}^l_\delta W(G)$.

The matrix $s = (s_{ij})$ consisting of the coefficients in (2.34) is called the scattering matrix. (We recall that the waves $U_1, \ldots, U_T$ are outgoing and $U_{T+1}, \ldots, U_{2T}$ incoming.)

Theorem 2.24. *The matrix* s *is unitary, i.e.* $s^* = s^{-1}$. *In addition,* $s^{-1} = t = (t_{jk})$ *where* t_{jk} *are the coefficients in (2.33).*

"Radiation conditions" provide a choice of solution (perhaps, with certain arbitrariness) and indicate the (type of) asymptotics of the desired solution. We call radiation conditions intrinsic if the corresponding asymptotics at infinity is a combination of outgoing waves. The following theorem justifies the statement of the problem with radiation conditions of such a type.

Let $U_1, \ldots, U_{2T}$ be the basis in Theorem 2.22. We choose representatives $u^j \in U_j, j = 1, \ldots, T$, arbitrarily, denote the linear hull $\mathcal{L}(u^1, \ldots, u^T)$ by Θ and consider the restriction $\mathbf{A}$ of $A(-\delta)$ to the space $\Theta + \mathcal{D}^l_\delta W(G)$. The mapping $\mathbf{A} : \Theta + \mathcal{D}^l_\delta W(G) \to \mathcal{R}^l_\delta W(G)$ is continuous.

Theorem 2.25. *Let* $z^1, \ldots, z^d$ *be a basis in* $\ker A(\delta), \{f, g\} \in \mathcal{R}^l_\delta W(G)$, *and* $(f, z^j)_G + (g, Qz^j)_{\partial G} = 0,\ j = 1, \ldots, d$. *Then*

1) There exists a solution $u \in \Theta + \mathcal{D}^l_\delta W(G)$ *of the equation* $\mathbf{A} = \{f, g\}$ *determined up to an arbitrary term in* $\mathcal{L}(z^1, \ldots, z^d)$.

2) The inclusion

$$v \equiv u - c_1 u^1 - \ldots - c_T u^T \in \mathcal{D}^l_\delta W(G)$$

holds where $c_j = i(f, \zeta^j) + i(g, Q\zeta^j),\ j = 1, \ldots,, T$, *and the functions* $\zeta^1, \ldots, \zeta^T$ *belong to* $\ker A(-\delta)$ *and satisfy (2.33).*

3) For the solution u *the inequality*

$$||v; \mathcal{D}^l_\delta W(G)|| + |c_1| + \ldots + |c_T| \leq \text{const}\,(||\{f, g\}; \mathcal{R}^l_\delta W(G)|| + ||\rho_\delta v; L_2(G)||) \tag{2.35}$$

holds. The solution u^0 *subject to the additional conditions* $(u^0, \zeta^j)_G = 0,\ j = 1, \ldots, d$, *is unique, and the estimate (2.35) is valid with the right-hand side* const $||\{f, g\}; \mathcal{R}^l_\delta W(G)||$.

§3. Boundary Value Problems in Domains with Edges

3.1. Statement of the Problem. Model Problems. Let G be a domain in R^n with compact closure $\overline{G}$ bounded by an $(n-1)$-dimensional manifold ∂G. We suppose that a subset $M \subset \partial G$ is selected for which the following conditions are satisfied: 1)M is a smooth d-dimensional submanifold of R^n without boundary; 2) $\partial G \backslash M$ is a smooth submanifold of R^n; 3) for every point

$x^0 \in M$ there is a neighborhood $\mathcal{U}$ in R^n and a diffeomorphism $\kappa : \mathcal{U} \to R^n$ such that $\kappa(\mathcal{U} \cap G) = B_1^n(O) \cap D$, where $B_1^0(O) = \{x \in R^n : |x| < 1\}, D = K \times R^d$, and K is an open $(n-d)$-dimensional cone cutting out an open set Ω on S^{n-d-1} with smooth boundary $\partial\Omega$.

A scalar differential operator

$$\mathcal{P}(x, D_x) = \sum_{|\gamma| \le m} p_\gamma(x) D_x^\gamma$$

with coefficients in $C^\infty(\overline{G} \setminus M)$ is called admissible if the representations

$$p_\gamma(x) = r^{-q+|\gamma|} p_\gamma^0(r, \omega, z), \quad p_\gamma^0 \in C^\infty(\overline{R}_+ \times \overline{\Omega} \times R^d)$$

are valid in a neighborhood $\mathcal{U}$ of every point $x^0 \in M$ where $\kappa(x) = (y, z), z \in R^d, y \in R^{n-d}$, and (r, ω) are spherical coordinates of y.

Let $\mathcal{L}(x, D_x)$ and $\mathcal{B}(x, D_x)$ be $k \times k$ and $m \times k$ matrices consisting of admissible differential operators; as usual, $\operatorname{ord} \mathcal{L}_{ij} = s_i + t_j$, $\operatorname{ord} \mathcal{B}_{ij} = \sigma_q + t_j$.

Completing $C_0^\infty(\overline{G} \setminus M)$ with respect to the norm

$$||u; V_\beta^l(G, M)|| = (\sum_{|\gamma| \le l} ||\rho^{\beta - l + |\alpha|} D_x^\gamma u; L_2(G)||^2)^{1/2}, \tag{3.1}$$

we introduce the space $V_\beta^l(G, M)$ for $\beta \in C^\infty(\overline{G}), l = 0, 1, \dots$ Here ρ is a smooth positive function on $\overline{G} \setminus M$ equivalent to the distance from M (near the edge M). Denote by $V_\beta^{l+1/2}(\partial G, M)$ the space of traces on $\partial G \setminus M$ of functions in $V_\beta^{l+1}(G, M)$ and by $\mathcal{D}_\beta^l V(G, M)$ and $\mathcal{R}_\beta^l V(G, M)$ the spaces defined in analogy with (1.21) when $V_\beta^s(K)$ is replaced by $V_\beta^s(G, M)$.

The operator

$$\mathcal{A} = \{\mathcal{L}, \mathcal{B}\} : \mathcal{D}_\beta^l V(G, M) \to \mathcal{R}_\beta^l V(G, M) \tag{3.2}$$

is continuous.

The principal part $\mathcal{P}^0$ of the above operator $\mathcal{P}$ is the operator on D defined by

$$\mathcal{P}^0(y, D_y, D_z) = \sum_{|\gamma| \le m} r^{-m+|\gamma|} p_\gamma^0(0, \omega, z) D_{(y,z)}^\gamma$$

(the diffeomorphism κ is subject to the requirements $\kappa(x^0) = 0, \kappa'(x^0) = 1$). By $\{\mathcal{L}^0, \mathcal{B}^0\}$ we denote the operator of the boundary value problem on D consisting of the principal parts of entries of $\mathcal{L}$ and $\mathcal{B}$. The problem $\{\mathcal{L}, \mathcal{B}\}$ is called elliptic on M if $\{\mathcal{L}^0, \mathcal{B}^0\}$ is elliptic on $D \setminus M_0$, where $M_0 = O \times R^d$ is the "edge of the wedge". We assume that $\{\mathcal{L}, \mathcal{B}\}$ is an elliptic problem on $\overline{G} \setminus M$ and on M.

We introduce the "model" problems in the wedge and cone. Let us consider the model problem in the wedge

$$\begin{aligned} \mathcal{L}^0(y, D_y, D_z) u(y, z) &= f(y, z), \quad (y, z) \in D, \\ \mathcal{B}^0(y, D_y, D_z) u(y, z) &= g(y, z), \quad (y, z) \in D \setminus M, \end{aligned} \tag{3.3}$$

where the operator $\{\mathcal{L}^0, \mathcal{B}^0\}$ is the principal part of $\{\mathcal{L}, \mathcal{B}\}$ at some point $x^0 \in M$. The norm in $V^l_\beta(D, M_0)$ is defined by (3.1) but with $D, M_0, r = |y|$ instead of G, M, ρ, the function β being replaced by a number β.

The operator

$$\mathcal{A}^0 = \{\mathcal{L}^0, \mathcal{B}^0\} : \mathcal{D}^l_\beta V(D, M_0) \to \mathcal{R}^0_\beta V(G, M_0) \tag{3.4}$$

is continuous

Let us apply the Fourier transform $\mathcal{F}_{z\to\xi}$ to (3.3) and obtain the following family of boundary value problems in a cone

$$\begin{aligned} \mathcal{L}^0(y, D_y, \xi)\hat{u}(y, \xi) &= \hat{f}(y, \xi), \quad y \in K, \\ \mathcal{B}^0(y, D_y, \xi)\hat{u}(y, \xi) &= \hat{g}(y, \xi), \quad y \in \partial K \setminus O, \end{aligned} \tag{3.5}$$

where $\hat{u}(y, \,\cdot\,)$ is the Fourier transform of the function $z \mapsto u(u, z)$, etc. Using new variables we can eliminate the dependence on $|\xi|$ of the operator of the problem (3.5). Set

$$\begin{aligned} &\eta = |\xi|y, \quad \theta = \xi|\xi|^{-1}, \quad U_j(\eta, \xi) = |\xi|^{t_j}\hat{u}_j(y, \xi), \\ &F_i(\eta, \xi) = |\xi|^{-s_i}\hat{f}_i(y, \xi), \quad G_q(\eta, \xi) = |\xi|^{-\sigma_q}\hat{g}_q(y, \xi). \end{aligned} \tag{3.6}$$

The equations (3.5) then take the form

$$\begin{aligned} \mathcal{L}^0(\eta, D_\eta, \theta)U(\eta, \xi) &= F(\eta, \xi), \quad \eta \in K, \\ \mathcal{B}^0(\eta, D_\eta, \theta)U(\eta, \xi) &= G(\eta, \xi), \quad \eta \in \partial K \setminus O. \end{aligned} \tag{3.7}$$

We call the problem (3.7) model in the cone K and denote its operator by $A(\theta), \theta \in S^{d-1}$.

Now we introduce the spaces where the operator $A(\theta)$ acts. Let $E^l_\beta(K)$ be the completion of $C_0^\infty(\overline{K} \setminus O)$ with respect to the norm

$$||U; E^l_\beta(K)|| = \Big(\sum_{|\alpha|\le l} ||\,|\eta|^\beta(1 + |\eta|^{|\alpha|-l})D^\alpha_y U; L_2(K)||\Big)^{1/2}$$

and $E^{l-1/2}_\beta(\partial K)$ the corresponding space of traces on $\partial K \setminus O$. The norms $||w; V^l_\beta(D, M_0)||$ and

$$\left(\int_{R^d} |\xi|^{2(l-\beta)-n+d}||W(\,\cdot\,, \xi); E^l_\beta(K)||^2\, d\xi\right)^{1/2}$$

are equivalent; here $W(\eta, \xi) = \hat{w}(|\xi|^{-1}\eta, \xi)$. Consequently the norms $||v; V^{l-1}_\beta(\partial D, M_0)||$ and

$$\left(\int_{R^d} |\xi|^{2(l-\beta)-n-d}||V(\,\cdot\,, \xi); E^{l-1/2}_\beta(\partial K)||^2\, d\xi\right)^{1/2}$$

are equivalent too.

The mapping

$$A(\theta) : \mathcal{D}^l_\beta E(K) \to \mathcal{R}^l_\beta E(K) \tag{3.8}$$

is continuous.

3.2. Properties of the Model Problems.

Theorem 3.1. *The operator (3.4) is an isomorphism if and only if (3.8) is an isomorphism for all $\theta \in S^{d-1}$.*

Note that the operator (3.4) is Fredholm if and only if it is an isomorphism.

Since the matrices $\mathcal{L}^0(y, D_y, D_z)$ consist of differential operators admissible near the vertex of the cone, one can define the operator pencil $\lambda \mapsto \Upsilon(\lambda)$ (see (1.37)). It is clear that the ellipticity of the original operator on the edge provides the ellipticity of the operator pencil.

Theorem 3.2. *The operator (3.8) is Fredholm if and only if the line $R + i(\beta - l + (n-d)/2)$ is free from the spectrum of the pencil Υ.*

We present a few properties of the model problems.

Proposition 3.3. *The inclusion $\chi U \in \mathcal{D}^s_\gamma E(K)$ holds for every solution $U \in \mathcal{D}^l_\beta E(K)$ of the problem (3.7); here χ is an arbitrary function in $C^\infty(\overline{K})$ equal to zero near the vertex of the cone, $s \geq l$ and $\gamma \in R$. (In other words, solutions of the homogeneous problem (1.11) are smooth on $\overline{K} \setminus O$ and vanish more rapidly than any degree of $|\eta|$ as $|\eta| \to 0$.)*

Proposition 3.4. *Let the closed strip between the lines $R + i(\beta_1 - l_1 + (n-d)/2)$ and $R + i(\beta_2 - l_2 + (n-d)/2)$ contain no eigenvalues of the operator pencil $\lambda \mapsto \Upsilon(\lambda)$ defined for the problem (3.7). If $\{F, G\} \in \mathcal{R}^{l_1}_{\beta_1} E(K) \cap \mathcal{R}^{l_2}_{\beta_2} E(K)$ then every solution U of the problem (3.7) in $\mathcal{D}^{l_1}_{\beta_1} E(K)$ belongs to $\mathcal{D}^{l_2}_{\beta_2} E(K)$.*

Proposition 3.5. *Let each of the lines $R + i(\beta_1 - l + (n-d)/2)$ and $R + i(\beta_2 - l + (n-d)/2)$ contain eigenvalues of Υ while the strip between these lines is free from the spectrum of Υ. Assume that for some $\beta \in (\beta_1, \beta_2)$ the operator (3.8) is an isomorphism. Then the operator (3.8) is an isomorphism for any $\beta \in (\beta_1, \beta_2)$, but not for $\beta \notin (\beta_1, \beta_2)$.*

From this proposition (and Theorem 3.1) it follows that there exists at most one interval (β_1, β_2) such that the operators (3.8) and (3.4) implement an isomorphism for $\beta \in (\beta_1, \beta_2)$. However, it may turn out that there are no such intervals. A similar situation arises for the Neumann problem.

Proposition 3.6. *Let the closed strip between the lines $R + i(\beta_1 - l_1 + (n-d)/2)$ and $R + i(\beta_2 - l_2 + (n-2)/2)$ contain no eigenvalues of the pencil Υ. Assume that the operator (3.8) has trivial kernel and cokernel for $l = l_1$ and $\beta = \beta_1$. Further, let $\{f, g\} \in \mathcal{R}^{l_1}_{\beta_1} V(D, M_0) \cap \mathcal{R}^{l_2}_{\beta_2} V(D, M_0)$. Then a solution $u \in \mathcal{D}^{l_1}_{\beta_1} V(D, M_0)$ of the problem (3.3) belongs to $\mathcal{D}^{l_2}_{\beta_2} V(D, M_0)$.*

3.3. Fredholm Property of the Problem in G. Now we turn to the operator (3.2). As in Sect.3.1, to every point $x^0 \in M$ we associate the operator

$$A(\theta, x^0) : \mathcal{D}^l_{\beta(x^0)} E(K(x^0)) \to \mathcal{R}^l_{\beta(x^0)} E(K(x^0)) \tag{3.9}$$

(in contrast to Sect.3.1 we indicate the dependence on $x^0 \in M$ in the notation). Suppose that for any point $x^0 \in M$ there exists a number $\beta_0(x^0)$ such that the operator (3.9) is an isomorphism for $\beta(x^0) = \beta_0(x^0)$. Then by Proposition 3.5 this property of the mapping (3.9) remains valid for all $\beta(x^0) \in (\beta_-(x^0), \beta_+(x^0))$. We suppose that this interval can not be expanded. The functions $x^0 \to \beta_\pm(x^0)$ are continuous.

Theorem 3.7. *Let the above conditions be fulfilled and let β be an arbitrary function in $C^\infty(\overline{G})$ with $M \ni x \mapsto \beta(x) \in (\beta_-(x), \beta_+(x))$. Then the operator (3.2) is Fredholm.*

This theorem can be proved by constructing regularizers patched by means of a partition of unity from the "local" regularizers for the model problems in the wedge D and space R^n. If the conditions of Theorem 3.7 are not satisfied then the operator (3.2) is not Fredholm.

Thus, to get a correct statement of the problem in the scale V^s_β (providing the Fredholm property for the operator (3.2)) one has to ensure that the operator (3.9) is an isomorphism. In general, it is not an easy matter. As was said, for some problems one cannot find a number β such that the operator (3.8) (or, what is the same, (3.9)) is an isomorphism. This sends us in search of a new scale of spaces, supplementary conditions on solutions near the edge, and so on. The situation is now better understood for the Dirichlet and Neumann problems in the case of self-adjoint elliptic systems ((Maz'ya, Plamenevskij (1977a)), (Maz'ya, Plamenevskij (1983)), (Nazarov (1988)), (Nazarov, Plamenevskij (1991a))), and for the second order equations with general boundary conditions ((Komech (1973)),(Maz'ya, Plamenevskij (1975a)),(Eskin (1985))).

3.4. Asymptotics of Solutions of the Dirichlet Problem for the Laplace Operator in a Three-dimensional Domain G with Edges on the Boundary. The decisive simplifying assumptions here are the ones about coefficients being constant and about the invariance along the edge of the opening of the dihedral angle formed by two tangent planes. These assumptions make it possible to use the Fourier transform.

3.4.1. Solvability of the Problem. Consider the Dirichlet problem

$$-\Delta u(x) = f(x), \quad x \in G; \quad u(x) = g(x), \quad x \in \partial G \setminus M. \tag{3.10}$$

in a bounded domain $G \subset R^3$ having a smooth edge on the boundary ∂G. Suppose that the opening α of the dihedral angle formed by the tangent planes does not change along the edge. Hence after straightening the boundary near a point of the edge the Laplace operator is rewritten in the form

$$\Delta_x = \Delta_y + \partial_z^2 + \mathcal{L}(y, z, \partial_y, \partial_z) \tag{3.11}$$

where (y, z) are coordinates on the wedge $D = K \times R, K = \{y \in R^2 : r > 0, \theta \in (0, \alpha)\}, (r, \theta)$ are polar coordinates, and $\mathcal{L}$ is a second order differential operator with smooth coefficients. Moreover, $\mathcal{L}(0, z, \partial_y, \partial_z)$ is a first order operator. So the model problem in the wedge D with edge M_0 is

$$\begin{aligned} &-(\Delta_y + \partial_z^2)u(y, z) = f(y, z), \quad (y, z) \in D, \\ &u(y, z) = g(y, z), \quad (y, z) \in \partial D \setminus M_0. \end{aligned} \tag{3.12}$$

Applying the Fourier transform $F_{z \to \xi}$ and the change of variables

$$\begin{aligned} &\eta = |\xi| y, \quad U(\eta, \xi) = \hat{u}(|\xi|^{-1}\eta, \xi), \\ &F(\eta, \xi) = |\xi|^{-2}\hat{f}(|\xi|^{-1}\eta, \xi), \quad G(\eta, \xi) = \hat{g}(|\xi|^{-1}\eta, \xi) \end{aligned} \tag{3.13}$$

we convert (3.11) to

$$-(\Delta_\eta - 1)U(\eta, \xi) = F(\eta, \xi), \quad \eta \in K; \quad U(\eta, \xi) = G(\eta, \xi), \quad \eta \in \partial K \setminus O. \tag{3.14}$$

Proposition 3.8. *The operator of the problem (3.14) implements an isomorphism* $E_\beta^{l+2}(K) \to E_\beta^l(K) \times E_\beta^{l+3/2}(\partial K)$ *if and only if* $|\beta - l - 1| < \pi\alpha^{-1}$.

Proposition 3.9. *1) The operator* $A^0 : V_\beta^{l+2}(D, M_0) \to V_\beta^l(D, M_0) \times V_\beta^{l+3/2}(\partial D, M_0)$ *of the problem (3.12) is an isomorphism for* $|\beta - l - 1| < \pi\alpha^{-1}$. *This operator is not Fredholm for all other* β.

2) The assertion 1 remains valid if we replace the sets D, M_0 *by* G, M *and the operator* A^0 *by the operator* A *of the problem (3.10).*

3.4.2. Asymptotics of Solutions. Let $u \in V_\beta^{l+2}(G, M)$ be a solution of the problem (3.10) with right-hand side $\{f, g\} \in V_\gamma^{l+3/2}(\partial G, M)$ and let the exponents β and γ be subject to the conditions

$$-\pi\alpha^{-1} < \gamma - l - 1 + \pi\alpha^{-1} < 0, \quad l + 1 - \pi\alpha^{-1} < \beta \leq \gamma + 1. \tag{3.15}$$

Denote by ζ a smooth function on $\overline{G}$ that vanishes outside some small neighborhood of a point on the edge M. According to the inequality $\gamma \geq \beta - 1$ and the properties of the operator $\mathcal{L}$ in (1.2), we have $\zeta \mathcal{L} u \in V_\gamma^l(D, M_0)$. Therefore, in order to obtain the asymptotics of the solution of the problem (3.10) it suffices to find the asymptotics of the solution $u \in V_\beta^{l+2}(D, M_0)$ of the problem (3.12) with right-hand side $\{f, g\} \in V_\gamma^l(D, M_0) \times V_\gamma^{l+3/2}(\partial D, M_0)$.

Applying the change of variables (3.13) we proceed to the problem (3.14). The right-hand side $\{F, G\}$ belongs to $E_\gamma^l(K) \times E_\gamma^{l+3/2}(\partial K)$ for almost all $\xi \in R$ (see Sect.3.1.). If we obtain the asymptotic formula for U with remainder in $E_\gamma^{l+2}(K)$ then the inverse Fourier transform will give the asymptotics for u with remainder in $V_\gamma^{l+2}(D, M_0)$. We have to apply Theorem 1.11 (on the asymptotics of solutions in a cone) where the spaces $V_\beta^l(K)$ were present.

Note that $E^{l+2}_\beta(K) \subset E^l_\gamma(K) \subset V^l_\gamma(K)$ and $E^{l+3/2}_\gamma(\partial K) \subset V^{l+3/2}_\gamma(\partial K)$. So the function $U \in E^{l+2}_\beta(K) \subset V^{l+2}_\beta(K)$ satisfies

$$-\Delta_\eta U = F' \equiv F - U \quad \text{in} \quad K, \quad U = G \quad \text{on} \quad \partial K \setminus O$$

where $\{F', G\} \in V^l_\gamma(K) \times V^{l+3/2}_\gamma(\partial K)$. Since

$$-2\pi\alpha^{-1} < \gamma - l - 1 < -\pi\alpha^{-1} < \beta - l - 1 < \pi\alpha^{-1}$$

(see (3.15)), we have

$$U(\eta, \xi) = p(\xi)|\eta|^{\pi/\alpha} \sin(\pi\theta\alpha^{-1}) + W(\eta, \xi) \,. \tag{3.16}$$

Moreover,

$$\begin{aligned}&|p(\xi)| + ||W; V^{l+2}_\gamma(K)|| \le c\,(||\{F', G\}; V^l_\gamma(K) \times V^{l+3/2}_\gamma(\partial K)|| + ||\{F', G\};\\ &V^l_\beta(K) \times V^{l+3/2}_\beta(\partial K)||) \le c\,(||\{F, G\}; E^l_\gamma(K) \times E^{l+3/2}_\gamma(\partial K)|| +\\ &+ ||U; E^{l+2}_\gamma(K)||).\end{aligned} \tag{3.17}$$

The remainder W is not in $E^{l+2}_\gamma(K)$ because of the behavior at infinity of the first term on the right in (3.16). We correct it by rewriting (3.16) as

$$U(\eta, \xi) = p(\xi)\chi(|\eta|)|\eta|^{\pi/\alpha} \sin(\pi\theta\alpha^{-1}) + V(\eta, \xi) \tag{3.18}$$

where $\chi \in C^\infty_c(R)$ and $\chi = 1$ near zero. Let ζ_0 and ζ_∞ be functions in $C^\infty(R)$ with $\zeta_0\chi = \zeta_0$ and $\zeta_\infty(1-\chi) = \zeta_\infty$. Combining (3.16) and (3.18) we obtain $\zeta_0 V = \zeta_0 W$ and $\zeta_\infty V = \zeta_\infty U$. Hence $V \in E^{l+2}_\gamma(K)$ and by (3.17) we have

$$\begin{aligned}&||V; E^{l+2}_\gamma(K)|| \le c\,(||\zeta_0 W; V^{l+2}_\gamma(K)|| + ||(1-\zeta_0)U; E^{l+2}_\gamma(K)|| + |p(\xi)|) \le\\ &\le c\,(||\{F, G\}; E^l_\gamma(K) \times E^{l+3/2}_\gamma(\partial K)|| + ||U; E^{l+2}_\beta(K)||).\end{aligned} \tag{3.19}$$

We return to the coordinates y to obtain the equality

$$\hat{u}(y, \xi) = \chi(|\xi|r)\hat{k}(\xi)r^{\pi/\alpha} \sin(\pi\theta\alpha^{-1}) + \hat{v}(y, \xi) \tag{3.20}$$

where $\hat{k}(\xi) = |\xi|^{\pi/\alpha}\rho(\xi)$ and $\hat{v}(y, \xi) = V(|\xi|y, \xi)$. After application of the inverse Fourier transform the relation (3.20) is rewritten as

$$u(y, z) = k(r, z)r^{\pi/\alpha} \sin(\pi\theta\alpha^{-1}) + v(y, z). \tag{3.21}$$

The multiplier k is defined by

$$k(z, z) = (2\pi)^{-1/2} \int_R \exp(i\xi z)\chi(|\xi|r)\hat{k}(\xi)\, d\xi \equiv (\mathcal{P}k)(r, z). \tag{3.22}$$

Denote by X the Fourier preimage of the function $\xi \mapsto \sqrt{2\pi}\chi(|\xi|)$. Then we obtain the following representation for the operator $\mathcal{P}$:

$$(\mathcal{P}k)(z,z) = \int_R X(r^{-1}\tau)k(z-\tau)r^{-1}\,d\tau = \int_R X(t)k(z-rt)\,dt.$$

If $k \in C_0^\infty(R^n)$ then $(\mathcal{P}k)(0,z) = k(z)\int X(t)\,dt = k(z)$; thus, $\mathcal{P}$ is an extension operator from the edge M_0 to the wedge. The inequalities

$$\begin{aligned} &||\mathcal{P}k - \chi k;\, V^0_{-s-1}(D,M_0)|| + ||\partial_z^j \partial_r^j \mathcal{P}k;\, V^0_{i+j-s-1}(D,M_0)|| \le \\ &\le c||\,|\xi|^s \hat{k};\, L_2(R)||, \quad i,j = 0,1,\dots,\ i+j>0 \end{aligned} \tag{3.23}$$

hold for $s = l+1-\gamma-\pi\alpha^{-1} \in (0,1)$. From (3.17) and (3.19) it follows that

$$\begin{aligned} &||\,|\xi|^s\hat{k}; L_2(R)|| + ||v; V^{l+2}_\gamma(D,M_0)|| \le c\,(||\{f,g\}; V^l_\gamma(D,M_0)\times \\ &\times V^{l+3/2}_\gamma(\partial D, M_0)|| + ||u; V^{l+2}_\beta(D,M_0)||)\,. \end{aligned} \tag{3.24}$$

Theorem 3.10. *Let $u \in V^{l+2}_\beta(D,M_0)$ be a solution of the problem (3.12) with right-hand side $\{f,g\} \in V^l_\gamma(D,M_0)\times V^{l+3/2}_\gamma(D,M_0)$ and let the condition (3.15) be fulfilled. Then the asymptotic formula (3.21) and estimate (3.24) are valid. The coefficient k in (3.21) is defined by (3.22), and the function k belongs to $H^s(R)$, $s = l+1-\gamma+\pi\alpha^{-1} \in (0,1)$.*

We have to check the inclusion $k \in H^s(R)$ only. By (3.24) it remains to estimate the norm of k in $L_2(R^n)$. Rewrite (3.21) in the form

$$u(y,z) = \chi(r)k(r)r^{\pi/\alpha}\sin(\pi\theta\alpha^{-1}) + w(y,z) \tag{3.25}$$

where according to (3.23) and (3.24) we have $w = (\mathcal{P}k - \chi k)r^{\pi/\alpha}\sin(\pi\theta\alpha^{-1}) + v$ and $w \in V^0_{\gamma-l-2}(D,M_0)$. Integrate (3.25) over $\Xi = \{(y,z) \in D : 1 < |y| < 2\}$ and obtain

$$\begin{aligned} &c_1||k; L_2(R)||^2 \le ||u-v; L_2(\Xi)||^2 \le c_2(||u; V^0_{\beta-l-2}(D,M_0)||^2 + \\ &+ ||w; V^0_{\gamma-l-2}(D,M_0)||^2). \end{aligned}$$

Theorem 3.10 is thus proven. Note that the formula (3.25) is simpler than (3.21) and can also be treated as an asymptotic formula (near the edge the remainder w decreases more rapidly than the first term). However the function w does not in general possess the same smoothness as the solution u far from the edge, because the coefficient k is not sufficiently smooth. This fact forces us to introduce the "smoothing-outside-the-edge" operator $\mathcal{P}$. If, for instance, the right-hand side $\{f,g\}$ vanishes near the edge then $k \in C^\infty(R)$, $w \in V^{l+2}_\gamma(D,M_0)$ and the formula (3.25) can replace (3.21) entirely.

3.4.3. On the Calculation of the Coefficient k in (3.25). If $s = l+1-\gamma-\pi\alpha^{-1} > 1/2$ then the inclusion $k \in H^s(R)$ implies the continuity of k. Assume that the latter inequality is satisfied as well as the conditions of Theorem 3.10. Let us treat a point x^0 of the edge M_0 as the vertex of the cone cutting out the domain $\Omega = \{\omega = (\omega_1,\omega_2) \in S^2 : \omega_1 \in (0,\pi), \omega_2 \in (0,\alpha)\}$ on the sphere

S^2 where ω_1 is the latitude and ω_2 is the longitude of ω. The asymptotics (3.25) can be rewritten in the form

$$u(x) = \chi(x)k(x^0)\rho^{\pi/\alpha}\Phi(\omega) + w(x) \tag{3.26}$$

where $\rho = |x - x_0|$, $\lambda = -i\pi/\alpha$ is an eigenvalue and $\Phi(\omega) = (\sin\omega_1)^{\pi/\alpha} \times \times \sin(\pi \leq -mega_2\alpha^{-1})$ is an eigenfunction of the corresponding operator pencil in Ω. The remainder $w(x)$ is $o(\rho^{\pi/\alpha})$ as $x \to x^0$ along directions nontangent to the edge. Finally, χ is a cut-off function. In other words, the formula (3.26) can be viewed as asymptotics in the cone with vertex x^0. Therefore, to find the coefficient $k(x^0)$, we make use of the same reasoning as in Theorem 2.6 to obtain the formula

$$k(x^0) = (f, \zeta(x^0; \cdot))_G - (g, \partial_\nu \zeta(x^0; \cdot))_{\partial G} \tag{3.27}$$

where $x \mapsto \zeta(x^0; x)$ is a solution of the homogeneous problem (3.10) with asymptotics of the form

$$\zeta(x^0; x) = \chi(x)a\rho^{-1-\pi\alpha}\Phi(\omega) + \eta(x).$$

The normalizing multiplier a is defined by

$$a = 2(\alpha + 2\pi)^{-1}\pi^{-1/2}\Gamma(\pi\alpha^{-1} + 3/2)\Gamma(\pi\alpha^{-1} + 1)^{-1}.$$

3.5. Asymptotics of Solutions of the General Problem in a Domain G with Edge. Let G be the same domain as in Sect.3.1 and $\mathcal{U}$ a neighborhood of a point x^0 on the edge M. We consider the elliptic problem

$$\begin{aligned} \mathcal{L}(x, D_x)u(x) &= f(x), \quad x \in G \\ \mathcal{B}(x, D_x)u(x) &= g(x), \quad x \in \partial G \setminus M \end{aligned}$$

with operators $\mathcal{L}$ and $\mathcal{B}$ defined in Sect.3.1. We multiply u by a cut-off function $\chi \in C_c^\infty(\mathcal{U})$ and introduce local coordinates (y, z) in the wedge D. Then we obtain the problem

$$\begin{aligned} \mathcal{L}(y, z, D_y, D_z)u(y, z) &= f(y, z), \quad (y, z) \in D \\ \mathcal{B}(y, z, D_y, D_z)u(y, z) &= g(y, z), \quad (y, z) \in \partial D \setminus M_0. \end{aligned} \tag{3.28}$$

Thus the matter reduces to the description of the asymptotics of solutions of the problem (3.28).

To every point $z \in M_0$ we associate the elliptic pencil $\lambda \mapsto \Upsilon(\lambda, z)$ corresponding to the operator $\{\mathcal{L}^0(y, z, D_y, 0), \mathcal{B}(y, z, D_y, 0)\}$ in K (see (1.27)); here $D = K \times R^d$, while $\mathcal{L}^0$ and $\mathcal{B}^0$ are the principal parts of $\mathcal{L}$ and $\mathcal{B}$ in (3.28) at the point z. As in Sect.3.3 we denote by $(\beta_-(z), \beta_+(z))$ the interval where the operator $A(\theta) \equiv A(\theta, z)$ is an isomorphism for each point (see (3.8)). The existence of such an interval is supposed, and $\beta_\pm$ are continuous functions on R^d with $\beta_\pm(z) = \beta_\pm^0 = \text{const}$ for $|z| > \rho > 0$. We consider a solution $u \in \mathcal{D}_\beta^l V(D, M_0)$ of the problem (3.28) with right-hand side $\{f, g\} \in \mathcal{R}^l_\gamma V(D, M_0)$. The β and γ are functions in $C^\infty(R^d)$ which are constant for $|z| > \rho > 0$ and satisfy $\gamma(z) < \beta(z) < \beta_+(z)$ for all $z \in R^d$.

Denote by $\lambda_1(z), \dots, \lambda_N(z)$ the eigenvalues of the pencil $\lambda \mapsto \Upsilon(\lambda, z)$ situated between the lines $R + i(\beta(z) - l + (n-d)/2)$ and $R + i(\gamma(z) - l + (n-d)/2)$. Let $\{\phi_\nu^{(h,j)}, j = 1, \dots, J_\nu, h = 0, \dots, \kappa_{j\nu} - 1\}$ be a canonical system of Jordan chains corresponding to $\lambda_\nu(z), \nu = 1, \dots, N$.

Suppose that the following conditions are fulfilled:

1. The above lines are free from the spectrum of the pencil $\lambda \mapsto \Upsilon(\lambda, z), z \in R^d$.

2. The numbers N, J_ν and $\kappa_{\nu j}$ do not depend on $z \in R^d$.

3. The Jordan chains may be chosen in such a way that the functions $z \mapsto \phi_\nu^{(h,j)}(\omega, z)$ are smooth on R^d for all $\omega \in \overline{\Omega}$.

Theorem 3.11. *Let $\beta(z) - \gamma(z) \in (0, 1)$ for all $z \in R^d$ and $D_z^\alpha\{f, g\} \in \mathcal{R}^l_\gamma V(D, M_0)$ for all multi-indices α. Suppose that the hypotheses 1-3 are satisfied. Then for a solution $u \in \mathcal{D}^l_\beta V(D, M_0)$ of the problem (3.28) we have the representation*

$$u(y, z) = \sum_{\nu=1}^{N} \sum_{\tau=1}^{J_\nu} \sum_{\sigma=0}^{\kappa_{\tau\nu}-1} c_\nu^{(\sigma,\tau)}(z) u_\nu^{(\sigma,\tau)}(y, z) + v(y, z) \tag{3.29}$$

where $v \in \mathcal{D}^l_\gamma V(D, M_0), c_\nu^{(\sigma,\tau)}$ are smooth functions on M_0, and $u_\nu^{(\sigma,\tau)}$ are defined by (1.35) with $r = |y|$, while λ_ν and $\{\phi_\nu^{h-j}\}$ are eigenvalues and Jordan chains of the pencil $\lambda \mapsto \Upsilon(\lambda, z)$.

If we drop the restriction $\beta - \gamma < 1$ the asymptotic formula (3.29) becomes more complicated (and more precise) because of lower terms. In general, the form of these terms does not coincide with that of parts of the asymptotics in a cone. The lower terms may contain differentiation in z. The representation (3.29) is a generalization of (3.25). If we do not impose any additional smoothness in z of the right-hand side of the problem (3.28) then the coefficients cease to be smooth. In such a case formulas of the type (3.21) containing the smoothing extensions of the coefficients given on the edge to D turn out to be more natural. Similar formulas were discussed in (Nikishkin (1979)), (Maz'ya, Rossman (1988)) and (Nazarov, Plamenevskij (1994)). Expressions for the coefficients $c_\nu^{(\sigma,\tau)}$ in (3.29) (like the equality (3.27)) and analogous results were given in (Maz'ya, Plamenevskij (1976)), (Maz'ya, Rossman (1988)) and (Nazarov, Plamenevskij (1994)).

3.6. Elliptic Problems on Manifolds with Intersecting Edges. We consider boundary value problems in domains whose boundaries contain edges of various dimensions intersecting at nonzero angles. For example, a polyhedron is a domain of such a type. The corresponding class $\mathcal{D}$ of manifolds may be described by induction. We introduce function spaces with weighted norms generalizing the spaces $V^s_\beta(G, M)$. Then we define collections of model problems whose operators are similar to $\Upsilon(\lambda)$ and $A(\theta)$ (compare with Sect.3.1). The original boundary value problem is Fredholm provided that the model problems are uniquely solvable.

3.6.1. Maps of Class D. Let $\Pi = R^{d_0} \times \overline{R_+} \times \ldots \times R^{d_q} \times \overline{R_+}, d_j \geq 0, j = 0, \ldots, q$. The points in Π are denoted by $(\zeta_0, p_0, \ldots, \zeta_q, p_q)$. Let μ_j be a multi-index of dimension d_j and ν_j a nonnegative integer. We set $M_k = (\mu_0, \ldots, \mu_k), N_k = (\nu_0, \ldots, \nu_k), M = M_q, N = N_q$; $s(M, N) = (|M_0| + |N_0| + \ldots + |M_{q-1}| + |N_{q-1}|, 0)$, $\zeta_j = (\zeta_{1,j}, \ldots, \zeta_{d_j,j}) \in R^{d_j}$, $\zeta_{ij} \in R; Z_k = (\zeta_0, \ldots, \zeta_k)$, $R_k = (\rho_0, \ldots, \rho_k)$, $Z = Z_q$, $R = R_q$. A function f is said to be in the class $\mathcal{D}$ if the functions $R^{s(M,N)} D^M_z D^N_R f$ are continuous on Π for all M and N. To make it more clear we write

$$R^{s(M,N)} D^M_Z D^N_R f = \rho_0^{|\mu_0|+\nu_0} \rho_1^{|\mu_0|+\nu_0+|\mu_1|+\nu_1} \ldots \rho_{q-1}^{|\mu_0|+\nu_0+\ldots+|\mu_{q-1}|+\nu_{q-1}} \times$$
$$\times D^{\mu_0}_\zeta D^{\nu_0}_{\rho_0} \ldots D^{\mu_q}_{\zeta_q} D^{\nu_q}_{\rho_q} f.$$

Let $\mathcal{W}_1$ and $\mathcal{W}_2$ be open subsets of Π. By definition, the map $\sigma : \mathcal{W}_1 \ni (\zeta_0, \rho_0, \ldots, \zeta_q, \rho_q) \mapsto (z_0, r_0, \ldots, z_q, r_q)$ is a diffeomorphism of the class $\mathcal{D}$ if the following conditions are satisfied:

1^0. The functions $z_j = z_j(\zeta_0, \rho_0, \ldots, \zeta_q, \rho_q)$ and $r_j = r_j(\zeta_0, \rho_0, \ldots, \zeta_q, \rho_q)$ are continuous.

2^0. The functions $r_j \rho_j^{-1}$ and $r_j^{-1} \rho_j$ are in $\mathcal{D}$.

3^0. The entries of the matrix

$$\left(\frac{r_1 \ldots r_{k-1}}{\rho_{-1} \ldots \rho_{k-1}} \frac{\partial(z_k, r_k)}{\partial(\zeta_j, \rho_j)} \right)^q_{k,j=0} \tag{3.30}$$

with $r_{-1} = \rho_{-1} = 1$ are in $\mathcal{D}$ and its determinant does not vanish (in the case $d_k = 0$ the variables ζ_k and z_k do not occur in (3.30)).

One can check that the class $\mathcal{D}$ is closed under composition of maps.

3.6.2. Manifolds of the Class $\mathcal{D}_n$. This class of n-dimensional manifolds is defined by induction. Denote by $\mathcal{D}_0$ the class of finite collections of points with the discrete topology and let us say that a manifold Ω in $\mathcal{D}_0$ has class $\mathcal{D}$ differential structure. Assume the classes $\mathcal{D}_k$, $0 \leq k \leq n-1$, of the manifolds having class $\mathcal{D}$ differential structure to be defined and introduce $\mathcal{D}_n$.

Let Ω be a Hausdorff compact topological space and let $\{\mathcal{U}_j\}$ be a finite open covering of Ω. Suppose that for each j a diffeomorphism κ_j maps $\mathcal{U}_j$ onto some open subset, with compact closure, of a product of the form $R^d \times K^{n-d}$, where $d = 0, \ldots, n-1$ and $K^{n-d} = R_+ \times \Omega'$, $\Omega' \in \mathcal{D}_{n-d-1}$.

In what follows it will sometimes be convenient to write $d_0, \zeta_0, \ldots, \mathcal{U}_{j,0}$ instead of $d, \zeta, \rho, \alpha, \kappa_j, \mathcal{U}_j$. We denote the set $\Omega' \in \mathcal{D}_{n-d_0-1}$ by Ω_1. For this set, a covering $\{\mathcal{U}'_j\}$ and maps $\{\kappa'_j\}$ are defined similarly to those for Ω. To every point $\alpha_0 \in \Omega_1$ there corresponds the triple $(\zeta_1, \rho_1, \alpha_1)$. By continuing this process we arrive at a C^∞ manifold $\Omega_q, d_q \geq 0$. Each point $\alpha_q \in \Omega_q$ has a neighborhood homeomorphic to some open subset of the half-space

$R^{d_q} \times R_+$. Therefore, to α_q we associate coordinates (ζ_q, ρ_q). Eventually, a point $x \in \mathcal{U}_j$ acquires the local coordinates $(\zeta_0, \rho_0, \ldots, \zeta_q, \rho_q)$, and an atlas $\{\mathcal{V}^{j_0,\ldots,j_q}, \sigma^{j_0,\ldots,j_q}\}$ proves to be defined on Ω, where $\{\mathcal{V}^{j_0,\ldots,j_q}\}$ is a family of open sets and

$$\sigma^{j_0,\ldots,j_q} = \tilde{\kappa}^{(q)}_{j_q} \circ \ldots \circ \tilde{\kappa}^{(0)}_{j_0} : \mathcal{V}^{j_0,\ldots,j_q} \to \Pi$$

with $\tilde{\kappa}^{(0)}_{j_0} = \kappa^{(0)}_{j_0}$ and $\tilde{\kappa}^{(h)}_{j_h}(\zeta_0, \rho_0, \ldots, \zeta_{h-1}, \rho_{h-1}, \alpha_{h-1}) = (\zeta_0, \rho_0, \ldots, \zeta_{h-1}, \rho_{h-1}, \kappa^{(h)}_{j_h}(\alpha_{h-1}))$.

We describe now the connections between local coordinates on intersections of coordinate neighborhoods. Let $\{\mathcal{V}_1, \sigma_1\}$ and $\{\mathcal{V}_2, \sigma_2\}$ be local charts such that $\mathcal{V}_1 \cap \mathcal{V}_2 \neq \emptyset$. Denote by $(\zeta_0, \rho_0, \ldots, \zeta_{q_1}, \rho_{q_1})$ and $(z_0, r_0, \ldots, z_{q_2}, r_{q_2})$ the coordinates in $\mathcal{V}_1$ and $\mathcal{V}_2$, respectively. First we consider the case $q_1 = q_2 = q$ and $z_j, \zeta_j \in R^{d_j}, j = 1, \ldots, q$. We require that the homeomorphism

$$\sigma_2 \circ \sigma_1^{-1} : \sigma_1(\mathcal{V}_1 \cap \mathcal{V}_2) \to \sigma_2(\mathcal{V}_1 \cap \mathcal{V}_2) \tag{3.31}$$

defines a class $\mathcal{D}$ diffeomorphism $(\zeta_0, \rho_0, \ldots, \zeta_q, \rho_q) \mapsto (z_0, r_0, \ldots, z_q, r_q)$. We come now to the general case. Let $r_i \geq \delta > 0$ on $\mathcal{V}_1 \cap \mathcal{V}_2$ for $i = 0, \ldots, k_0 - 1$ and let k_{r_0} vanish at some points of $\overline{\mathcal{V}_1 \cap \mathcal{V}_2}$. Set $\tilde{z}_0 = (z_0, r_0, \ldots, z_{k_0-1}, r_{k_0-1}, z_{k_0})$, $\tilde{r}_0 = r_{k_0}$. Further, we suppose that $r_i \geq \delta > 0$ on $\mathcal{V}_1 \cap \mathcal{V}_2$ for $k_0 < i < k_1$ while r_{k_1} vanishes at some points of $\overline{\mathcal{V}_1 \cap \mathcal{V}_2}$. Set $\tilde{z}_1 = (z_{k_0+1}, r_{k_0+1}, \ldots, z_{k_1})$ and $\tilde{r}_1 = r_{k_1}$. We can continue this process to rewrite the coordinates $(z_0, r_0, \ldots, z_{q_1}, r_{q_1})$ in the form $(\tilde{z}_0, \tilde{r}_0, \ldots, \tilde{z}_{p_1}, \tilde{r}_{p_1})$ where $\tilde{r}_{p_1} = r_{q_1}$. A similar procedure (with possibly some numbers $l_0, l_1, \ldots$ instead of $k_0, k_1, \ldots$) rewrites the coordinates $(\zeta_0, \rho_0, \ldots, \zeta_{q_2}, \rho_{q_2})$ as $(\tilde{\zeta}_0, \tilde{\rho}_0, \ldots, \tilde{\zeta}_{p_2}, \rho_{p_2})$. Assume that $p_1 = p_2 = p$ and the vectors $\tilde{\zeta}_j, \tilde{z}_j$ have the same number $\tilde{d}_j$ of coordinates, $j = 1, \ldots, p$. We require that the mapping (3.31) is a diffeomorphism $(\tilde{\zeta}_0, \tilde{\rho}_0, \ldots, \tilde{\zeta}_p, \tilde{\rho}_p) \mapsto (\tilde{z}_0, \tilde{r}_0, \ldots, \tilde{z}_p, \tilde{r}_p)$ of the class $\mathcal{D}$.

Any atlas subject to the above conditions will be called admissible. We shall say that two classes are equivalent if and only if their union is an admissible atlas. The set of equivalent admissible atlases defines the class $\mathcal{D}_n$ differential structure. The manifolds with $\mathcal{D}_n$ structure constitute the class $\mathcal{D}_n$. A function f given on $\Omega \in \mathcal{D}_n$ is said to be in $\mathcal{D}(\Omega)$ if $f \circ \sigma^{-1} \in \mathcal{D}$ for every chart $\{\mathcal{V}, \sigma\}$.

3.6.3. Stratification of the Manifold $\Omega \in \mathcal{D}_n$. Let $x^h = (\zeta_0^h, \rho_0^h, \ldots, \zeta_q^h, \rho_q^h)$, $h = 1, 2$, be two points of Ω with coordinates written in some local system. We say that these points are equivalent if $\zeta_j^1 = \zeta_j^2, 0 \leq j \leq k, \rho_j^1 = \rho_j^2 > 0, 0 \leq j \leq k-1$ and $\rho_k^1 = \rho_k^2 = 0$ for some k. Identifying equivalent points we obtain the quotient space $\Omega^{\cdot}$. Denote by π the projection $\Omega \to \Omega^{\cdot}$. The point $x^{\cdot} \in \Omega^{\cdot}$ is said to be in $\mathcal{M}^d \subset \Omega^{\cdot}$ if in some local coordinate system a representative of this point has coordinates $(\zeta_0, \rho_0, \ldots, \zeta_q, \rho_q)$, where $\rho_j > 0$ for $j < k$ and $\rho_k = 0$ while $d = d_0 + \ldots + d_k + k, \zeta_j \in R^{d_j}$. The connected components of $\mathcal{M}^d$ are called d-dimensional strata. The strata of codimension one in Ω will have a special name: the faces of Ω. The set $\Omega \setminus (\mathcal{M}^0 \cup \ldots \cup \mathcal{M}^{n-1})$ is a (noncompact) C^∞ manifold with boundary; the faces of Ω form the boundary of this manifold. The union $\mathcal{M}^0 \cup \ldots \cup \mathcal{M}^{n-1}$ is called the boundary $\partial\Omega$ of Ω. The boundary of each stratum consists of strata of lower dimensions. The set of all strata whose boundaries contain the stratum T is called the star of T and denoted by $st(T)$. An open subset $\mathcal{U}$ of Ω is said to be servicing a d-dimensional stratum if there exists a homeomorphism

$$\kappa : \mathcal{U} \to R^d \times \overline{R}_+ \times \Omega_1, \quad \Omega_1 \in \mathcal{D}_{n-d-1}$$

such that the local coordinates defined by means of $(\mathcal{U}, \kappa)$ are compatible with the differential structure on Ω. For every point $x \in x^{\cdot} \in T$ there exists a neighborhood $\mathcal{U}$ servicing the stratum T. Denote by τ the projection $R^{d_0} \times \overline{R}_+ \times \Omega_1 \to \Omega_1$. If s is a stratum in $st(T)$ then the set $\pi_1 \circ \tau \circ \kappa(\mathcal{U} \cap \pi^{-1}(s))$, where $\pi_1 : \Omega \to \Omega_1^{\cdot}$ is the projection, proves to be a stratum in Ω_1 and there are no other strata in Ω_1. In what follows we assume that for each of manifolds $\Omega, \Omega_1, \ldots$ there exists a finite covering by neighborhoods, either disjoint with the boundary or servicing some stratum. We fix such a covering on every manifold. Let $T_0, \ldots, T_m$ be a sequence of strata, $T_k \in st(T_j)$ for $k > j$, and let a neighborhood $\mathcal{U}$ be servicing T_0 while the triple of projections $(\zeta_0, \rho_0, \alpha_0)$ is defined by means of $\mathcal{U}$. By deciphering consecutively this triple (in the same manner as was done in defining the coordinates $(\zeta_0, \rho_0, \ldots, \zeta_q, \rho_q)$) we associate to this sequence of strata the coordinate system $(\zeta_0, \rho_0, \ldots, \zeta_m, \rho_m, \alpha_m)$ where $\alpha_m \in \Omega_{m+1}$. The stratum T_k is determined by the conditions $\rho_j > 0$ for $j < k$ and $\rho_k = 0$.

3.6.4. Operators and Function Spaces. We define the class $O(p, \Omega)$ of differential operators $\mathcal{P}$ on $\Omega \setminus \partial\Omega$ having in any local coordinate system $(\zeta_0, \rho_0, \ldots, \zeta_q, \rho_q)$ the representation

$$\mathcal{P} = \sum_{|M|+|N| \le p} \mathcal{P}_{MN} R^{s(M,N)-p} D_Z^M D_R^N \tag{3.32}$$

where $p = (p, \ldots, p, 0) \in R^{q+1}$ and the coefficients $\mathcal{P}_{MN}$ belong to the class $\mathcal{D}$. In more detail, (3.32) has the form

$$\mathcal{P} = \sum_{|\mu_0|+\nu_0+\ldots+|\mu_q|+\nu_q \le p} \mathcal{P}_{\mu_0\nu_0\ldots\mu_q\nu_q}(\zeta_0, \rho_0, \ldots, \zeta_q, \rho_q) \rho_0^{|\mu_0|+\nu_0-p} \times$$
$$\times \rho_1^{|\mu_0|+\nu_0+|\mu_1|+\nu_1-p} \ldots \rho_{q-1}^{|\mu_0|+\nu_0+\ldots+|\mu_{q-1}|+\nu_{q-1}-p} D_{\zeta_0}^{\mu_0} D_{\rho_0}^{\nu_0} \ldots D_{\zeta_q}^{\mu_q} D_{\rho_q}^{\nu_q}$$

where $\mathcal{P}_{\mu_0\nu_o\ldots\mu_q\nu_q}$ are class $\mathcal{D}$ functions.

By definition, a differential operator $\mathcal{P}(\xi)$ belongs to the class $\tilde{O}(p, \Omega)$ with parameter $\xi \in R^\tau$ if $\mathcal{P}(\xi) = \sum \xi^\gamma \mathcal{P}_\gamma, |\gamma| \le p$, γ being a multi-index $(\gamma_1, \ldots, \gamma_\tau)$, and $\mathcal{P} \in O(p - |\gamma|, \Omega)$.

We come now to define function spaces. To every d-dimensional stratum T, $0 \le d \le n-2$, we associate a real number β_T and denote by B the collection of all β_T. Let $\mathcal{U}$ be a neighborhood servicing T, and let κ be a homeomorphism $\kappa : \mathcal{U} \to R^d \times \overline{R}_+ \times \Omega_1$. Assume that $S \in st(T)$, $\dim S \le n-2$ and denote by S_1

the stratum of Ω_1 generated by S. Set $\beta_{S_1} = \beta_S$ and $B_1 = \{\beta_{S_1} : S \in st(T)\}$. If only $st(T)$ contains $(n-1)$-dimensional strata then the set B_1 is empty.

We introduce the function space $V^s_B(\Omega)$ by induction on $dim\, \Omega$. Suppose that the space has already been defined for all $\Omega \in \mathcal{D}_k$ with $k < n$. In case of n-dimensional manifolds Ω, the norm in $V^s_B(\Omega), s = 0, 1, \ldots$, is obtained from local norms with the help of a partition of unity. Let $\mathcal{U}$ be a neighborhood servicing a d-dimensional stratum T. For functions u supported in $\mathcal{U}$ we introduce the norm

$$||u; V^s_B(\Omega)|| = (\int_{R^d} d\zeta_0 \int_0^\infty \rho_0^{2(\beta_T - s)+n-d-1} \sum_{|\mu|+\nu\le l} ||(\rho_0 D_{\zeta_0})^\mu (\rho_0 D_{\rho_0})^\nu \times$$

$$\times\, u \circ \kappa^{-1}; V^{s-|\mu|-\nu}_{B_1}(\Omega_1)||^2\, d\rho_0)^{1/2}.$$

If B is void (i.e. $\partial\Omega$ is smooth) then $V^s_B(\Omega)$ is the Sobolev space $H^s(\Omega)$. Let Γ be the collection of all faces (i.e. $(n-1)$-dimensional strata) of Ω. Denote by $V^{s-1/2}_B(\Gamma), s = 1, 2, \ldots$, the space of traces on Γ of functions in $V^s_\beta(\Omega)$. Also, we introduce the following norm in $V^s_B(\Omega)$ depending on a parameter $t \in R_+$:

$$||u; V^s_\beta(\Omega; t)|| = (\sum_{k=0}^{s} t^{2k} ||u; V^{s-k}_B(\Omega)||^2)^{1/2}.$$

Let $\mathcal{L}$ be a $k\times k$ matrix differential operator with entries $\mathcal{L}_{ij}$ in $O(s_i + t_j, \Omega)$. Denote by $\Gamma^{(1)}, \ldots, \Gamma^{(N)}$ the faces of Ω and by $\mathcal{B}^{(1)}, \ldots, \mathcal{B}^{(N)}$ $(m \times k)$ matrix differential operators with entries $\mathcal{B}^{(p)}_{ij} \in O(\sigma^{(p)}_i + t_j, \Omega)$. The operator $\{\mathcal{L}, \mathcal{B}^{(1)}, \ldots, \mathcal{B}^{(n)}\}$ of boundary value problem implements a continuous mapping

$$\{\mathcal{L}, \mathcal{B}\} : \mathcal{D}^l_B V(\Omega) \to \mathcal{R}^l_B V(\Omega) \tag{3.33}$$

where $l \ge \max\{1 + \max \sigma^{(p)}_q, 0\}$, and $\mathcal{D}^l_B V(\Omega), \mathcal{R}^l_B V(\Omega)$ are defined similary to (1.21).

The operator $\{\mathcal{L}(\xi), \mathcal{B}^{(1)}(\xi), \ldots, \mathcal{B}^{(N)}(\xi)\}$ of boundary value problem with parameter,

$$\{\mathcal{L}(\xi), \mathcal{B}(\xi)\} : \mathcal{D}^l_B V(\Omega, |\xi|) \to \mathcal{R}^l_B V((\Omega, |\xi|), \tag{3.34}$$

is bounded and its norm is majorized by a constant independent of ξ.

3.6.5. Fredholm Property of the Operator (3.33). Let $\mathcal{U}$ be a neighborhood servicing a d-dimensional stratum T, $0 \le d \le n-2, x \in \mathcal{U}$ and let $\kappa : \mathcal{U} \to R^d \times \overline{R}_+ \times \Omega_1$ be a map defining coordinates (ζ, ρ, α). In these coordinates an operator $\mathcal{P} \in O(p, \Omega)$ can be represented in the form

$$\mathcal{P} = \rho^{-p} \sum_{|\mu|+\nu\le p} \mathcal{P}_{\mu\nu}(\zeta, \rho)(\rho D_\zeta)^\mu (\rho D_\rho)^\nu$$

where $\mathcal{P}_{\mu\nu} \in O(p - |\mu| - \nu, \Omega)$. For every point $x^{(0)} = (\zeta^{(0)}, \rho = 0)$ we define the "model" operator on $R_+ \times \Omega_1$:

$$\dot{\mathcal{P}}(\zeta^{(0)}, 0; \rho\eta, \rho D_\rho) = \sum_{|\mu|+\nu\le p} \mathcal{P}_{\mu\nu}(\zeta^{(0)}, 0)(\rho\eta)^\mu(\rho\eta)^\mu(\rho D_\rho)^\nu.$$

To the operator $\mathcal{P}(\xi) \in \tilde{O}(p, \Omega)$ we associate the model one

$$\dot{\mathcal{P}}(\zeta^{(0)}, 0; \rho\xi, \rho\eta, \rho D_\rho) = \sum_{|\gamma|\le p} (\rho\xi)^\gamma \dot{\mathcal{P}}_\gamma(\zeta^0, 0; \rho\eta, \rho D_\rho).$$

The operator $\{\mathcal{L}, \mathcal{B}\}$ of the boundary value problem on Ω generates at every point $x^{(0)} \in \mathcal{M}^0 \cup \ldots \cup \mathcal{M}^{n-2}$ the model operator $\{\dot{\mathcal{L}}, \dot{\mathcal{B}}\}(\zeta^0, 0; \rho\eta, \rho D_\rho)$ of boundary value problem on $K^{n-d} = R_+ \times \Omega_1$, where d is the dimension of the stratum T containing $x^{(0)}$. Let $\{\dot{\mathcal{L}}, \dot{\mathcal{B}}\}(\zeta^0, 0; \rho\xi, \rho\eta, \rho D_\rho)$ stand for the model operator of boundary value problem with parameter.

Introduce the space $E^s_{B(T)}(K^{n-d})$ where $B(T) = \{\beta_T, B_1\}$ endowed with the norm

$$||v; E^s_{B(T)}(K^{n-d})|| = (\int_0^\infty \rho^{2(\beta_T - s)+n-d-1} \sum_{k_1+k_2\le s} \rho^{2k_1} ||(\rho D_\rho)^{k_2} v;$$
$$V^{s-k_1-k_2}_{B_1}(\Omega_1)||^2 \, dr)^{1/2}.$$

The operator

$$\{\dot{\mathcal{L}}, \dot{\mathcal{B}}\}(\zeta^0, 0; \rho\eta, \rho D_\rho) : \mathcal{D}^l_{B(T)} E(K^{n-d}) \to \mathcal{R}^l_{B(T)} E(K^{n-d}) \tag{3.35}$$

is continuous.

Proposition 3.12. *The operator (3.33) is Fredholm if and only if, for all $\eta \in R^d$ with $|\eta| = 1$, the mapping (3.35) is an isomorphism for any $x^{(0)} \in M^1 \cup \ldots \cup M^{n-2}$ and Fredholm for $x^{(0)} \in M^0$.*

Proposition 3.13. *The operator (3.34) is an isomorphism for large $|\xi|$ and*

$$||u; \mathcal{D}^l_B V(\Omega, |\xi|)|| \le c\,||\{\mathcal{L}(\xi), \mathcal{B}(\xi)\}u; \mathcal{R}^l_B V(\Omega, |\xi|)||$$

with constant c independent of ξ if and only if the following condition is satisfied: for all ξ, η such that $|\xi|^2 + |\eta|^2 = 1$ the mapping

$$\{\dot{\mathcal{L}}, \dot{\mathcal{B}}\}(\zeta^0, 0, \rho\xi, \rho\eta, \rho D_\rho) : \mathcal{D}^l_{B(T)} E(K^{n-d}) \to \mathcal{R}^l_{B(T)} E(K^{n-d})$$

is an isomorphism for any $x^{(0)} \in M^0 \cup \ldots \cup M^{n-2}$.

We define the pencil $\Upsilon(\lambda) = \{\dot{\mathcal{L}}, \dot{\mathcal{B}}\}(\zeta^0, 0, 0, \lambda)$ for the operator $\{\dot{\mathcal{L}}, \dot{\mathcal{B}}\}(\zeta^0, 0, \rho\eta, \rho D_\rho)$. Suppose that Υ satisfies the conditions : a) for large $|\lambda|, \lambda \in R$, the mapping $\Upsilon(\lambda) : \mathcal{D}^l_{B_1} V(\Omega, |\lambda|) \to \mathcal{R}^l_{B_1} V(\Omega_1)$ is an isomorphism; b) for the above λ the estimate

$$||u; \mathcal{D}^l_{B_1} V(\Omega_1, |\lambda|)|| \le c\,||\Upsilon(\lambda)u; \mathcal{R}^l_{B_1} V(\Omega_1, |\lambda|)||$$

holds with constant c independent of $|\lambda|$. Then $\lambda \mapsto \Upsilon(\lambda)^{-1}$ is a meromorphic operator-function. Its poles, with the possible exception of finitely many points, are situated inside a double angle containing the imaginary axis.

Proposition 3.14. *The operator (3.14) is Fredholm if and only if the conditions a) and b) are satisfied and the line $R + i(\beta_T - l + (n-d)/2)$ is free from the spectrum of the pencil Υ, where T is the stratum containing $x^{(0)}$.*

Consider the pencil $\Upsilon(\lambda)$ as an operator $\{\mathcal{L}(\xi), \mathcal{B}(\xi)\}$, setting $\xi \equiv \lambda \in R$. To obtain the model problem for $\Upsilon(\lambda)$ we define the operator of the type $\{\dot{\mathcal{L}}, \dot{\mathcal{B}}\}(\zeta^0, 0, r\theta, rD_r)$ at every point ζ^0 of a d-dimensional stratum of the manifold Ω_1, where $\theta = (\lambda(\lambda^2 + |\eta|^2)^{-1/2}, \eta(\lambda^2 + |\eta|^2)^{-1/2})$ and $\eta \in R^{\delta}$.

Proposition 3.15. *Let each operator $\{\dot{\mathcal{L}}, \dot{\mathcal{B}}\}(\zeta^0, 0, r\theta, rD_r)$ be an isomorphism of the form (3.35) for every stratum of dimension $\delta <$ $\dim \Omega_1 - 1$. Then the pencil Υ satisfies the conditions a) and b).*

By Propositions 3.12-3.15 the verification of the Fredholm property for the operator (3.33) (the invertibility of the operator (3.34) for large $|\xi|$) reduces to consecutively checking the triviality of the kernel and cokernel of every model problem and the absence of the spectrum of the pencils Υ (on manifolds of lower dimensions) on some lines in the complex plane.

This scheme is activated in the following way. Let T be a d-dimensional stratum of Ω and let $T_0, \dots, T_s = T$ be a chain of strata such that $T_k \in st(T_j)$ for $k > j$. In any coordinate system $(\zeta_0, \rho_0, \dots, \zeta_s, \rho_s, \alpha_s), \alpha_s \in \Omega_{s+1}$, the operator $\mathcal{P} \in O(p, \Omega)$ has the form

$$\mathcal{P} = \hat{\rho}_s^{-p} \sum_{|M_s|+|N_s| \le p} \mathcal{P}_{M_s,N_s}(Z_s, R_s)(\hat{R}_s D_{Z_s})^{M_s}(\hat{R}_s D_{R_s})^{N_s}$$

where $\hat{\rho}_j = \rho_j \dots \rho_0, \hat{R}_s = \{\hat{\rho}_0, \dots, \hat{\rho}_l\}, \mathcal{P}_{M_s,N_s} \in O(p - |M_s| - |N_s|, \Omega_{s+1})$. Generalizing the above definition of a model operator we associate to every collection $(Z_s^{(0)}, 0) = (\zeta^0, 0, \dots, \zeta_s^{(0)}, 0)$ the model operator on $R_+ \times \Omega_{s+1}$:

$$\dot{\mathcal{P}} = \sum_{|M_s|+|N_s| \le p} \mathcal{P}_{M_s,N_s}(Z_s^0, 0)(\rho\eta)^{H_s}(\rho D_\rho)^{\nu_s} \tag{3.36}$$

where $\eta \in R^{d_0 + \dots + d_s + s}, d_s = d$ and $H_s = \{\mu_0, \nu_0, \dots, \mu_{s-1}, \nu_{s-1}, \mu_s\}$. Now the operators $A_d(Z_s^0, 0, \rho\eta, \rho D_\rho)$ of the model boundary value problem on $R_+ \times \Omega_{s+1}$ are defined in a natural way. The entries of the corresponding matrix operators are expressions of the form (3.36).

If the original boundary value problem on Ω contains a parameter ξ then it gives rise to an operator of the form

$$A_d(Z_s^0, 0, \rho\xi, \rho\eta, \rho D_\rho) \equiv A_d(\rho\tau, \rho D_\rho)$$

on $R_+ \times \Omega_{s+1}$ where $\tau = (\xi, \eta)$.

By definition, a collection B of reals is admissible if it satisfies the following condition. To every $(n-2)$-dimensional stratum T there corresponds a number β_T such that the line $R + i(\beta_T + 1 - l)$ contains no eigenvalues of the pencil $\lambda \mapsto \Upsilon(\lambda) = A_{n-2}(0, \lambda)$ (this operator is associated to the stratum T). Suppose that $\ker A_{n-2}(\rho\tau, \rho D_\rho) = 0$ and $\operatorname{coker} A_{n-2}(\rho\tau, \rho D_\rho) = 0$ for all operators associated to the $(n-2)$-dimensional strata, where $|\tau| = 1$. Then, by Propositions 3.14 and 3.15, the operator pencils $\lambda \mapsto A_{n-3}(0, \lambda)$ satisfy the conditions a) and b) where the collection B_1 is defined uniquely by the exponents β_T corresponding to the $(n-2)$-dimensional strata. Assume that all the exponents for d-dimensional strata with $d > n - k$ have already been chosen so that the conditions a) and b) are fulfilled for the pencils $\lambda \mapsto A_{n-k}(0, \lambda)$, where the collection B_1 is determined in a unique way by the exponents of the $(n-k)$-dimensional strata. Moreover, let the exponents of the $(n-k)$-dimensional strata be chosen in such a way that the lines $R + i(\beta_T - l + n/2)$ contain no eigenvalues of the pencils $\lambda \mapsto A_{n-k}(0, \lambda)$ and the kernel and cokernel of every operator $A_{n-k}(\rho\tau, \rho D_\rho)$ are trivial, where $|\tau| = 1$. Then all the operators $A_{n-k}(0, \lambda)$ satisfy the conditions a) and b). If we are able to go through all dimensions $d = n-2, \dots, 0$ in this way then we obtain an admissible collection B.

Combining Propositions 3.12-3.14 we arrive at the following result.

Theorem 3.16. *1) if the collection B is admissible then the operator*

$$\{\mathcal{L}, \mathcal{B}\} : \mathcal{D}^l_B V(\Omega) \to \mathcal{R}^l_B V(\Omega)$$

is Fredholm.

2) If the collection B is admissible then the operator

$$\{\mathcal{L}(\xi), \mathcal{B}(\xi)\} : \mathcal{D}^l_B V(\Omega) \mapsto \mathcal{R}^l_B V(\Omega)$$

is an isomorphism for large $|\xi|$ and the estimate

$$||u; \mathcal{D}^l_B V(\Omega, |\xi|)|| \leq c\, ||\{\mathcal{L}(\xi), \mathcal{B}(\xi)\}u; \mathcal{R}^l_B V(\Omega, |\xi|)||$$

is valid.

Bibliography

The solvability of general elliptic boundary value problems in a cone was proved in (Kondrat'ev (1967)) (in the Hilbert spaces V^l_β) and (Maz'ya, Plamenevskij (1978a)) (in the classes $V^l_{p,\beta}$ and $\Lambda^{l,\alpha}_\beta$). The notions related to the spectrum of meromorphic operator-functions and Proposition 1.9 are taken from (Gohberg, Sigal (1971)); Proposition 1.8 on the special choice of Jordan chains appeared in (Maz'ya, Plamenevskij (1975b)). The asymptotics of solutions in a cylinder and cone were studied in (Kondrat'ev (1967)), (Agmon, Nirenberg (1963)), and the formulas for the coefficients in the asymptotics were given in (Maz'ya, Plamenevskij (1975b)). The estimates of fundamental solutions are taken from (Maz'ya, Plamenevskij (1979)).

The Fredholm property of boundary value problems in domains with conical points in the Hilbert spaces V^l_β was proved in (Kondrat'ev (1967)). The formulas for the coefficients in the asymptotics of solutions (Theorem 2.6) were obtained in (Maz'ya, Plamenevskij (1977a)). Theorem 2.5 on the index is essentially taken from the same paper (Maz'ya, Plamenevskij (1977a)). (The analogous representations of the coefficients in the asymptotics of solutions for some problems in crack mechanics were found in (Bueckner (1970)).) The weighted spaces with nonhomogeneous norms and the elliptic problems in such spaces were studied in (Kondrat'ev (1967)), (Maz'ya, Plamenevskij (1978e)), (Solonnikov (1979a)) and others. Sect.2.7 contains results from (Nazarov, Plamenevskij (1991b)), (Nazarov, Plamenevskij (1990b)); see also (Nazarov, Plamenevskij (1994)).

An extensive literature is devoted to the spectrum of the special operator pencils (defining the asymptotics of solutions near vertices of cones). We refer to (Kozlov, Maz,ya (1988)), (Maz'ya, Nazarov (1986)), (Maz'ya, Nazarov (1989)), (Maz'ya, Nazarov, Plamenevskij (1983)), (Maz'ya, Plamenevskij (1981)); in connection with elasticity theory we mention (MoF (1988)), (Parton, Perlin (1981)).

We turn now to problems in domains with edges of positive dimensions. The Fredholm property of the Dirichlet problem for second order equations in domains with smooth edges was proved in (Kondrat'ev (1970)). The Fredholm property of general elliptic problems (under the further assumptions on the triviality of the kernels and cokernels of model operators) were established in (Maz'ya, Plamenevskij (1973a)), (Maz'ya, Plamenevskij (1978b)) for the weighted spaces generated by L_2- and L_p-norms. The estimates in the weighted Hölder classes were obtained in (Maz'ya, Plamenevskij (1978c)), (Maz'ya, Plamenevskij (1978d)), (Solonnikov (1979b)). The Dirichlet problem was considered in (Maz'ya, Plamenevskij (1977b)), (Maz'ya, Plamenevskij (1978b)), (Maz'ya, Plamenevskij (1983)). The Neumann problem (without the above assumption on the model operators) in spaces with nonhomogeneous norms were studied in (Zajonchkovskij, Solonnikov (1983)), (Nazarov (1988)). The convolution operator with homogeneous function in R^n in nonhomogeneous weighted spaces was considered in (Plamenevskij, Taschian (1990)) (with the description of the kernels and cokernels of the model operators).

Various problems for second order equations were analysed in (Komech (1973)), (Maz'ya, Plamenevskij (1971)), (Maz'ya, Plamenevskij (1975a)), (Eskin (1985)). In these papers some conditions providing the triviality of the kernels and cokernels of model operators were obtained. The related topic (for equations of arbitrary order) was discussed in (Kozlov (1989)). The papers (Birman, Solomjak (1987)), (Maz'ya, Morozov, Plamenevskij (1979)), (Maz'ya, Plamenevskij (1973b)), (Maz'ya, Plamenevskij (1983)), (Maz'ya, Plamenevskij, Stupyalis (1979)), (Solonnikov (1979a)), (Solonnikov (1979b)) were devoted to various problems of the mechanics of continua.

Section 3.6 is based on the paper (Maz'ya, Plamenevskij (1977b)). Special problems for the classical equations in domains with intersecting edges were studied in (Maz'ya (1973)), (Maz'ya, Plamenevskij (1980)), (Maz'ya, Plamenevskij (1983)), (Fichera (1975)) and (Grisvard (1985)). The Neumann problem for self-adjoint elliptic systems was investigated in (Nazarov, Plamenevskij (1991a)).

The asymptotic expansions of solutions near a smooth edge with remainder having worse differential properties than the solution (i.e. the formulas of type (3.25)) were derived in (Kondrat'ev (1977)), (Maz'ya, Plamenevskij (1973a)). The asymptotics with "smooth" remainder (i.e. the formulas of type (3.21)) for solutions of second order equations was obtained in (Nikishkin (1979)). The asymptotic formulas of such a kind for solutions of general elliptic problems were established in (Maz'ya, Rossman (1988)). The related topics were discussed in (Nazarov, Plamenevskij (1994)).

A few words on the monographs devoted to elliptic problems in domains with nonsmooth boundary. The classical boundary value problems for second order equations were considered in (Grisvard (1985)). Elliptic problems in the classical (nonweighted) Sobolev spaces were studied in (Dauge (1988)). Numerical solutions were discussed in (Kufner, Sänding (1987)). The monograph (Nazarov, Plamenevskij (1994)) is devoted to the general theory of elliptic boundary value problems in domains with piecewise smooth boundary.

References[1]

Agmon, S., Nirenberg, L. (1963): Properties of solutions of ordinary differential equations in Banach space. Commun. Pure Appl. Math. *16*, 121–239. Zbl. 117,100

Agranovich, M.S., Vishik, M.I. (1964): Elliptic problems with a parameter and parabolic problems of general type. Usp. Mat. Nauk *19*(3), 53–161. English transl.: Russ. Math. Surv. *19*(3) (1964), 53–157. Zbl. 137,296

Birman, M.S., Solomyak, M.Z. (1987): L_2-theory of the Maxwell operator in arbitrary domains. Usp. Mat. Nauk *42*(6), 61–76. English transl.: Russ. Math. Surv. *42*(6) (1987), 75–96. Zbl. 635.35075

Bueckner, H.F. (1970): A novel principle for the computation of stress intensity factors. Z. Angew. Math. Mech. *50*, 529–546. Zbl. 213,266

Dauge, M. (1988): Elliptic boundary value problems on corner domains. Smoothness and asymptotics of solutions. Lecture Notes in Mathematics *1341*, Springer. Zbl. 638.35001

Eskin, G. (1985): Boundary-value problems for second-order elliptic equations in domains with corners. Proc. Sympos. Pure Math. *43*, 105–131. Zbl. 574.35029

Fichera, G. (1975): Asymptotic behaviour of the electric field and density of the electric charge in the neighborhood of singular points of a conducting surface. Usp. Mat. Nauk *30*(3), 105–124. English transl.: Russ. Math. Surv. 30(3) (1975), 107–127. Zbl. 328.31008

[1] For the convenience of the reader, references to reviews in Zentralblatt für Mathematik (Zbl.), compiled by means of the MATH database, and Jahrbuch über die Fortschritte der Mathematik (Jbuch) have, as far as possible, been included in this bibliography.

Gohberg, I.C., Sigal, E.I. (1971): An operator generalization of the logarithmic residue theorem and the theorem of Rouche. Mat. Sb., Nov. Ser. *84*, 607–629. English transl.: Math. USSR, Sb. *13* (1971), 603–625. Zbl. 254.47046

Grisvard, P. (1985): Elliptic Problems in Nonsmooth Domains. Pitman. Zbl. 695.35060

Komech, A.I. (1973): Elliptic boundary value problems on manifolds with a piecewise smooth boundary. Mat. Sb., Nov. Ser. *92*, 89–134. English transl.: Math. USSR, Sb. *31* (1973), 91–135. Zbl. 286.35027

Kondrat'ev, V.A. (1967): Boundary value problems for elliptic equations in domains with conical or angular points. Tr. Mosk. Mat. O.-va *16*, 209–292. English transl.: Trans. Moscow Math. Soc. *16* (1967), 227–313. Zbl. 162,163

Kondrat'ev, V.A. (1970): The smoothness of solution of Dirichlet's problem for second-order elliptic equations in a region with piecewise-smooth boundary. Differ. Uravn. *6*, 1831–1843. English transl.: Differ. Equations *6* (1970), 1392–1401. Zbl. 209,411

Kondrat'ev, V.A. (1977): Singularities of a solution of Dirichlet's problem for a second-order elliptic equation in a neighborhood of an edge. Differ. Uravn. 13, 2026–2032. English transl.: Differ. Equations *13* (1977), 1411-1415. Zbl. 379.35020

Kondrat'ev, V.A., Oleinik, O.A. (1983): Boundary value problems for partial differential equations in non-smooth domains. Usp. Mat. Nauk *38*(2), 3–76. English transl.: Russ. Math. Surv. *38*(2) (1983), 1–86. Zbl. 523.35010

Koplienko, L.S., Plamenevskij, B.A. (1983): A radiation principle for periodic problems. Differ. Uravn. *19*, 1713–1723. English transl.: Differ. Equations *19* (1983), 1273–1281. Zbl. 543.35027

Kozlov, V.A. (1989): The strong zero theorem for an elliptic boundary value problem in an angle. Mat. Sb. *180*, 831–849. English transl.: Math. USSR, Sb. *67* (1990), 283–302. Zbl. 695.35005

Kozlov, V.A., Maz'ya, V.G. (1988): Spectral properties of the operator bundles generated by elliptic boundary value problems in a cone. Funkts. Anal. Prilozh. *22*(2), 38–46. English transl.: Funct. Anal. Appl. *22* (1988), 114–121. Zbl. 672.35050

Kufner, A., Sändig, A.-M. (1987): Some Applications of Weighted Sobolev Spaces. Teubner, Leipzig. Zbl. 662.46034

Maz'ya, V.G. (1973): On the oblique derivative problem in a domain with edges of different dimensions. Vestn. Leningr. Univ., Mat. Mekh. Astron. *7*, 34–39. English transl.: Vestn. Leningr. Univ. Math. *6* (1979), 148–154. Zbl. 257.35032

Maz'ya, V.G., Morozov, N.F., Plamenevskij, B.A. (1979): On nonlinear bending of a plate with a crack. Differential and integral equations. Boundary value problems (I.N. Vekua Memorial Collection), Tbilisi, 145–163. English transl.: Am. Math. Soc. Transl. (Ser. 2) *123* (1984), 125–139. Zbl. 451.73030

Maz'ya, V.G., Nazarov, S.A. (1986): The vertex of a cone can be nonregular in the Wiener sense for a fourth-order elliptic equation. Mat. Zametki *39*, 24–28. English transl.: Math. Notes *39* (1986), 14–16. Zbl. 604.35016

Maz'ya, V.G., Nazarov, S.A. (1989): Singularities of solutions of the Neumann problem at a conical point. Sib. Mat. Zh. *30*(3), 52–63. English transl.: Sib. Math. J. *30* (1989), 387–396. Zbl. 701.35021

Maz'ya, V.G., Nazarov, S.A., Plamenevskij, B.A. (1983): On the singularities of solutions of the Dirichlet problem in the exterior of a slender cone. Mat. Sb., Nov. Ser. *122*, 435–456. English transl.: Math. USSR, Sb. *50* (1985), 415–437

Maz'ya, V.G., Plamenevskij, B.A. (1971): Problems with oblique derivatives in regions with piecewise smooth boundaries. Funkts. Anal. Prilozh. *5*(3), 102–103. English transl.: Funct. Anal. Appl. *5* (1971), 256–258. Zbl. 232.35027

Maz'ya, V.G., Plamenevskij, B.A. (1973a): Elliptic boundary value problems in a domain with piecewise smooth boundary. Proc. Sympos. Continuum Mechanics and Related Problems of Analysis. Tbilisi, Mecnieraba *1*, 171–181 (Russian). Zbl. 283.35037

Maz'ya, V.G., Plamenevskij, B.A. (1973b): The asymptotic behavior of solutions of the Navier-Stokes equations near edges. Dokl. Akad. Nauk SSSR *210*, 803–806. English transl.: Sov. Phys., Dokl. *18* (1973/1974), 379–381. Zbl. 295.35006

Maz'ya, V.G., Plamenevskij, B.A. (1975a): On boundary value problems for a second order elliptic equation in a domain with edges. Vestn. Leningr. Univ., Mat. Mekh. Astron. *1*, 102–108. English transl.: Vestn. Leningr. Univ. Math. *8* (1980), 99–106. Zbl. 296.35029

Maz'ya, V.G., Plamenevskij, B.A. (1975b): On the coefficients in the asymptotic expansion of the solution of elliptic boundary value problems in a cone. Zap. Nauchn. Semin. LOMI *52*, 110–127. English transl.: J. Sov. Math. *9* (1978), 750–764. Zbl. 351.35010

Maz'ya, V.G., Plamenevskij, B.A. (1976): On the coefficients in the asymptotics of solutions of elliptic boundary value problems near the edge. Dokl. Akad. Nauk SSSR *229*, 33–36. English transl.: Sov. Math., Dokl. *17* (1976), 970–974. Zbl. 355.35032

Maz'ya, V.G., Plamenevskij, B.A. (1977a): The coefficients in the asymptotics of solutions of elliptic boundary value problems in domains with conical points. Math. Nachr. *76*, 29–60 English transl.: Am. Math. Soc., Transl. (Ser. 2) *123* (1984), 57–88. Zbl. 359.35024

Maz'ya, V.G., Plamenevskij, B.A. (1977b): Elliptic boundary value problems on manifolds with singularities. Probl. Mat. Anal. *6*, 85–142 (Russian). Zbl. 453.58022

Maz'ya, V.G., Plamenevskij, B.A. (1978a): Estimates in L_p and in Hölder classes and the Miranda-Agmon maximum principle for solutions of elliptic boundary value problems in domains with singular points on the boundary. Math. Nachr. *81*, 25–82. English transl.: Am. Math. Soc., Transl. (Ser. 2) *123* (1984), 1–56. Zbl. 371.35018

Maz'ya, V.G., Plamenevskij, B.A. (1978b): L_p-estimates of solutions of elliptic boundary value problems in domains with edges. Tr. Mosk. Mat. O.-va *37*, 49–93. English transl.: Trans. Moscow Math. Soc. *37* (1980), 49–97. Zbl. 441.35028

Maz'ya, V.G., Plamenevskij, B.A. (1978c): Schauder estimates of solutions of elliptic boundary value problems in domains with edges of the boundary. Proc. Semin. S.L. Sobolev *2*, Novosibirsk, 69–102. English transl.: Am. Math. Soc., Transl. (Ser. 2) *123* (1984), 141–169. Zbl. 423.35021

Maz'ya, V.G., Plamenevskij, B.A. (1978d): Estimates of Green's functions and Schauder estimates for solutions of elliptic boundary value problems in a dihedral angle. Sib. Mat. Zh. *19*, 1065–1082. English transl.: Sib. Math. J. *19* (1978), 752–764. Zbl. 408.35014

Maz'ya, V.G., Plamenevskij, B.A. (1978e): Weighted spaces with nonhomogeneous norms and boundary value problems in domains with conical points. Elliptische Differentialgleichungen (Meeting, Rostock, 1977), Wilhelm-Pieck-Univ., Rostock, 161–190. English transl.: Am. Math. Soc., Transl. (Ser. 2) *123* (1984), 89–107. Zbl. 429.35031

Maz'ya, V.G., Plamenevskij, B.A. (1979): Asymptotic behavior of the fundamental solutions of elliptic boundary value problems in domains with conical points. Probl. Mat. Anal. *7*, 100–145 English transl.: Sel. Math. Sov. *4* (1985), 363–397. Zbl. 417.35014

Maz'ya, V.G., Plamenevskij, B.A. (1980): A problem on the motion of a fluid with a free surface in a container with piecewise smooth walls. Dokl. Akad. Nauk SSSR *250*, 1315–1318. English transl.: Sov. Math., Dokl. *21* (1980), 317–319. Zbl. 444.35067

Maz'ya, V.G., Plamenevskij, B.A. (1981): On the properties of solutions of three-dimensional problems of elasticity theory and hydrodynamics in domains with isolated singular points. Din. Sploshnoj. Sredy *50*, 99–121. English transl.: Am. Math. Soc., Transl. (Ser. 2) *123* (1984), 109–123. Zbl. 561.73020

Maz'ya, V.G., Plamenevskij, B.A. (1983): The first boundary value problem for classical equations of mathematical physics in domains with piecewise smooth boundaries. Z. Anal. Anwend. *2*, 335–359. Zbl. 532.35065

Maz'ya, V.G., Plamenevskij, B.A., Stupyalis, L.I. (1979): The three-dimensional problem of steady-state motion of a fluid with a free surface. Differ. Uravn. Primen. *23* 1–155. English transl.: Am. Math. Soc., Transl. (Ser. 2) *123* (1984), 171–286. Zbl. 431.76027

Maz'ya, V.G., Rossman, J. (1988): Über die Asymptotik der Lösungen elliptischer Randwertaufgaben in der Umgebung von Kanten. Math. Nachr. *138*, 27–53. Zbl. 672.35020

Maz'ya, V.G., Rossman, J. (1984): Über die Lösbarkeit und die Asymptotik der Lösungen elliptischer Randwertaufgaben in Gebieten mit Kanten. Preprint Akadem. Wiss. DDR, P-MATH.-31/84, 1–44. Zbl. 547.35042

MoF (1988): Mechanics of Fracture and Strength of Materials (in 4 volumes). Vol. 2, Naukova Dumka, Kiev (Russian)

Nazarov, S.A. (1988): Estimates near an edge for the solution of the Neumann problem for an elliptic system. Vestn. Leningr. Univ., Mat. Mekh. Astron. *21*, 37–42. English transl.: Vestn. Leningr. Univ. *21* (1988), 52–59. Zbl. 684.35021

Nazarov, S.A., Plamenevskij, B.A. (1991a): Neumann problem for self-adjoint systems in a domain with piecewise smooth boundary. Tr. Leningr. Mat. O.-va *1*, 174–211. English transl.: Am. Math. Soc., Transl. (Ser. 2) *155* (1993), 169–206. Zbl. 778.35033

Nazarov, S.A., Plamenevskij, B.A. (1991b): Radiation principles for self-adjoint elliptic systems. Probl. Mat. Fiz. *13*, 192–245 (Russian)

Nazarov, S.A., Plamenevskij, B.A. (1990): On radiation conditions for self-adjoint elliptic problems. Dokl. Akad. Nauk SSSR *311*, 523–535. English transl.: Sov. Math., Dokl. *41* (1990), 274–277. Zbl. 725.35027

Nazarov, S.A., Plamenevskij, B.A. (1994): Elliptic Problems in Domains with Piecewise Smooth Boundary. W. de Gruyter&Co, Berlin New York

Nikishkin, V.A. (1979): Singularities of the solutions to the Dirichlet problem for a second-order equation in a neighborhood of an edge. Vestn. Mosk. Univ., Ser. I Mat. Mekh. 1979, No. *2*, 51–62. English transl.: Moscow Univ. Math. Bull. *34*(2) (1979), 53–64. Zbl. 399.35049

Parton, V.Z., Perlin, P.I. (1981): Methods of Mathematical Elasticity Theory. Nauka (Russian). Zbl. 506.73005

Pazy, A. (1967): Asymptotic expansions of solutions of ordinary differential equations in Hilbert space. Arch. Rat. Mech. Anal. *24*(2), 193–218. Zbl. 147,123

Plamenevskij, B.A., Tashchiyan, G.M. (1990): Convolution operator in weighted spaces. Probl. Mat. Anal. *11*, 208–237. English transl.: J. Sov. Math. *64*(6) (1993), 1363–1381

Rice, J.R. (1972): Some remarks on elastic crack-tip stress fields. Int. J. Solids Struct. *8*(6), 751–758. Zbl. 245.73003

Solonnikov, V.A. (1979a): Solvability of a problem on the plane motion of a heavy viscous incompressible capillary liquid partially filling a container. Izv. Akad. Nauk SSSR, Ser. Mat. *43*, 203–236. English transl.: Math. USSR, Izv. *14* (1980), 193–221. Zbl. 411.76019

Solonnikov, V.A. (1979b): Solvability of the three-dimensional problem with a free boundary for the stationary Navier-Stokes system. Zap. Nauchn. Semin. LOMI *84*, 252–285. English transl.: J. Sov. Math. *21* (1983), 427–450. Zbl. 414.35062

Zajaczkowski, W., Solonnikov, V.A. (1983): The Neumann problem for second-order elliptic equations in domains with ribs on the boundary. Boundary value problems of mathematical physics and related questions in the theory of functions. Zap. Nauchn. Semin. LOMI *127*, 7–48. English transl.: J. Sov. Math. *27* (1984), 2561–2586

Author Index

Subject Index

Printing: Saladruck, Berlin
Binding: Buchbinderei Lüderitz & Bauer, Berlin